Science and Technology of Fast Ion Conductors

NATO ASI Series

Advanced Science Institutes Series

A series presenting the results of activities sponsored by the NATO Science Committee, which aims at the dissemination of advanced scientific and technological knowledge, with a view to strengthening links between scientific communities.

The series is published by an international board of publishers in conjunction with the NATO Scientific Affairs Division

A	**Life Sciences**	Plenum Publishing Corporation
B	**Physics**	New York and London
C	**Mathematical**	Kluwer Academic Publishers
	and Physical Sciences	Dordrecht, Boston, and London
D	**Behavioral and Social Sciences**	
E	**Applied Sciences**	
F	**Computer and Systems Sciences**	Springer-Verlag
G	**Ecological Sciences**	Berlin, Heidelberg, New York, London,
H	**Cell Biology**	Paris, and Tokyo

Recent Volumes in this Series

Series B: Physics

Science and Technology of Fast Ion Conductors

Edited by

Harry L. Tuller

Massachusetts Institute of Technology
Cambridge, Massachusetts

and

Minko Balkanski

Université Pierre et Marie Curie
Paris, France

Plenum Press
New York and London
Published in cooperation with NATO Scientific Affairs Division

Proceedings of a NATO Advanced Study Institute,
which was the International School of Materials
Science and Technology Twelfth Course on
Fast Ion Conductors,
held July 1–15, 1987,
in Erice, Italy

Library of Congress Cataloging in Publication Data

International School of Materials Science and Technology (1987: Erice, Italy)
   Science and technology of fast ion conductors / edited by Harry L. Tuller and
Minko Balkanski.
     p.   cm. — (NATO ASI series. Series B, Physics: v 199)
  "Proceedings of a NATO Advanced Study Institute, which was the Interna-
tional School of Materials Science and Technology Twelfth Course on Fast Ion
Conductors, held July 1–15, 1987, in Erice, Italy"—T.p. verso.
  "Published in cooperation with NATO Scientific Affairs Division."
  "Held within the program of activities of the NATO Special Program on Con-
densed Systems of Low Dimensionality, running from 1983 to 1988 as part of the
activities of the NATO Science Committee."
  Includes bibliographies and index.
  ISBN-13:978-1-4612-7842-9   e-ISBN-13:978-1-4613-0509-5
  DOI: 10.1007/978-1-4613-0509-5

  1. Superionic conductors—Congresses. 2. Electrolytes—Conductivity—
Congresses. I. Tuller, Harry L. II. Balkanski, Minko, 1927-   . III. North Atlantic
Treaty Organization. Scientific Affairs Division. IV. Special Program on Condens-
ed Systems of Low Dimensionality (NATO). V. Title. VI. Series.
QD561.I564 1987   89-33399
530.4′1—dc20   CIP

© 1989 Plenum Press, New York
Softcover reprint of the hardcover 1st edition 1989

A Division of Plenum Publishing Corporation
233 Spring Street, New York, N.Y. 10013

SPECIAL PROGRAM ON CONDENSED SYSTEMS OF LOW DIMENSIONALITY

This book contains the proceedings of a NATO Advanced Research Workshop held within the program of activities of the NATO Special Program on Condensed Systems of Low Dimensionality, running from 1983 to 1988 as part of the activities of the NATO Science Committee.

Other books previously published as a result of the activities of the Special Program are:

PREFACE

The rediscovery of fast ion conduction in solids in the 1960's stimulated interest both in the scientific community in which the fundamentals of diffusion, order-disorder phenomena and crystal structure evaluation required re-examination, and in the technical community in which novel approaches to energy conversion and chemical sensing became possible with the introduction of the new field of "Solid State Ionics."

Because of both the novelty and the vitality of this field, it has grown rapidly in many directions. This growth has included the discovery of many new crystalline fast ion conductors, and the extension to the fields of organic and amorphous compounds. The growth has involved the extension of classical diffusion theory in an attempt to account for carrier interactions and the development of sophisticated computer models. Diffraction techniques have been refined to detect carrier distributions and anharmonic vibrations. Similar advances in the application of other techniques such as NMR, Raman, IR, and Impedance Spectroscopies to this field have also occurred.

The applications of fast ion conducting solid electrolytes have also developed in many directions. High energy density Na/S batteries are now reaching the last stages of development, Li batteries are being implanted in humans for heart pacemakers, and solid state fuel cells are again being considered for future power plants. The proliferation of inexpensive microcomputers has stimulated the need for improved chemical sensors--a major application now being the zirconia auto exhaust sensor being sold by the millions each year. Other applications such as electro-catalysts, electrochromic displays, etc., are also being examined. Many of the new developments require thin films and structures of low dimensionality, which provides additional challenges to the technologist.

This Advanced Study Institute was formulated to synthesize concepts applicable to the characterization, interpretation, discovery, and application of fast ion conducting materials. This activity was initiated by a review of classical diffusion theory and the modifications needed when applied to fast ion conductors. Of special interest was a detailed analysis which examined diffusion--structure correlations--utilizing recent neutron diffraction data which provided detailed mappings of ion distributions amongst various sites in the crystals. A direction which will certainly be more aggressively pursued in the future, i.e., static and dynamic simulations of transport, was examined in some detail with impressive results. It appears that these calculations can now be relied upon in a predictive manner which should speed our understanding of complex systems and enable discovery of new optimized materials.

Since structure plays a key role in controlling transport, a number of key characterization techniques were described in some detail including x-ray and neutron diffraction, nuclear magnetic resonance, and raman and IR spectroscopies. New developments in these fields and their applicability to fast ion conductors were emphasized.

Initially, almost all work in this field was focused on crystalline inorganic materials. Several lectures examined in some detail those cation and anion conductors which have become the "classic" systems in this field and the extensions to newer materials. A greater effort in this Institute, however, was placed on the new emerging materials systems including polymeric and amorphous systems both of which stimulated a great deal of interest and discussion.

Although applications of fast ion conductors usually involve highly aggressive enviroments such as in batteries and fuel cells, the chemical and electrochemical stability of these materials are rarely discussed in detail. In this Institute, this subject was ivestigated in some detail from a number of standpoints considering both thermodynamic and kinetic issues and as applied to crystalline and amorphous solids. Redox stability was investigated in relation to the generation of mixed ionic and electonic conductors.

An extended description of the theory and analysis of complex impedance spectroscopy served to emphasize the necessity of using frequency dependent measurements to deconvolute the roles of bulk and interfacial effects. Unfortunately, this lecture was not made available for publication in these proceedings. Along these lines, an analysis of polycrystalline and multiphase materials was presented. Such materials allow for additional property manipulation by control of interfacial chemistry.

The relevance and utility of the above concepts and issues were then directed to applications which focused on advanced battery, fuel cell, and sensor designs. In this area, we drew broadly on the combined expertise of the lecturers and participants.

Due to the rapid expansion of the field, international symposia held on these subjects have been, by their nature. large and have not in general provided individuals entering the field with a means of developing a proper perspective. For this reason, we have, in this Advanced Research Institute, concentrated on the presentation of tutorials to provide our participants with the proper background to understand the latest developments as well as the tools necessary to forge forward on their own. We thank our invited lecturers for their substantial efforts towards achieving these goals.

Harry L. Tuller

CONTENTS

APPLICATIONS

CONTRIBUTED PAPERS

COMPUTER MODELLING OF SUPERIONICS

C.R.A. Catlow

Department of Chemistry
University of Keele
Staffs ST5 5BG, U.K.

(1)   INTRODUCTION

Computer modelling methods are now used extensively in solid state chemistry and physics; and fast ion conductors have represented a particularly successful field for application of these techniques. Both the techniques and their application have been reviewed several times in recent years[1]-[7]. In this present article we therefore concentrate on those aspects that are of particular relevance to superionics, and on a number of relatively recent applications. We hope to show that modelling methods can now be used in a predictive sense in studying superionics, and that they can yield unique information on the mechanisms responsible for fast ion conduction.

(2)   AIMS AND METHODOLOGY

Computer modelling methods have three main aims: first to predict the structures and properties of materials; secondly to describe the behaviour of defects in solids, where the term defect is used in its widest possible sense to embrace any discontinuity from the three dimensional periodicity and includes for example the surfaces of solids; thirdly modelling methods are used to elucidate the detailed mechanisms of atomic migration − a feature that is of particular importance in the study of superionic conductors.

Most calculations in the first two categories use static simulation procedures, that is methods in which no explicit account is taken of the thermal motions of the atoms in the material. The detailed study of mechanism requires, however, that these kinetic effects be included explicitly. Dynamical simulation methods will therefore be discussed later in this chapter after the static methods have been described. In every case, the simulation methods are based on effective interatomic potentials which describe the forces acting between the atoms in the solid. Potential model models thus play a central role in the field of simulations as they must contain the essential physics of the system simulated. They will be described in detail at the end of this section.

(2.1)   Perfect Lattice Simulations

In these methods we must specify a unit cell structure which is then repeated infinitely in order to generate the infinite periodic system. In modern work, the unit cell may be very large and complex containing several hundred atoms. Static calculations on perfect structures have three main components:
(i)   Lattice summations of both Coulomb and short−range terms. The former are slowly converging owing to the r^{-1} dependence of the electrostatic interactions. They must be taken to infinity and not truncated. This can be achieved by use of the Ewald procedure which transforms the Coulomb sums from a slowly converging series in real space to a rapidly converging reciprocal space summation: a good discussion is given by Tosi[8]. In contrast, the short range terms due to overlap repulsive, and covalent and dispersive

1

attractive forces can be readily handled in real space, as they may be safely truncated beyond a 'cut-off' distance normally taken as between 5 and 10Å· The sum of both types of term will give the lattice energy of the solid, E, knowledge of which is of considerable value in predicting the stability of inorganic materials.

(ii)   <u>First and Second Derivatives</u> of the lattice energy with respect to particle coordinates and with respect to the six components of the strain tensor which describe the deformation of the unit cell. Evaluation of these derivatives is straightforward (if complex) and is discussed in detail by Catlow and Norgett[9]. Having evaluated these quantities we may then calculate the <u>elastic</u> and <u>dielectric</u> properties of the material and the <u>phonon dispersion curves</u>. From the latter we may obtain the entropy at constant volume of the material by integrating over the Brillouin zone, that is by writing:

$$ S = NKT \left[\frac{\partial \ln Q}{\partial T} \right]_V + NK \ln Q \qquad (1a) $$

$$ \text{where } \ln Q = \int_0^\infty dk \sum_n \ln [1 - \exp\left[\frac{-h\nu nk}{KT} \right]] \qquad (1b) $$

where the sum is over all n vibrational frequencies.
The Helmholtz free energy, A, of the solid may then be obtained by writing:

$$ A = E - TS \qquad (2) $$

(iii)   <u>Energy Minimisation</u> in which we adjust iteratively atomic coordinates, and/or the cell dimensions until we have generated the minimum energy, or more ambitiously free energy, configuration. Calculations in which only coordinates are varied are referred to as 'constant volume' calculations in contrast to the 'constant pressure' minimisations which also vary cell-dimensions. The most efficient minimisation procedures involve the use of second derivatives of the energy function: a widely used procedure is the <u>Newton-Raphson</u> method in which the atomic coordinates (in the vector $\underline{x}_{j+1}$) in the (j+1)th iteration are related to those in the jth iteration by:

$$ \underline{x}_{j+1} = \underline{x}_j - \underline{H}_j \underline{g}_j \qquad (3) $$

where the Hessian matrix $\underline{H}_j = \underline{W}_j^{-1}$, in which the component matrix elements W_{ij} are the second derivatives $(\frac{\partial^2 E}{\partial x_i \partial x_j})$; g_i is a vector of gradients, i.e. of first derivatives of the energy function. Such methods are rapidly converging but require storage of the Hessian which for the largest most complex crystal structures may exhaust the cpu memories even of modern supercomputers. For this reason, alternative minimisation methods which require calculation and storage of only the first derivatives of the energy are used in some studies. A favoured procedure is the 'conjugate gradient' method, the memory requirements of which are modest, but at the expense of much slower convergence.

Energy minimisation is a simple but very effective technique which now has a wide range of scientific applications, including studies of protein conformation on the one hand, and the applications to inorganic materials that are our main concern in this article. The latter area is now advancing rapidly, with recent studies of highly complex solids for example zeolites[10] and ternary transition metal oxides[11]. The methods could unquestionably have a predictive role in the study of superionics. Their major limitation (in addition to these inherent to the approximations arising from the use of static lattice methods and, of course, those arising from the potential models employed) is the requirement for a reasonable initial approximation to the final minimised structure. In studying a new or hypothetical structure this may pose serious difficulties. However, we consider that this problem may be at least partially overcome by the combination of energy minimisation methods with simpler structure prediction procedures, e.g. the distance least squares (DLS) method[12] in which trial structures can be generated on the basis of existing information on bond lengths in inorganic crystals. In the opinion of the present author, such a combination of techniques offers an exciting future for structure prediction studies.

2

We should also emphasise that recent work, especially of Parker and Price[13] has demonstrated the viability of free energy minimisatioan studies. This opens up the prospect of detailed predictions of phase diagrams of inorganic materials.

Finally, we recall that the techniques embrace calculations of properties as well as structures. Again, it is clear from several studies that elastic and dielectric properties of materials can be accurately modelled. Such calculations, in addition to their intrinsic interest and predictive value, may also be used as a check on the interatomic potentials used in the modelling calculations.

(2.2) Defect Calculations

A Energies

Simulation studies of defects is now a standard and well developed field. The basis of the method is the Mott–Littleton procedure[14] using the two–region strategy summarised in fig. (1). In the inner region, atomic coordinates are adjusted (generally using the Newton Raphson method discussed above) until they are at zero force. The polarisation, $\underline{P}$, of the outer region at a point $\underline{r}$ is calculated from the relationship:

$$\underline{P}(r) = \frac{1}{4\pi} \frac{q\underline{r}}{r^3} (1 - \epsilon_o^{-1}) \tag{4}$$

where q is the effective charge of the defect and ϵ_o is the static dielectric constant of the solid. In practice an interface region IIa is used between region I and the remainder of region II which extends to infinity. We note that a modified form of equation (4) is used in studies of non cubic materials[1][15].

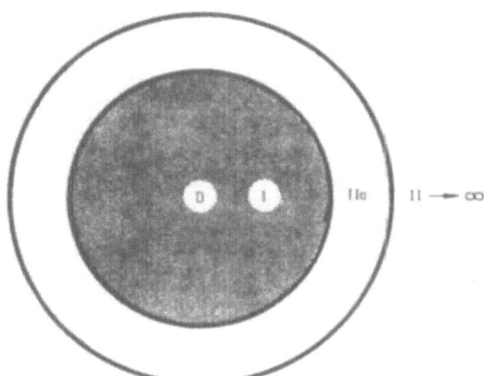

Figure 1. Two region strategy for defect calculations

The mathematical development of the theory of defect energies is described extensively elsewhere[1]. And it is now clear from several studies (see e.g. the reviews in refs (1)–(5)) that provided good interatomic potentials are available, and that provided a sufficiently large sized region I is used, accurate values can be calculated for the energies of formation, migration and interaction of defects; typical results will be presented in section (3). Regarding the size of the inner region, 100–150 ions is usually sufficient, although a larger region may be needed for non cubic materials. Particular care must, however, be taken when performing calculations of defect cluster or migration energies, which may depend on the differences between two large numbers. In these cases it is necessary to make the inner regions used in the two calculations strictly comparable; thus in the calculation of an activation energy, exactly the same region I must be used to calculate the energies of the saddle points and the ground states of the defects. Given such precautions, however, excellent results may be achieved; and the methods are now quite routine for the study of defects in polar solids.

B   Entropies

There are two main types of contribution to the entropy of a defective crystal. The first are purely configurational terms arising from the orientational or site degeneracy of the defects. In both cases, the resulting entropy can be written as a simple logarithmic expression, which in the case of the orientational degeneracy is given by:

$$S = K \log_e(N_R) \tag{5}$$

where N_R is the number of distinct orientations of the defect. In the case of site degeneracy, the resulting entropy term is included automatically in the mass action treatment of defect equilibria.

The more difficult term to evaluate is that due to vibrational effects; i.e. that arising from the perturbation of the vibrations of the surrounding lattice ions by the presence of the defects. Earlier work was based on Greens function methodologies; a good review is given by Jacobs[16]. An important landmark in the development of the theory was the publication of the paper by Gillan and Jacobs[17], which demonstrated the viability of the embedded crystallite method for calculating defect entropies. The approach is similar in some respects to that used in the defect energy calculations. The vibrational contribution to the defect entropy may be written as:

$$S_{vib} = K \ln \left\{ \frac{\prod_{i=1}^{3N'} \omega_i'}{\prod_{i=1}^{3N} \omega_i} \right\} + 3K(N'-N)\{1 - \ln(\frac{\hbar}{KT})\} \tag{6}$$

where the primes indicate perturbed vibrational frequencies; the unprimed frequencies refer to the normal perfect lattice modes. The values of the perturbed frequencies are evaluated in a region surrounding the defect and compared with the unperturbed values. Difficulties are encountered in the convergence of the calculated entropy as the crystallite is expanded; extrapolation procedures are used to overcome these problems. The method has been used in a number of studies in the last three years by Harding and coworkers[18]-[21]; and Harding has developed an efficient computer program (SHEOL) for undertaking defect entropy calculations. The results obtained to date are encouraging with good agreement being obtained between calculated and experimental entropies of Frenkel pair formation in CaF_2[18][19]. Harding's work does, however, suggest that there is rather stronger sensitivity of calculated defect entropies to interatomic potential parameters than is the case for the defect energy terms.

C   Super-cell Calculations

The methods described in the previous sections base their approach on a embedding the defect in an infinite and otherwise perfect crystal. An alternative procedure is to set up a defect supercell, i.e. to repeat the defect and a surrounding region of the lattice infinitely, and to carry out a perfect lattice calculation on the resulting structure. As the size of the supercell increases, the results should converge to those obtained by the techniques described earlier. And that this is the case has been shown by detailed studies of Allen et al[22]. Supercell methods also provide a convenient alternative approach for evaluating defect entropies[23]. Moreover, the method provides additional information to that given by the calculations on isolated defects, in that it yields energies of interaction between the defects and the defect separation may be adjusted by changing the size of the supercell. This latter feature was explored by Cormack et al[24], who used supercell methods to study the interaction between the planar defects which form in the non-stoichiometric reduced oxides, TiO_{2-x} and WO_{3-x}. It is likely that the method will be applied to a greater extent in the future.

D   Calculations of Defect Migration Parameters

In studies of superionics we are, of course, predominantly interested in the transport phenomena. It was shown thirty years ago by Vineyard[25], that the rate of atomic transport in solids could be treated by a method based on absolute rate theory and which gives the frequency of atom jumps, ν, as:

4

$$\nu = \nu_0 \exp(-g_{ACT}/KT) \qquad (7)$$

where ν_0 is a pre–exponential factor and g_{ACT} is the free energy of activation of the migration process which is normally effected by defects. Techniques of the type summarised above may be used to calculate g_{ACT} by performing calculations on the energies and entropies of the ground state and proposed saddle points for the migration mechanism. In a high symmetry structure such as that of NaCl or CaF_2, identification of the saddle point may be straightforward; it is far more difficult in the case of lower symmetry structures where an extensive search of the potential energy surface may be needed.

One point we should stress is that the use of equation (7) implies the validity of the 'hopping model' of defect transport. This model implies that the time, τ_j, associated with the jump process is such that

$$\tau_j \ll \tau_R$$

where τ_R is the residence time of the defect at a particular site. It is generally a sufficient condition for the validity of the hopping model that:

$$E_{ACT} \gg KT$$

which holds in the vast majority of ionic conductors, but not necessarily in superionics. In the case where the conditions fails, recourse must be had to other simulation techniques, in particular, these based on the molecular dynamics technique discussed in section (2.3).

E   Constant Volume vs. Constant Pressure Parameters

In all the techniques discussed above we have assumed implicity that the defect parameters are being calculated at constant volume or, more strictly, constant lattice parameter. Of course experimentally, constant pressure parameters are measured. The relationship between the two can, however, be straight forwardly[26] written as:

$$h_p = u_v + \frac{T\beta}{\kappa_T} V_p \qquad 8(a)$$

and

$$s_p = s_v + \frac{\beta V_p}{\kappa_T} \qquad 8(b)$$

where the subscripts p and v refer to constant pressure and volume quantities respectively and where β is the expansivity of the solid and κ_T the isothermal compressibility. V_p is the volume of formation of the defect, defined as

$$V_p = (\frac{\partial g_p}{\partial p})_T = -\kappa_T V (\frac{\partial g_p}{\partial V})_T \qquad (9)$$

where V is the unit cell volume
Substitution of equation (9) into equations (8) eliminates the dependence on κ_T. Moreover, the volume derivative of g may be readily evaluated by performing calculations as a function of lattice parameter. We should also note in this context the relationship derived by Gillan[27], who showed that:

$$g_p = f_u \qquad (10)$$

where f_u is the Helmholtz constant volume free energy of defect formation.

Several points should be noted about the above relationships. First, h_p and s_p are clearly temperature dependent. This arises implicitly through the dependence of the defect parameter on the temperature dependent lattice parameter and explicitly as in equation (8a). However, experimental analyses are generally made assuming temperature independent parameters. Harding[26] has shown that the enthalpy deduced from such analyses is most likely to correspond to the calculated value of u_v at OK. This approximate relationship which arises from cancellation of terms does not hold for entropy terms. A second point which has emerged from work of Harding and coworkers[18]–[21] and Mackrodt's group[22], is that it is important in studying high temperatures defect properties for the interatomic potentials correctly to reproduce the lattice expansion of the crystal. Failure to ensure this may result in misleading results. A good case study of the temperature dependence of defect parameters is provided by the silver halides; a recent study is reported in reference (21) and earlier work is presented in reference (28).

5

(2.3)   Dynamical Simulation Studies

We argued above that, for superionic systems, static lattice methods may become inappropriate.   And for systems with low activation energies or possibly at very high temperatures, the molecular dynamics (MD) technique becomes more suitable.   The basis of this method is simple in concept, and the essential features of the simulations are as follows:

(i)   An ensemble of particles in a 'simulation box' is specified and periodic boundary conditions are applied;   thus in standard applications an infinite periodic system is simulated and surface effects are not included.   The ensemble may contain anything from 100 to several thousand particles with the computer time increasingly rapidly with the number of ions of simulated ions.   In applications to crystalline solids, the 'simulation box' will normally be a super-cell of the basic unit cell.

(ii)   All particles are assigned positions and velocities.   In the case of crystalline solids the former will normally be close to the crystallographically determined coordinates (if available).   The latter are assigned in accordance with the target temperature.

(iii)   The simulation proceeds by a succession of time steps, Δt, where Δt is typically $10^{-14}-10^{-15}$ sec, i.e. much shorter than the period of an atomic vibration.   After each time step the positions and velocities of the ions are updated using the classical equations of motion.   This in the limit of infinitely small, Δt, we could write

$$x_i(t+\Delta t) \;=\; x_i(t) + v(t)\Delta t \tag{11a}$$

$$v_i(t+\Delta t) \;=\; v_i(t) + \frac{f_i}{m_i}\Delta t \tag{11b}$$

where f_i is the force acting on the ith particle which is of mass m_i;   the forces must be evaluated from knowledge of the interatomic potentials.   For a finite time step, the use of the above equations, will lead to errors due to the variation of f and v during the period Δt.   As a result, more sophisticated updating algorithms must be employed involving higher powers of Δt.   An example is the Beeman[29] alogorithm which updates x and v as follows:

$$x_i(t + \Delta t) = x_i(t) + v_i(t)\Delta t + \frac{\Delta t}{6m_i} \{4f_i(t) - f_i(t-\Delta t)\} \tag{12a}$$

$$v_i(t + \Delta t) = v_i(t) + \frac{\Delta t}{6m_i} \{2f_i(t + \Delta t) + 5f_i(t) - f_i(t-\Delta t) \tag{12b}$$

For other algorithms we refer to Sangster and Dixon[30] and to Allen and Tildesley[31].

(iv)   In the initial stages of the simulation the system is 'equilibrated', i.e. it is allowed to evolve until equipartition has been achieved between potential and kinetic energy, and there is a Maxwellian distribution of velocities.   During this stage of the simulation, which may take several thousand time steps, it is generally necessary to scale the velocities of the ions regularly in order to counteract the drift in the temperature of the simulation.   A good indication that equilibrium has been attained is the observation that the temperature remains constant within the statistical limits owing to the use of a finite number of particles.

(v)   After equilibration has been achieved, a production run follows, in which the simulation is run for several thousand time steps, corresponding to 5–100 psec of 'real time'.   The coordinates and velocities of the successive time steps are stored on tape or disc.   They contain a detailed record of the time evolution of the simulation, analysis of which can yield the following information.

a) Structural properties via radial distribution functions $g_{\alpha\beta}(r)$, (i.e. the probability that if we have a particle of type α at position O we have a second particle within a given volume element at a distance r away).   Fig (2) illustrates calculated r.d.fs for Li...Li pairs in superionic Li_3N.

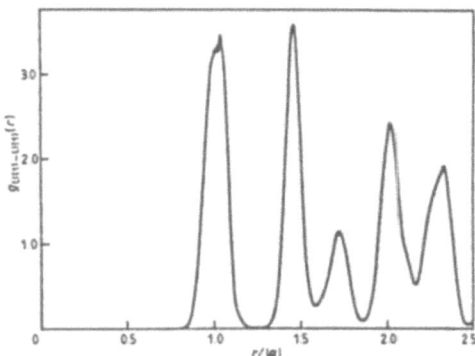

Figure 2. Li(1) - Li(1) Radical distribution function. 300 K.

b) <u>Diffusion coefficients</u>, D_α, which may be obtained from the Einstein like relationship for the dependence of the root mean square displacement $<r_\alpha^2>$ of particle α on time t:

$$<r_\alpha^2> = 6D_\alpha t + B_\alpha \tag{13}$$

where B_α represents the mean square amplitude of vibrational, i.e. non–diffusive motion and can be related to crystallographic temperature factors. Fig (3) illustrates data for the case of F^- migration in superionic CaF_2 obtained in a simulation study of Dixon and Gillan[32]. The linear increase of $<r^2>$ with t is a clear indication of diffusive motion.

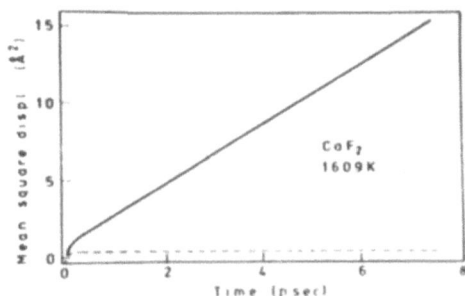

Figure 3. Plot of mean square displacement vs. time for F^- (full–line) and Ca^{2+} (dotted line) in CaF_2.

c) <u>Dynamical correlation functions</u> including the velocity autocorrelation function (see reference 31) and the van Hove correlation function G(r,t). The latter is of particular importance; it is defined as the probability that if we have a particle at r=0 at t=0, that we have a second particle at r at time t. The van Hove self correlation function $G_s(r,t)$ indicates that the particle must be the same in both cases. As we shall see in the chapter on X–ray and neutron scattering, the space/time Fourier transform of this function is proportional to the neutron scattering cross section of the system.

d) <u>Ion migration mechanisms</u> which can be elucidated and displayed by examining successive timesteps in the configuration. Several examples will be given in section (3).

It should be apparent from the above discussion that the MD technique is one of very considerable power in the study of the dynamics of condensed matter. Its range of applications are, however, limited for the following reasons:

(i)   The short time scale of the simulations. It is rare for the 'real time' of the simulation to be greater than 100 psec. If it is desired to use the simulation to model ion transport it is necessary that a sufficient number of particle jumps occur during this period. In effect this requires that the material be a superionic conductor.

(ii)   Interpretation of the results in term of defect models may be difficult; and if defect energies are required it is necessary to run the simulations at several temperatures.

(iii)   The use of periodic boundary conditions removes surfaces from the simulation. Defects in Schottky disordered crystals cannot therefore be generated naturally; and in such systems the defects must be introduced into the simulation.

(iv)   The simulations are computationally expensive requiring often several hours of c.p.u. time even on a modern supercomputer. Moreover, the computational requirements can increase by up to an order of magnitude if the interatomic potential models are extended to include ionic polarisation (see below).

In practice, static simulation methods of the types described in sections 2A and B are by far the most effective in describing 'normal', 'non-superionic' systems and in obtaining simple information on defect parameters even in superionics. M.D. methods do, however, yield a wealth of detailed dynamical information which is of great value in understanding superionics.

(2.4)   Monte-Carlo Methods

Application of these techniques is ideal for systems in which there are several different types of jump mechanism; and indeed the most extensive applications have been made to alloys systems by Murch and coworkers[33][34]. They are also well suited to the study of certain classes of superionic, especially heavily disordered systems. The methods which are essentially a computational statistical sampling approach, proceed like MD by a succession of steps or 'moves'; but in the case of MC techniques there is no time dimension and the successive moves do not represent a time sequence. The simulations proceed as follows:

(i)   As with MD a simulation box is set up; several thousand particles are usually included, and periodic boundary conditions are applied.

(ii)   Each 'move', in the simulation involves selecting a defect at random, and a jump direction also at random. The type of jump is then checked against a 'look-up table' and either its frequency or its activation energy is noted. In the latter case the frequency is then taken as proportional to the Boltzmann factor:

$$\nu_i \ \alpha \ \exp(-E_{ACT}/KT) \tag{14}$$

and if entropies are known g_{ACT} could be substituted for E_{ACT}. We note that in most applications, normalised frequencies are used, i.e. the values are scaled so that the maximum is unity. Only relative rather than absolute values of the diffusion coefficient are therefore calculated.

(iii)   A random number R in the range 0-1 is generated. If $\nu_i > R$ the jump proceeds; if $\nu_i < R$ the jump fails. In either case the simulation proceeds by selecting another defect at random.

(iv)   After an initial equilibrium period involving several thousand moves, the production run follows. Several thousand more moves are generated and stored for subsequent analysis.

(v)   In studies of electrical conductivity, a field, E is applied across the simulation box.

Analysis of the results yield diffusion coefficients via the Einstein-Stokes equation: while conductivities may be calculated straight forwardly by following the drift of charge under the action of the applied field. More subtle information may also be obtained. For example, Murch has defined the 'conductivity correlation factor', f_I, as:

$$f_i \ = \ \frac{2KT<X>}{nq_i a^2 E} \tag{15}$$

where $<X>$ is the mean drift of ions of change q_i under the action of the field E, effected by n jumps of length a. The coefficient represents the efficiency of the defect jump in effecting bulk conductivity. We will find this quantity of value in our discussion of applications in section (3).

8

(2.5)   Interatomic Potentials

The development of potentials for polar solids, which underwrites the whole field of simulation studies discussed in this article, has proceeded rapidly in the last ten years. There are now useful potential models for a wide range of halide and oxide materials; and recent extensions of the formalism used in describing short range potentials, allow semi-covalent materials to be modelled. Most of the potential models reported in the literature have the following main features:

(i)   The use of charged ionic species. Fully ionic, i.e. integral charges, are commonly but not necessarily employed.

(ii)   The description of the short-range interactions that come into play when atomic change clouds overlap in terms of simple analytical pair potentials. A favoured form is the Buckingham potential:

$$V(r) = Ae^{-r/\rho} - Cr^{-6} \tag{14}$$

where the attractive r^{-6} term represents the effects of dispersive forces or of other weakly attractive terms. Recently[35][36], we have extended our description of short-range interaction to include many-body terms modelled by a bond-bending formalism in which we add to the two-body functions of the type described above, a term $E(\theta)$, where

$$E(\theta) = \frac{1}{2}K_B(\theta-\theta_o)^2 \tag{17}$$

which is applied to selected bond angles whose equilibrium bond angle is θ_o. K_B is a force constant. The inclusion of such terms has proved to be of particular value in modelling silicates, where the bond-bending terms are applied to O ^{Si} O bond angles and where θ_o is the tetrahedral angle. Other many-body terms, for example the 'triple-dipole' model[37] are currently being investigated for other materials.

(iii)   In defect calculations it is essential to include a representation of ionic polarisation which is generally described using the shell model[38] — a simple mechanical model of a polarisable ion, which describes polarisation in terms of the displacement of a massless shell representing the polarisable valence shell electrons from a core in which the mass of the ion is concentrated and which simulates the nucleus and core electrons. Core and shell are connected by an harmonic spring. Despite its crude nature, the shell model has been successful in describing lattice dynamical and properties of solids. And the importance of including a description of polarisability in defect calculations should be stressed. Defects in polar solids polarise the surrounding lattice ions. If a reliable defect energy is to be calculated this polarisation must be correctly modelled.

(iv)   The potential models must be parameterised. Thus for example the parameters A, ρ and C must be fixed in the description of the short-range interactions; and the shell model includes two variable parameters — the shell charge and the spring constant. Two strategies are available for determining the variable parameters: first theoretical methods may be used: interatomic potentials may be obtained by calculating the energies of pairs or clusters of ions as a function of the interatomic spacing. The resulting energy surface may then be fitted to potential functions. A variety of theoretical methods may be used, including simpler electron gas methods of the type developed by Wedepohl[39] and Gordon & Kim[40], and which have been extensively applied to ionic crystals by Mackrodt and coworkers[41][42]. Increased use is being made of ab-initio Hartree-Fock methods which promise a higher degree of accuracy. Again Mackrodt et al[43] have made made important contributions in this field.

The alternative parameterisation procedure is to 'fit' the variable parameters to the known properties of model compounds. The method is the only one currently available for determining shell model parameters, and has been widely used by the present author and coworkers. It is, however, inherently limited by the need for suitable model compounds, and by the inevitable fact that empirical fitting procedures yield information only about potentials at interatomic spacings close to those in the perfect lattice; extrapolation of the resulting potentials to those spacings observed in the defective lattice are of questionable validity, although we note that empirical potentials have performed well in defect studies.

In concluding this section we attempt to present a necessarily subjective summary of the present states of interatomic potentials for a range of polar solids. Our classification presented in table (1) groups materials into these for which potentials are 'well developed', those which are 'good' and those described as 'usable'. The first category have been shown to yield accurate defect energies (and in some cases entropies); the available potentials also yield accurate physical properties for the perfect solids. The second category are those for which reliable perfect and defect lattice properties can be calculated but for which it is known that the interatomic potentials involve simplifications. In the third category, there are known be inadequacies in the models, which are nevertheless useful when used with caution.

TABLE 1

Status of interatomic potentials for some ionic and semi-ionic solids

Classification	Materials
1. Reliable and well developed	NaCl + alkali halides; MgO + alkaline earth oxides; CaF_2 + alkaline earth fluorides
2. Good	NiO + low valence binary transition metal oxides; MgF_2, MnF_2, LaF_3; $BaTiO_3$; UO_2 + fluorite oxides.
3. Usable	SiO_2 + silicates; TiO_2 WO_3, Nb_2O_5, AgCl, ZnSe

It is clear that future work in the simulation field must concentrate on improving the quality and reliability of the interatomic potential functions. The role of theoretical methods will be crucial in this endeavour.

(3)   APPLICATIONS

(3.1)   Static Lattice Calculations

We recall that static methods may be applied to both perfect and defective structures. In the former area the most interesting applications have concerned silicates and aluminosilicates. Successful simulation studies have been reported on a range of structures[44], including most recently the complex porous structures adopted by zeolites[45]. There is no doubt that the techniques can be used to model the structures and properties of solid electrolytes; and it should be noted that currently available potential models successfully model the structures of e.g. β-Al_2O_3[46], and that useful applications of static lattice techniques have been made to the complex structural problems posed by the δ phase of Bi_2O_3[47].

Most applications of static lattice techniques have, however concerned defect properties. We have reviewed the field previously[3] and wish here to emphasise three points. The first is the possibility of accurate and routine calculations of defect energies using the techniques described in section (2.1A). Table (2) summarises calculated and experimental defect formation and migration energies for a number of simple ionic compounds; in general the agreement is excellent. Secondly, we note the useful insight yielded by the calculations on defect parameters, as an example of which we take the work of Butler et al[48] which has recently been refined by Murray[49]. This work examined the interesting problem posed by the oxygen ion conductor CeO_2 doped with a range of rare-earth oxides. The system had been investigated experimentally by Gerhardt-Anderson and Nowick[50], and the basic defect structure is well understood: the substitutional rare earth ions are compensated by mobile oxygen ion vacancies in the fluorite structured host; and at low dopant concentrations, with which we are concerned here, the dopants and vacancies form the simple pair clusters illustrated in fig (4). Analysis of the conductivity data suggested, moreover, that the binding energies were strongly sensitive to the radius of

10

the dopant ion – a result which was initially surprising as it had generally been considered that the binding energies were largely electrostatic in origin. As shown in fig (5) calculations[48,49], largely reproduced the experimental data, and that showed that a large component of the binding energy was due to elastic effects arising from the mismatch between the dopant and host ionic radii, hence the largest energy is obtained for Sc^{3+}

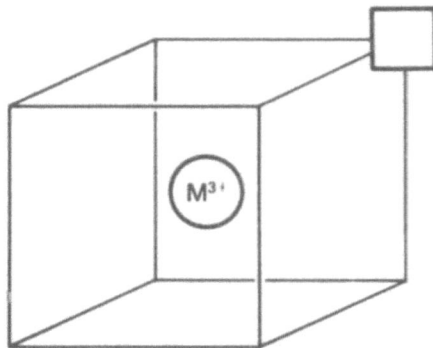

Figure 4.   Cluster comprising one trivalent cation substitutional and one anion vacancy in doped fluorite oxides.

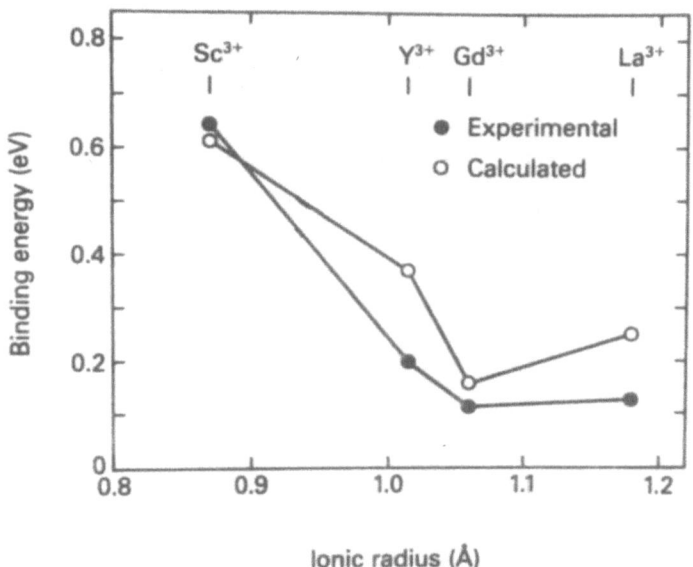

Ionic radius (Å)

Figure 5.   Calculated and experimental binding energies for the cluster shown in Fig. 4, for a variety of dopants.

which has the smallest radius and the greatest mismatch. The simulation studies therefore confirmed the interpretation of the experimental data and provided useful insight into the nature of the factors controlling the unusual behaviour of these systems that had been revealed by experiment.

TABLE 2

Defect energy calculations

Crystal	Process	Calculated Energy* (eV)
NaCl	Schottky pair formation	2.4-2.7 (2.3-2.7)
	Cation vacancy migration	0.66 (0.7-0.8)
CaF_2	Anion Frenkel pair formation	2.6-2.7 (2.6-2.7)
	Anion interstitial activation	0.91 (0.9-1.0)
MgO	Scottky pair formation	7.5-7.7 (5-7)
	Cation vacancy activation	1.8-2.2 (2.0-2.3)
NiO	Schottky pair formation	6-7
	Cation vacancy activation	1.86 (1.5)

*Experimental values are given in brackets. Detailed discussions and references are given in ref (5).

Our third type of application relates to guidance and to complex defect models. A good illustration is provided by the high temperature superionic fluorites. As discussed elsewhere in this book, all fluorite structured compounds show a transition to a superionic state within a few hundred degrees of their melting point. The transition to the superionic state is known to be accompanied by the generation of anion disorder. One model suggested from neutron scattering studies[51] has proposed the formation of interstitial clusters of the type shown in fig. (6). In these clusters pairs of interstitials are stabilised

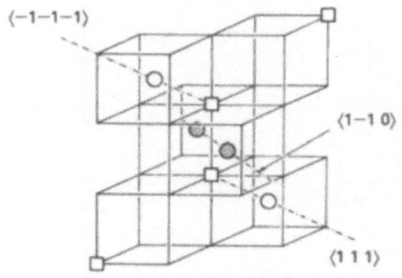

Figure 6. Proposed interstitial-vacancy cluster in high-temperature fluorite.

by relaxations of neighbouring lattice ions; the vacancies corresponding to the interstitials are located in the vicinity of the interstitial cluster. Static defect calculations found that such clusters were stable and strongly bound, thereby supporting the interpretation of the experimental data given by Hutchings and coworkers[51]. We should note that this interpretation remains controversial; but the value of the static defect calculations in providing guidance in interpreting complex defect structures is demonstrated.

Static defect calculations have now been extensively applied to superionics; and other system examined recently include β-Al_2O_3[46], Bi_2O_3[47], LaF_3[53], $RbBiF_4$[54], stabilised zirconia[55][56] and PbF_2[57]. The techniques are now quite standard and generally applicable to solid state electrolytes.

(3.2)   Molecular Dynamics Simulations

MD techniques have been used in studying several superionics. Early work of Vashista and Raman[58] investigated AgI; the superionic properties of this material were simulated effectively using simple, ionic model pair potentials. Moreover, structural properties were in good agreement with the experimental, crystallographic data on the superionic phase of this material (see the chapter of Wuensch in this book for further details). Vashishta and coworkers[59] have also reported useful MD studies on the silver chalcogenides. Again, these studies use relatively simple rigid ion potentials with which, perhaps surprisingly, they are able to describe the superionic properties of these materials.

One of the most detailed and comprehensive MD simulations of superionic behaviour was that undertaken by Dixon and Gillan[32][60] on the high temperature fluorite structured superionics. The work again employed rigid ion potentials which had, however, been reparameterized in order to ensure that the experimental Frenkel energies were reproduced. The studies which have been reviewed and expanded by Gillan[6][7] first had important structural consequences: they showed that in the superionic phase interstitials do not occupy the body-centre position of the cubic interstitial site. This is in line with the neutron diffraction studies[51]. They also verified that transport takes place by a hopping mechanism – predominantly of vacancies. In addition, recently Gillan[61][62] has shown that by calculating the dynamical structure factor $S(Q,\omega)$ it is possible to reproduce the observed diffuse quasi-elastic scattering measured by Hutchings and coworkers[51] on the high temperature fluorites, although the interpretation of these data in terms of defect models remains controversial.

These and other studies therefore emphasise that MD techniques yield valuable information on structural and dynamic properties of superionics, and that the predictions can be directly tested and verified experimentally. In the remainder of this section we concentrate on more qualitative applications to the elucidation of mechanism where straight forward applications of the techniques could give unique insight. We will consider two particular systems that we have recently studied: first the superionic lithium conductor Li_3N and secondly the mixed cation fluoride ion conductor $RbBiF_4$.

A   Lithium Nitride

Li_3N is one of the best Li^+ ion conductors. It has a surprising and unique crystal structure which is shown in fig (7). Hexagonal nets of lithium ions are arranged in layers of composition Li_2N. The layers are connected by the remaining 'bridging' lithium ions. As would be expected the conductivity is anisotropic being greater parallel than perpendicular to the layers[63]. A detailed simulation study of Li_3N was reported by Wolf et al[64][65]. The work successfully reproduced the superionic properties of the material and established the following qualitative features concerning structure and mechanism in the material.

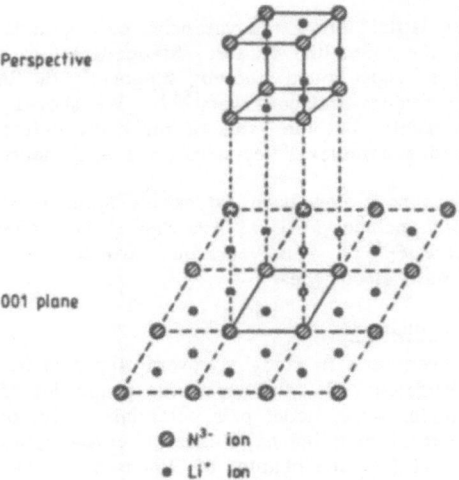

Perspective

001 plane

⊘ N³⁻ ion
• Li⁺ ion

Figure 8. Schematic plan of a migration event involving one Li(1) ion and five Li(2) ions.

    (i)   The material is essentially a normal ordered crystal structure with relatively low levels of thermally generated disorder. This contrasts with other intrinsic superionics e.g. the high temperature fluorites and AgI, where there is known to be extensive disordering of one sublattice.

    (ii)   It does nevertheless seem to be relatively easy to excite Li^+ ions from the Li_2N layers into the interlayer region. The simulations suggested than at ~100°C, ~1% of the Li^+ ion had been excited from the layer sites.

    (iii)   The resulting creation of vacancies on the layers allows highly correlated Li^+ migration mechanisms to occur: a typical example is shown in fig (8), where six lithium ions move in a concerted process. A whole range of related mechanisms were discovered by Wolf in a detailed examination of the results of the simulation study. The operation of such mechanisms, together with the relative ease of vacancy creation, explains why the material is such an effective superionic.

    The MD studies also revealed the mechanisms for ion transport perpendicular to the Li_2N layers. Typical mechanisms are shown in fig (9) which shows the Z coordinates of $4Li^+$ ions as a function of time. We note that one ion   oscillates in the inter−layer region; after ~2 psec it displaces the bridging Li^+ ion which rapidly reoccupies a site in one of the Li_2N layers. This process, as well as effecting transport in the Z direction also causes exchange of layer and bridging Li^+ ions − a process which had been shown to occur by NMR studies. Once again the simulations show that the key to understanding the superionic properties of Li_3N are concerted mechanisms involving several lithium ions. The same theme will emerge in our study of the F^- ion conductor discussed next.

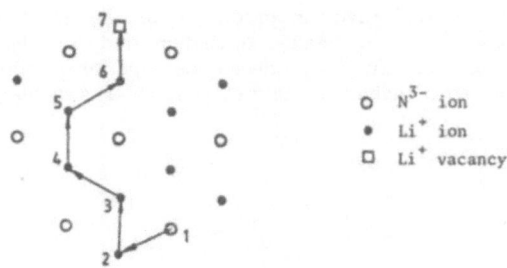

○ N³⁻ ion
• Li⁺ ion
□ Li⁺ vacancy

Figure 7. The structure of Li_3N.

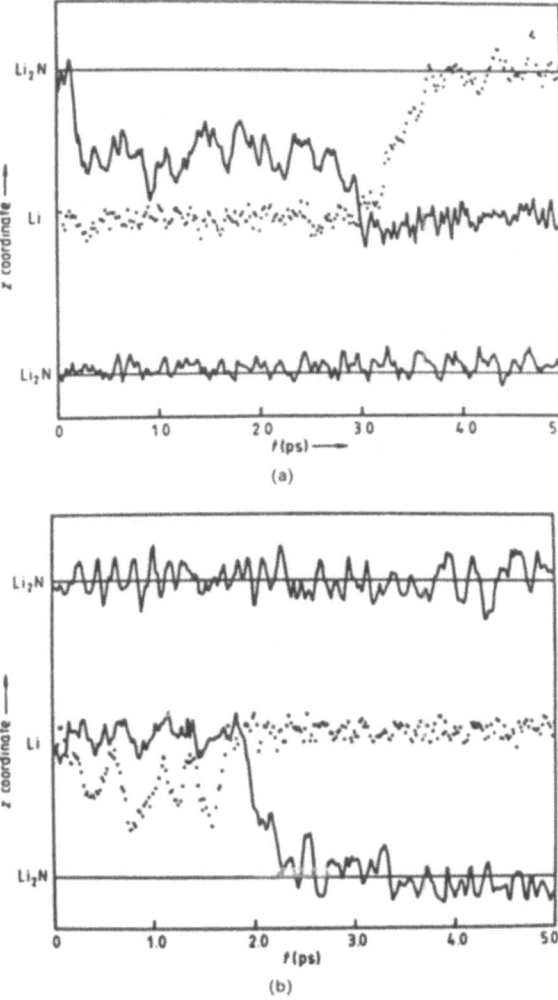

Figure 9. Migration events parallel to the c axis in Li_3N. (a) A Li^+ ion in the Li_2N layer moves into the intergap region and subsequently displaces a bridging Li^+ ion (dotted line), which then enters the Li_2N layer. (b) A Li^+ "interstitial" in the intergap region (dotted line) initially oscillates and then displaces a bridging Li^+ ion. Both graphs (a) and (b) show the z-coodinate of three ions plotted against time at 400K.

B   RbBiF$_4$

Several studies of Reau, Matar and coworkers[66][67] have established that this cation disordered, fluorite structured material is a good F^- ion conductor – considerably better than for example PbF_2. Static lattice calculations of Cox[54][68] have suggested that one of the key features in providing superionic behaviour in the material is the relative ease of exciting F^- ions into interstitial sites from lattice positions surrounding which there is a local excess of Rb^+; experimental evidence for this proposal has been obtained from EXAFS studies[68]. As a result, the system contains a high concentration of interstitials and unlike most fluuorite structured systems, the interstitials appear to be the more mobile species. M.D. studies have confirmed the essential features of the static simulations and have, moreover, shown that the interstitial migration mechanisms once more involve a highly correlated process.

If we examine individual F^- ions we find that they move regularly between lattice and interstitial positions. A typical example is shown in fig (10): an ion is seen to oscillate around an interstitial position then to move and occupy a regular lattice site, before moving out again into an intersitital configuration. Moreover, the movement of ions between lattice and intersitital sites is a highly correlated process as shown in fig (11) which displays the trajectories of 3 ions one of which originally occupies an interstitial site. We note that the three ions move together effecting exchange of F^- between lattice and interstitial positions. Cox[69] has shown that even more complex concerted processes can occur involving at least six F^- ions. And correlated motion is again seen as the key to understanding superionicity in this material.

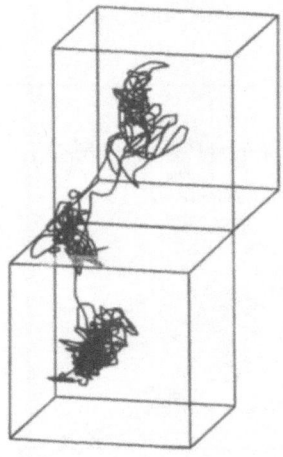

Figure 10. Trajectory for F^- ion in $RbBiF_4$.

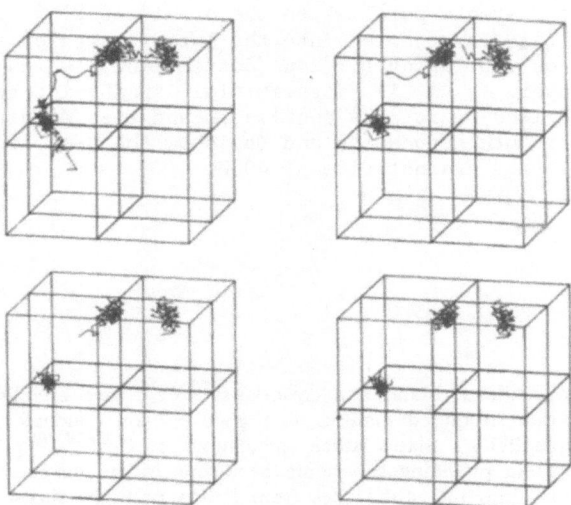

Figure 11. Correlated motion of three migrating F^- ions in $RbBiF_4$.

Many other examples and studies could be used to illustrate our themes of the value of MD calculation in providing insight concerning ion migration mechanisms. The growth of computer power is certain to lead to increased use of these techniques in solid state chemistry.

(3.3)   Monte–Carlo Methods

The application of this powerful technique to solid electrolytes has to date been limited. A useful study is reported by Murch et al[70], which reveals the potential of the technique. Murch et al were concerned with the problem posed by the fluorite structured oxygen ion conductors CeO_2/Y_2O_3 which has already been discussed in section (3.1). However, in contrast to the previous discussion which was concerned with dilute systems, the present study was concerned with the transport properties of these materials at high dopant concentrations. Work of Wang et al[71] had established the remarkable fact that the variation of conductivity with dopant concentration exhibits a maximum at a vacancy concentration of ~5%. The data, which are illustrated in fig (12) show that the conductivity decreases dramatically, by several orders of magnitude after the maximum.

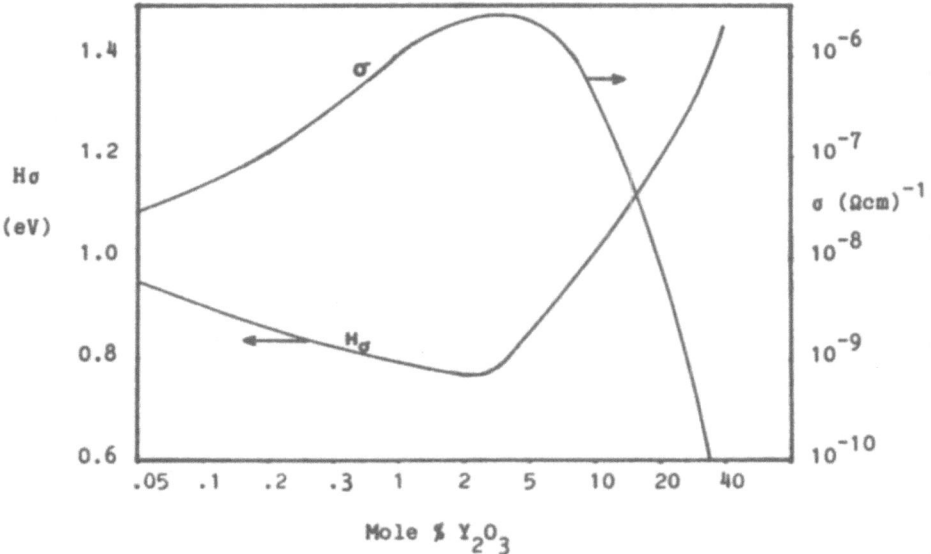

Figure 12.   Variation in Conductivity and Arrhenius Energy in Y_2O_3 Doped CeO_2 (after Wang et al. (71).

Murch et al[70] attempted to model this system by the following procedure.

(i)   They set up a large M.C. simulation box into which dopants and vacancies were distributed at random.

(ii)   A preliminary set of static defect calculations were performed which examined the migration energies for vacancy jumps for every possible distribution of dopant ions in nearest–neighbour cation sites around a pair of anion sites.

(iii)   In each move in the M.C. procedure they examined the dopant environment which was matched with one of those used in static defect calculations. Normalised frequencies were calculated as described in section (2.4).

(iv)   The simulation proceeded as summarised in section (2.4). An initial equilibration period was followed by a production run. The simulations were performed at two temperatures, and the results in terms of conductivities, Arrhenius energies and conductivity correlation factors, f_I, are summarised in figs. (13)–(15).

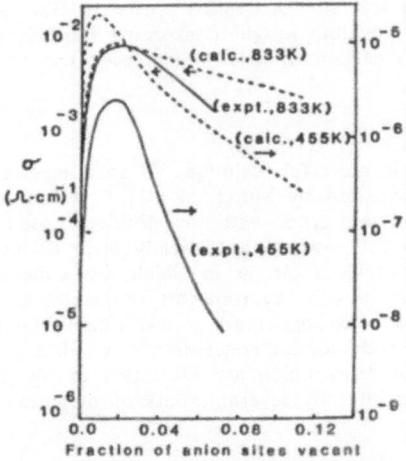

Figure 13.   Calculated and Experimental Conductivities (after ref 70)

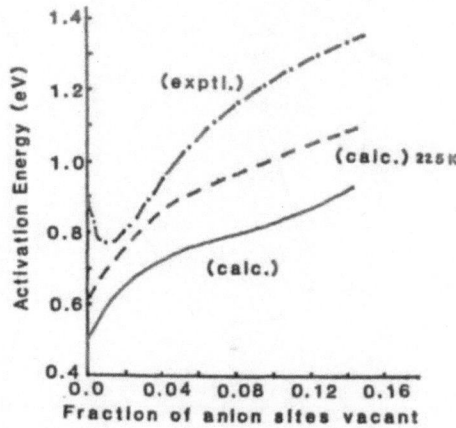

Figure 14.   Calculated and Experimental Arrhenius Energies

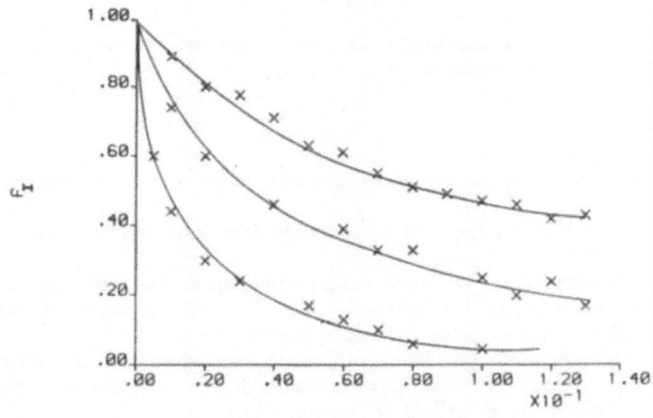

Fraction of anion sites vacant

Figuer 15.   Calculated Variation of f_I with Vacancy Concentration for 3
temperatures: 1200 K; 833 K; and 455 K (lowest curve).

The results are satisfactory at the qualitative level in that they reproduce the observed maximum in the conductivity. The calculated behaviour for the Arrhenius energy is less satisfactory, as although the increase in the energy at high dopant concentration is reproduced; the initial decrease at low x is not found in the simulations. Perhaps the most interesting feature of the calculations is the rapid decrease in f_I at higher values of x. This suggests a possible explanation of the remarkable experimental results. The conductivity decreases with dopant concentration at high dopant level because the vacancy jumps become decreasingly effective in causing bulk conductivity; the vacancies increasingly perform 'ineffective' jumps in which they oscillate between sites or jump around a dopant or group of dopant ions.

We also note that Murch et al[70] performed a study of an ordered dopant system. The results demonstrated a dramatic reduction in the conductivity indicating that ageing which may lead to dopant ordering might be expected to have pronounced deleterious effects on the ionic conductivity of dopant fluorite oxides.

Finally, we should point out that the behaviour of rare earth doped CeO_2 is not unique. Stabilised zirconias show similar variations of conductivity with dopant concentration; the application of MC methods to these systems should be encouraged.

(4)   SUMMARY AND CONCLUSIONS

We hope to have shown in this chapter that modelling methods are now a major technique in the study of solid state ionics. The methods are well established and with growth in super–computer power their range and scope will expand. Progress in the field will be controlled first by developments in technique, of which the most notable will be the use of Langevin dynamics which allows the simulations to focus on the properties of the mobile ions, and quantum dynamics which enable quantum effects of, for example, protons to be described in terms of 'necklaces' of classical particles. Secondly, our ability to develop increasingly accurate and realistic interatomic potentials will control the extent to which accurate simulations can be performed on new types of material.

One area that has not been considered in this chapter concerns the simulations of amorphous materials. Work of e.g. Soules[72] has shown that by performing M.D. simulations on melts followed by a simulated quench it is possible to obtain useful simulation of glass structures. In view of the importance of amorphous systems as ionic conductors simulation studies of these systems should clearly be encouraged.

ACKNOWLEDGEMENTS

I would like to thank Paul Cox for permission to quote his unpublished work. Much of the work described in this chapter has been supported by grants from the Science and Engineering Research Council and from the Theoretical Physics Division, AERE, Harwell.

REFERENCES

(1)   C.R.A. Catlow and W.C. Mackrodt (eds) 'Computer Simulation Solids', Lecture Notes in Physics vol 166 (Springer, Berlin), 1982.
(2)   W.C. Mackrodt, in 'Transport in Non–Stoichiometric Compoundds' (eds G. Petot Ervas, Hj. Matzke, C Monty), North Holland, Amsterdam, 1984.
(3)   C.R.A. Catlow, Solid State Ionics, 8, 89 (1983).
(4)   C.R.A. Catlow, Ann. Rev. Mat. Sci. 16, 517 (1986).
(5)   C.R.A. Catlow in 'Defect Physics : Modern Techniques' (eds A.V. Chadwick and M. Terenzi) (Plenum) (1986).
(6)   M.J. Gillan, in 'Solid State Ionics '83' (eds M. Kleitz, B. Sapoval and D. Ravaire), North–Holland, Amsterdam, 1983.
(7)   M.J. Gillan, Physics B131, 157 (1985).
(8)   M. Tosi in Solid State Physics (eds F. Seitz and S. Turnbull) p 161.

(9)   C.R.A. Catlow and M.J. Norgett, UKAEA Report – AERE-M 2936 (1976).

(10)   M.J. Sanders, C.R.A. Catlow and J.V. Smith, J. Phys. Chem. **88**, 2796 (1984).

(11)   G.V. Lewis and C.R.A. Catlow, J. Phys. Chem. Solids **47**, 896 (1986).

(12)   W.M. Meier and M. Villiger, Z. Kristrogr. **129**, 411 (1969).

(13)   S.C. Parker and G.D. Price, Ann. Rev. Solid State Chem. **1** – in press.

(14)   N.F. Mott and M.J. Littleton, Trans. Farad. Soc. **34**, 485 (1938).

(15)   C.R.A. Catlow, R. James, W.C. Mackrodt and R.F. Stewart, Phys. Rev. **B25**, 1006 (1982).

(16)   P.W.M. Jacobs in 'Computer Simulation of Solids', Lecture Notes in Physics vol **166** (eds C.R.A. Catlow and W.C. Mackrodt), Springer, Berlin, (1982).

(17)   M.J. Gillan and P.W.M. Jacobs, Phys. Rev. **B28**, 759 (1983).

(18)   J.H. Harding, Phys. Rev. B. **32**, 6861 (1985).

(19)   J.H. Harding, Physica **B131**, 13 (1985).

(20)   R.A. Jackson, A.D. Murray, J.H. Harding and C.R.A. Catlow, Phil. Mag. **53**, 27 (1985).

(21)   C.R.A. Catlow, J. Corish, J.H. Harding and P.W.M. Jacobs, Phil. Mag. **A55**, 481 (1987).

(22)   N. Allen, W.C. Mackrodt and M. Leslie, in Adv. in Ceramics vol. 23 (eds C.R.A. Catlow and W.C. Mackrodt), 1987.

(23)   J.H. Harding and A.M. Stoneham, Phil. Mag., **B43**, 705 (1981).

(24)   A.N. Cormack, P.W. Tasker, R. Jones and C.R.A. Catlow, J. Solid State Chem., **44**, 174 (1982).

(25)   G. Vineyard, J. Phys. Chem. Solids, **3**, 157 (1957).

(26)   J.H. Harding in 'Defect Physics : Modern Techniques', (eds A.V. Chadwick and M. G. Terenzi)(Plenum)(1986).

(27)   M.J. Gillan, Phil. Mag. **A43**, 301 (1981).

(28)   C.R.A. Catlow, J. Corish and P.W.M. Jacobs, J. Phys. C. **72**, 3433 (1979).

(29)   D. Beeman, J. Comput. Phys. **20**, 130 (1976).

(30)   M.J. Sangster and M. Dixon, Adv. Phys., **25**, 247 (1976).

(31)   M.J. Allen and D. Tildesley, 'Computer Simulation of Liquids' Oxford University Press (1987).

(32)   M.J. Gillan and M. Dixon, J. Phys. C. **13**, 1901 (1980).

(33)   G.E. Murch, in 'Diffusion in Crystalline Solids (eds G.E. Murch and A.S. Nowick) Academic Press, New York (1984).

(34)   G.E. Murch, Phil. Mag. **A46**, 575 (1984).

(35)   M.J. Sanders, M. Leslie and C.R.A. Catlow, J. Chem. Soc. Chem. Commun., 1271 (1984).

(36)   C.R.A. Catlow, C.M. Freeman and R.L. Royle, Physica **B131**, 1 (1985).

(37)   B.M. Axilrod and E. Teller, J. Chem. Phys. **11**, 299 (1943).

(38)   B.G. Dick and A.W. Overhauser, Phys. Rev. **B112**, 90 (1958).

(39)   P.T. Wedepohl, Proc. Phys. Soc. **92**, 79 (1967).

(40)   R.G. Gordon and Y.S. Kim, J. Chem. Phys. **56**, 3122 (1972).

(41)   W.C. Mackrodt and R.F. Stewart, J. Phys. C. **12**, 431 (1979).

(42)   W.C. Mackrodt and R.F. Stewart, J. Phys. C. **12**, 5015 (1979).

(43)   W.C. Mackrodt, R.F. Stewart, J.C. Campbell, I.M. Hillier, J. Phys. (Paris) **41** C7, 64 (1980).

(44)   C.R.A. Catlow and A.N. Cormack, Int. Rev. Phys. Chem. – in press.

(45)   R.A. Jackson and C.R.A. Catlow, Molecular Simulation – in press.

(46)   C.R.A. Catlow and J.R. Walker, J. Phys. C., **15**, 6151 (1982).

(47)   P.W.M. Jacobs in 'Advances in Ceramics', vol 23 (eds C.R.A. Catlow and W.C. Mackrodt) (1987).

(48)   V. Butler, C.R.A. Catlow, B.E.F. Fender and J.H. Harding, Solid State Ionics. **8**, 109 (1983).

(49)   A.D. Murray, PhD Thesis, University of London (1985).

(50)   R. Gerhardt-Anderson and A.S. Nowick. Solid State Ionics **5**, 547 (1981).

(51)   M.T. Hutchings, K. Clausen, M.H. Dickens, W. Hayes, K.J. Kjems, P.G. Schnabel and C. Smith, J. Phys. **C17**, 3903 (1984).

(52)   C.R.A. Catlow and W. Hayes, J. Phys. C. (1982).

(53)   W.M. Jordan and C.R.A. Catlow, Radiation Effects – in press.

(54)   P. Cox and C.R.A. Catlow – to be published.

(55) A.N. Cormack – to be published.

(56) C.R.A. Catlow, A.V. Chadwick, A.N. Cormack, G.N. Greaves, M. Leslie and L.M. Moroney, Journal of the Materials Research Society – in press.

(57) C.R.A. Catlow, S.F. Matar, J.M. Réau and P. Hagenmuller, J. Solid State Chem. (1985).

(58) P. Vashishta anad A. Rahman in 'Fast Ion Transport in Solids' (eds P. Vashishta, J.N. Mundy and G.K. Shenoy) p 527 North Holland; Amsterdam, (1979).

(59) P. Vashishta, Solid State Ionics. 18/19, 3 (1986).

(60) M.Dixon and M.J. Gillan, J. Phys. C13, 1919 (1980).

(61) M.J. Gillan, J. Phys. C. 19, 3391 (1980).

(62) M.J. Gillan – to be published.

(63) A. Rabenau, Solid State Ionics 6, 277 (1982).

(64) M.L. Wolf, J.R. Walker and C.R.A. Catlow, J. Phys. C. 17, 6623 (1984).

(65) M.L. Wolf and C.R.A. Catlow, J. Phys. C. 17, 6635 (1984).

(66) S. Matar, J.M. Réau, G. Demazeau, J. Leuat and P. Hagenmuller, Mat. Res. Bull. 15, 1295 (1980).

(67) S. Matar and J.M. Réau, C.R. Acad. Sci. Paris, 294, 649 (1982).

(68) P.A. Cox, PhD Thesis, University of Keele (1988).

(69) P.A. Cox and C.R.A. Catlow – to be published.

(70) G.E. Murch, A.D. Murray and C.R.A. Catlow, Solid State Ionics 18, 196 (1986).

(71) P.Y. Wang, D.S. Park, J. Griffiths and A.S. Nowick, Solid State Ionics 2, 95 (1981).

(72) T.F. Soules, J. Non. Cryst. Solids 49, 29 (1982).

CRYSTALLINE ANIONIC FAST ION CONDUCTION

I. Riess

Physics Department
Technion - IIT
Haifa 32000, Israel

ABSTRACT

Propagation of an ion in a solid involves the penetration of a rather large species through a structure, which, from the point of view of the ion, is densely populated. One would, therefore, expect small ion (cation) motion to be more likely than large ion (anion) motion. While this is indeed the trend, fast anionic conduction in crystalline solids can be observed at elevated temperatures where conductivities of the order of 1 $ohm^{-1}-cm^{-1}$ are measured.

The most intensively investigated anionic conductors are those conducting oxygen and fluorine ions. However, conduction of other anions, in particular halogen ions, has been observed.

We discuss the following topics: The various conduction mechanisms, crystalline structures and defect structures that enable the conduction of anions. Special emphasis will be on the fluorite structure. Typical examples of fast O^{2-} and F^- conductors will be discussed in detail.

Data collected from a range of different experimental methods for the same material or group of materials will be presented. This will constitute the main part of the lecture.

We shall close the presentation with short references to a) application of fast anion conductors, and b) "high temperature" superconductor oxides, the crystalline and defect structure of which are rather close to some of the oxide fast ionic conductors that will be discussed.

INTRODUCTION

Conditions for Fast Ionic Conduction

For ionic conduction in crystalline solids to occur, ions have to move through a rather dense matrix of other ionic species of comparable size. To enable this, two conditions must be fulfilled:

1. an empty site exists in the "forwards" direction, into which a conducting ion can move, and

2. the propagation of this ion from site to site is not impeded.

If these conditions are fulfilled, a continuous conduction path is formed.

One would expect, a-priori, that small ions, i.e. cations, might have a better chance to fulfill these conditions. Indeed, most fast ion conductors are cationic. However, fast anionic conduction, via the smaller O^{2-} and F^- anions, has been observed. The conductivity of these anions can reach that found in fast cation conductors, namely $\sim 1 \ \Omega^{-1} cm^{-1}$, though at somewhat higher temperatures.

Before going on with a general discussion of anionic conduction, let us consider in detail an example that will clarify the meaning and importance of conditions (1) and (2). We consider the fluorite (CaF_2) structure which is the structure we shall be most interested in. The fluorite cubic f.c.c. unit cell is shown in Fig. 1a. Possible interstitial sites are indicated at the center of the cubic cell, I_1 (and its equivalent, on the center of the cube edge, I_2). Figure 1a presents the (equilibrium) positions of the centers of the positive and negative ions. The Coulomb interaction energy of an interstitial anion at I_1 with its surroundings can be calculated, provided the interstitial ion is small enough to enter the site. To find out if the site I_1 is roomy enough, one has to take the ionic radii[1,2] into consideration. The position of the anions in the fluorite unit cell of CeO_2 is shown in Figs. 1b, 1c. It is clearly seen that due to the close packing of the oxygen ions the free space about I_1 is too small to accommodate another oxygen, O^{2-}, ion. In other words, forcing an O^{2-} ion into the I_1 position requires high activation energy. We would therefore expect that a perfect fluorite crystal will be a poor anion conductor.

The situation may be drastically changed if the anion sublattice structure is modified or if anion vacancies are introduced into the given fluorite structure. This can be achieved either by a collective thermal excitation and rearrangement of anions, resulting in an order-disorder transition, by reduction, or by doping with a lower valent cation, such as Gd_2O_3 in CeO_2. In either case the solid is no longer a perfect crystal. There is, however, another structure, related to the fluorite one, that of δ-Bi_2O_3. It has the cubic f.c.c. unit cell with six oxygen anions placed in the eight tetrahedra. Thus inherent in the crystalline structure are sites, formally interstitial ones, which serve

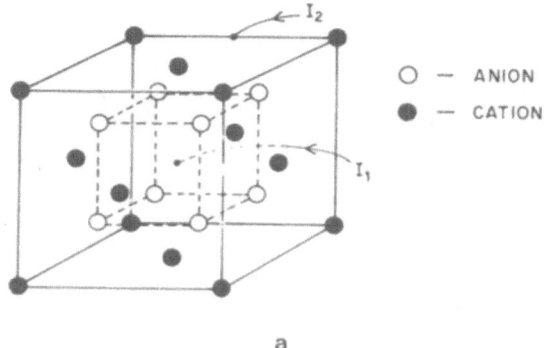

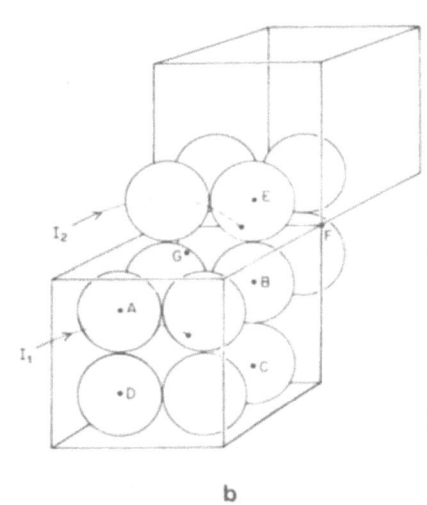

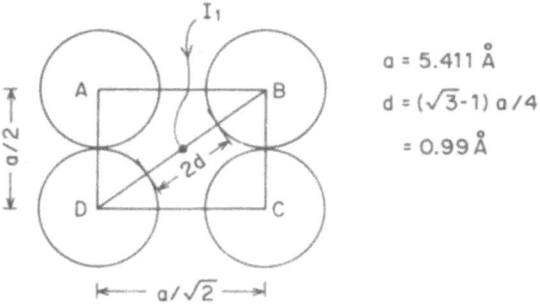

Fig. 1. (a) Cubic f.c.c. fluorite unit cell, (b) oxygen ions surrounding interstitial sites I_1 (I_2) (see text) in the fluorite CeO_2, and (c) cross section of the free space of the interstitial site I_1 (I_2) in CeO_2.

as quasi-oxygen vacancies. In either of the above mentioned ways, the first condition is fulfilled, i.e. the anion can find an empty site to move into.

The motion of the anion into the empty site may be impeded by a high potential barrier. We would anticipate that this motion will be difficult, if the anion has to squeeze through a narrow passage formed by other ions. Thus the crystalline structure and the polarizability of the ions involved will play an important role in determining the mobility of the anion. Let us consider the fluorite structure, Fig. 1b. An anion on site B (which is surrounded by four cations) is moving to occupy a vacancy assumed now at site E. This anion will not follow a straight line from B to E, mainly because of the cations in F and G. The path of minimum energy will pass closer to the point I_2. Obviously, the temperature plays an important role in overcoming the potential barrier, and the mobility of anions will be much enhanced at elevated temperatures.

Correlation Effects

Correlation may exist between the various ions. We have already mentioned one kind, which occurs in the order-disorder (Faraday) transition.

A correlated motion of mobile ions can occur in the so-called caterpillar-like motion in which an ion moves into a position being held by another ion, that leaves its position synchronously with the first ion. This yields one answer to the problem of site availability.

Another correlation effect enhances the mobility of the ion, if the motion of the ion is facilitated by the rotation of a group of other ions. This is analogous to the passing of a person through a rotating door. Some researchers suggest that the rotating ions may even push the moving ion and accelerate it, and therefore refer to this as the paddle wheel mechanism.[3]

Finally, for a high concentration of mobile ions mutual repulsion will result in a Debye-Hückel type relaxation process.[4]

The various correlation effects have been observed in cationic conductors, but to our best knowledge only the Faraday transition has yet been reported for anionic conductors.

Phase Transformations

Many of the solid ion conductors may exist in more than one phase, not all exhibiting fast ionic conduction. The transition from the non-conducting phase to the conducting one, associated with a jump in crystallographic dimensions, is of first order. We shall assume in this lecture that the solid is in the conducting phase except for the solid undergoing the Faraday transition. The latter will be discussed for a wider temperature range.

Summary of Conduction Mechanisms

1. <u>Conduction on free surfaces of chemically adsorbed ions</u>.
    Though of potentially high mobility, it is of little
    interest for macroscopic systems, since the contribution
    to the conductivity from an atomic monolayer is expected
    to be negligible.

2. <u>Conduction within subsurface space charge region</u>.
    Conduction within a few atomic layers or even a few
    hundred atomic layers may be affected by a local space
    charge and enhanced defect concentration,[5] but again will
    have little effect on the overall conductivity. This
    may, however, have an effect on interface reaction, in
    particular at electrodes, but this is outside the scope
    of our talk.

    On the other hand, on the grain boundaries between two
    insoluble solids, one a fast ion conductor and the other
    an insulator or both two different ion conductors, there
    may be a high concentration of charged defects and a
    corresponding deep, 0.1–1 μm, space charge and enhanced
    defect concentration, region.[6,7] The ionic conduction in
    this defect rich space charge region may be enhanced to
    an extent that it affects the total conductivity.

3. <u>Enhanced or reduced conduction in grain boundary layer
    0.1–1 μm thick having a composition different from the
    bulk due to inhomogeneous doping</u>. This may occur because
    of the tendency of the dopant material to concentrate
    closer to the grain boundary. For small grains of the
    size 1–10 μm this may have an effect on the total
    conductivity.[8] However, it is basically the bulk
    property of the grain boundary layer that counts. We
    shall therefore cover this topic when bulk ionic
    conduction is discussed, below.

4. <u>Bulk conductivity</u>

    a. <u>Thermal excitation of defects</u>. All solids will
        exhibit some ionic conduction at elevated
        temperatures due to thermal excitation of defects in
        particular Schottky and Frenkel defects. However,
        to fulfill conditions (1) and (2), for fast ionic
        conduction, one has to consider the following more
        restricted conduction mechanisms.

    b. <u>Faraday transition</u>. As mentioned before, an ordered
        crystal with poor anionic conduction may go through
        an order-disorder phase transition, in which the
        anions change position and move in a correlated way.
        This transition, though first order, is somewhat
        diffuse, taking place over a wide temperature range,
        close, but below the melting temperature, T_m. This
        is a correlated many body transition and not a
        transition of independent anions between two energy
        levels. For the latter the specific heat has a wide
        Schottky anomaly with a width of the order of T_m

contrary to the much narrower peak observed experimentally.[9] Another argument in favour of the cooperative model will be given later.

c. <u>Doping with aliovalent cations</u>. An aliovalent cation is one whose charge differs from that of the host cation. If it enters the solid substitutionally, a defect is created also in the anion sublattice. If the dopant ion has a lower valency, than that of the host cation, an anion vacancy is created. If the valency is higher, an interstitial anion will be introduced. The effect of doping depends on the host material, the number of equivalent sites available, mobility in interstitial sites, etc. However, for poor anionic conductors, such as the perfect fluorite structure below the Faraday transition, the doping with a low valence aliovalent cation, which introduces anion vacancies, enhances the ionic conductivity drastically.

d. <u>Change of stoichiometry</u>. Let MX represent an ionic solid. Under reducing condition the composition will change to $MX_{1-\delta}$ or $M_{1+\delta}X$. The different notation attempts to clarify whether the defects are on the anion sublattice ($MX_{1-\delta}$ with X vacancies) or on the cation sublattice. In $MX_{1-\delta}$ condition (1) is fulfilled and one expects to observe enhanced anionic conductivity at least in a poorly conducting MX material. Under oxidizing conditions the nonstoichiometric compound will be $MX_{1+\delta}$ or $M_{1-\delta}X$. In the first case anion interstitials are introduced, which may enhance the anionic conduction.

The defects existing in this kind of nonstoichiometric compound usually act as donors or acceptors. These introduce quasi-free electrons or holes, and the electronic conductivity may become predominant.

When the concentration of defects is high, but the temperature is not too high, the defects order with a superlattice periodicity and a new phase is formed. The ordered defects may be concentrated on regularly spaced one-dimensional strings, on two-dimensional crystal shear, or twinned planes and chemical layered segregation, or be distributed rather uniformly, but periodically, in the solid.

e. <u>More than one equivalent site per anion</u>. In those crystals, in which the number of equivalent sites exceeds the number of anions (e.g. δ-Bi_2O_3) and these sites are closely packed, one finds a low excitation energy of an anion into a nearby empty equivalent site, and a continuous conduction path exists. At low temperatures when the anions are fully ordered, some of these sites are occupied in a regularly spaced form. The empty sites are really not equivalent because of the anionic sublattice

28

formed. At elevated temperatures, on the other hand, one can expect all equivalent sites to be equally occupied. So the anions can diffuse within the equivalent sites. If, as in δ-Bi_2O_3, this structure exists only at elevated temperatures, one can observe only the highly conducting disordered structure of the ion conductor.[10]

    f.   <u>Correlated motions</u>. Caterpillar, rotating door (or paddle wheel) mechanisms, and Debye-Hückel type relaxations were mentioned before. They were observed in cation conductors and will not be treated further here.

    g.   <u>One- and two-dimensional conductors</u>. In certain ion conductors other ions are arranged so that tunnels or channels are available for filling and motion of small cations. Alternatively, there are ion conductors that have a layered structure (e.g. ß-Al_2O_3) in which a large spacing exists between some of the atomic layers enabling the incorporation and high mobility of small cations. It is less reasonable to expect that much more space can be found to accommodate and let pass the much larger anions. Yet one can find complementary structures, which will exhibit high anion conduction. We notice that the counterpart of a mobile ion is a mobile vacancy and the counterpart of a one-dimensional tunnel is a one-dimensional chain of ions. For the larger anions, in particular in compounds rich with anions, one would expect to find such one-dimensional chains (or two-dimensional sheets) of anions, in which an anion vacancy can be introduced and move. Vacancies could be introduced by doping with aliovalent ions. Experimentally, compounds such as YF_3 or LaF_3 (See Figs. 2 and 3) have anions in non-equivalent positions thus reducing drastically the density of chains of equivalent and mobile anions. The situation is altered only at elevated temperatures, when anions on different types of sites can participate in the anion conduction.

Fast Anion Conductors

Fast anionic conduction is found mainly in solids of the fluorite and fluorite related structures. It was also observed in solids with the perovskite, YF_3, tysonate and the simple cubic structures (see Figs. 1-5), and in stabilized Ta_2O_5.[79] The smaller anions O^{2-} and F^- [1,2] show the fastest conduction. Good anionic conduction is also found for the other halogens: Cl^-, Br^- and I^- and for S^{2-}.

The combination of the proper structure and anion results in an anionic conductivity, σ_i, of the order of 1 $\Omega^{-1}cm^{-1}$, which is of the same magnitude as that of fast cation conductors or good liquid electrolytes. The highest O^{2-} conductivity is found in δ-Bi_2O_3 and doped Bi_2O_3 [11,12] with $\sigma_{ion} \sim 1 \ \Omega^{-1}cm^{-1}$ at ~ 800°C and an activation energy, ΔH, of 0.47 eV.

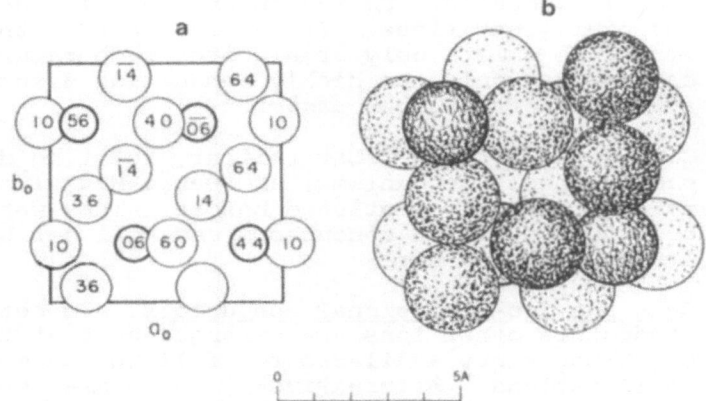

Fig. 2. a) A projection along the Co axis of the orthorhombic
           structure of YF$_3$. Fluorine ions are the larger
           spheres and b) a drawing of the same structure
           viewed along the Co axis. The fluorine ions are
           dotted. (From Ref. (10), Fig. VB, 9).

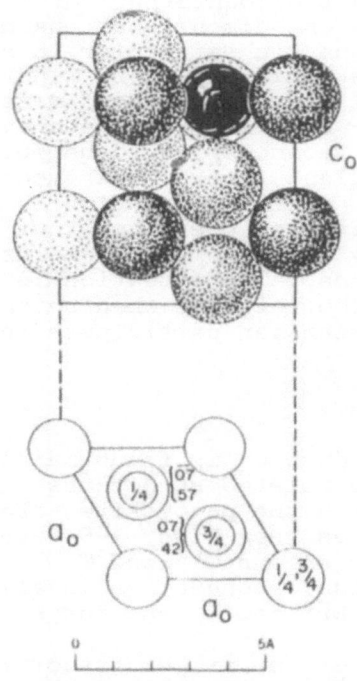

Fig. 3.   Two projections of the hexagonal cell of LaF$_3$. In
           the upper one the lanthanum atom is black. (From
           the Ref. (10), Fig. VB, 10).

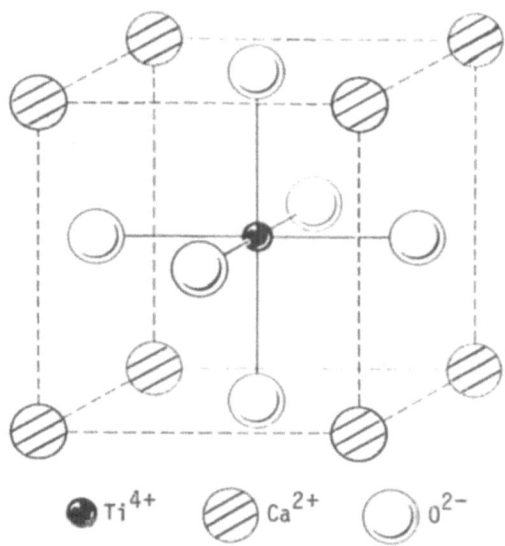

$\bullet$ Ti^{4+} Ca^{2+} O^{2-}

Fig. 4.   CaTiO$_3$ perovskite (idealized cubic) unit cell (from Ref. (2), Fig. 2.28).

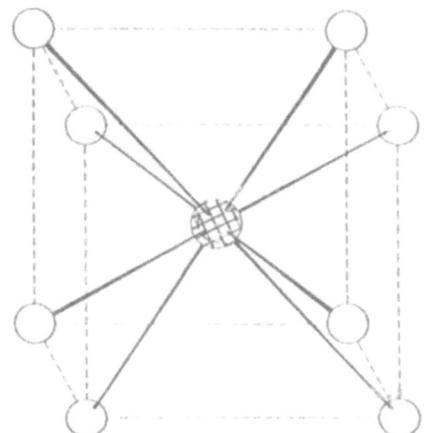

Fig. 5.   Cesium chloride, simple cubic unit cell.   (From Ref. (2), Fig. 2.26)

The fast F$^-$ conductors also have the fluorite structure. For RbBiF$_4$, KBiF$_4$, and Pb$_{0.75}$Bi$_{0.25}$F$_{2.25}$, σ_{ion} ~0.1 Ω^{-1}cm^{-1} at ~300°C, and ΔH ~ 0.39 eV (see Ref. 14, p. 328).

Literature

A number of good reviews were written in recent years, on solid electrolytes, including anionic conductors, and on defect chemistry. We name here a few of the more recent ones. A general text book on solid electrolytes was written by H. Rickert.[13] Reviews can be found in the books edited by Hagenmuller and van Gool[14] and by Geller.[15] Sorensen edited a book containing reviews specifically on oxide ionic

31

conductors[16] and Hagenmuller edited a book on fluorides.[17] Oxide solid electrolytes are discussed also by Kingery et al.[2] Experimental techniques used to investigate solid electrolytes were reviewed by Linford and Hackwood.[18] Reactions in solids were discussed by Schmalzried.[19] Finally, the extended book by Kroger[20] covers the chemistry of many imperfect crystals.

Scope

In the Introduction, we have presented general ideas concerning fast anion conductors. In the next sections we present experimental data to illustrate these ideas that will give us also an opportunity to mention the more interesting materials and experimental techniques used. We shall refer to results published in recent years (1980-87) thereby providing an up to date review of the literature. We shall close the presentation mentioning shortly the new "high temperature" superconductors which are closely related to perovskite oxide anion conductors, and then mention the possible applications of fast anion conductors.

ENHANCED IONIC CONDUCTION BY SURFACE INDUCED DEFECTS

If the surface has an excess of charged defects of one kind, a space charge is built up near the surface, with a high concentration of defects of opposite charge. Due to this high concentration, the ionic conductivity is enhanced. When the surface is not a free one, but the interface between two insoluble solids, the defect concentration at the surface is found, in many cases, to increase, which in turn increases the space charge region.[6,7] $SrCl_2$ and Al_2O_3 were recently reported to exhibit enhanced ionic conductivity as small particles of Al_2O_3 are added to $SrCl_2$.[21] The two materials do not form a solid solution and the enhancement is ascribed to the surface effect. It is estimated that the space charge region is 0.1-1 μm and that the ionic conductivity in the space charge region is two orders of magnitude higher than in the bulk.

INHOMOGENEOUS DOPING OF GRAIN BOUNDARIES

A dopant introduced into a polycrystalline solid may distribute non-uniformly. In polycrystalline CeO_2 (grain size ≤ 4 μm) doped with 10 mol% Gd_2O_3 (which yields the highest conductivity) the activation energy for ionic conduction, ΔH depends on temperature. Further, ΔH increases with increase in Gd_2O_3 concentration. The apparent temperature dependence of ΔH was interpreted with a model, in which each grain has slightly less than the nominal Gd_2O_3 concentration in its bulk, but has an excess by a factor 1.7 of Gd_2O_3 concentration in the grain boundary layer. The thickness of the layer was determined to be 0.15 μm. The higher ΔH seen at low temperatures is then that of the conductivity in the layer with excess dopant, while the lower ΔH seen at more elevated T is that of the bulk conductivity.[8]

THERMAL EXCITATION OF DEFECTS

Thermal excitation of Schottky and Frenkel defects results in a gradual increase in ionic conductivity with temperature. Examples: TlCl has the CsCl simple cubic structure. At elevated temperatures Schottky defects are excited. Ionic conduction is mainly due to Cl^- vacancies and to a lesser extent due to Tl^+ vacancies (t_{cl} - 0,9 at 500K). σ_{ion} - $5 \times 10^{-5} \Omega^{-1} cm^{-1}$ at 500K.[22] Doping with divalent cations reduces σ_{ion} at intermediate temperatures, - 300K, as the faster anion vacancies are replaced by the slower cation vacancies via the Schottky equilibrium. $PbCl_2$ and $PbBr_2$ form solid solutions having an orthorhombic structure (see Fig. 6) Schottky defect anion vacancies, facilitate the ionic conduction.[23,24] There is a sharp increase in σ_{ion} close to the melting temperature, which is probably due to thermal excitation of Frenkel defects.

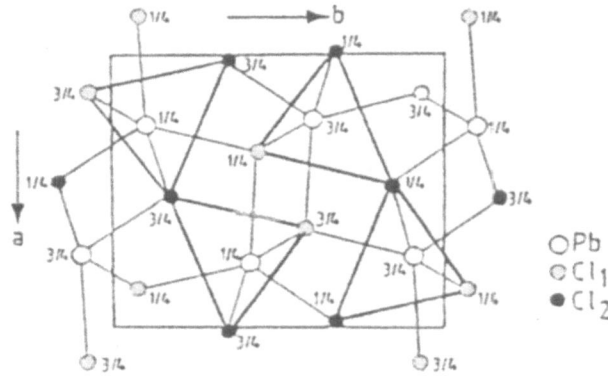

Fig. 6. C-axis projection of the orthorhombic $PbCl_2$ structure. In $PbCl|_{1-x}-PbBr_2|_x$ solid solution for x < 0.5, Br^- ions are on site type (2) only. (From Ref. (24), Fig. 1).

Ac conductivity measurements on Bi_2WO_6 (orthorhombic structure) yields thermally enhaced O^{2-} ionic conductivity that increases sharply above - 500°C.[25] σ_{ion} is strongly anisotropic with the highest conductivity being along the [001] direction: -0.1 $\Omega^{-1} cm^{-1}$ at 950°C. Electronic conduction sets in at these temperature for oxygen partial pressure, $P(O_2)$, below 10^{-2} atm.

ORDER-DISORDER TRANSITION, THE FARADAY TRANSITION

Specific heat, Cp, measurements show a peak at the cooperative Faraday transition, while thermal excitation of independent Frenkel defects results in a monotonic increase in C_p.[26] Cp for $SrCl_2$ (fluorite structure) is shown in Fig. 7. A similar behaviour is observed in βPbF_2, CaF_2, SrF_2, BaF_2 (fluorite structure). Doping these fluorides with UF_4

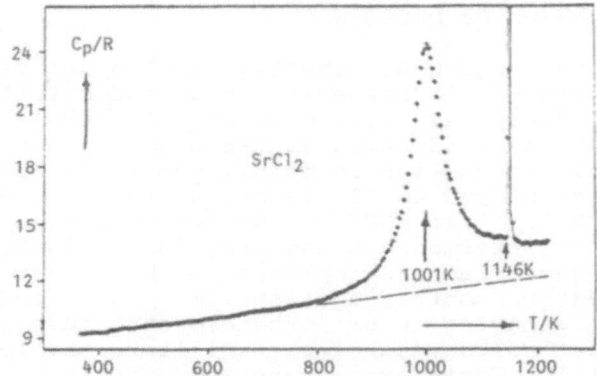

Fig. 7.   Specific heat of strontium chloride. Melting
temperature: 1146K. Faraday transition peak: 1001K.
(From Ref. 9, Fig. 1).

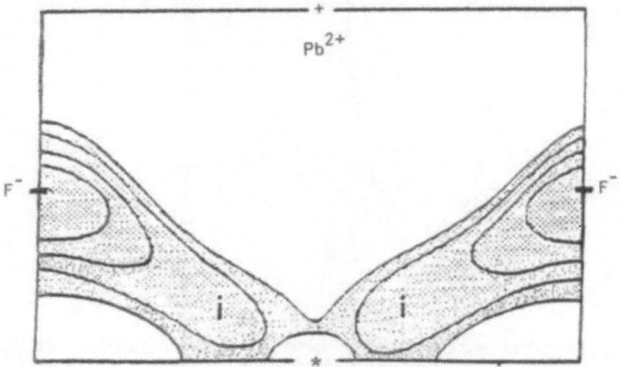

Fig 8.   Crystal field potential along the minimum energy
path in the fluorite structure. Model calculation.
* is the center of unit cell, octahedral site.
(From Ref. 29, Fig. 9).

introduces F^- interstitials, F_i, which lower the Faraday
transition temperature and enthalpy.[27,28] Above the Faraday
transition, anions flow in a highly defective anion
sublattice. Their concentration becomes almost constant and
the activation energy for conductivity reflects mainly the
small mobility activation energy. Eventually, this activation
energy vanishes.[29] The critical temperature T_c, at which the
$\ln \sigma_{ion}$ vs. T curve changes its slope, is approximately T_λ,
the peak temperature in Cp.

XRD measurements on single crystal β-PbF_2[29] reveal a
sharp change in lattice constant at T_c. Electron density
mapping of the F^- ions is shown in Fig. 8. This model
calculation fits rather well the experimental data for
β-PbF_2. Asymmetric "smearing" of F^- positions about the
regular fluorite anion (0.25, 0.25, 0.25) sites is attributed

to anharmonic vibrations. This was also observed by the
broading and lowering (Debye-Waller factor) of diffraction
lines in neutron scattering measurements on ß-PbF$_2$ and
SrCl$_2$.[30] Above T$_c$ at ~800K, the F$^-$ ions are displaced into
one of four equivalent sites (x,x,x) with x=0.25 + δ,
δ=0.045.[29] In addition, F$_i$ are found to diffuse through a
region with no a-priori potential minima. The path is shown
in Fig. 8. It does not pass through the cubic cell center.
The fraction of anions leaving the tetrahedral sites above T$_c$
is 40% in ß-PbF$_2$, but only 3.5% in SrCl$_2$.[30] ß-PbF$_2$, quenched
from temperatures (920K) above T$_c$, reveals a lower activation
energy ΔH=0,40 eV at ~300K[31] than annealed samples (ΔH = 0.70
eV).27 This is due to quenching of the highly defective high
temperature anion sublatice. The fact that ion motion occurs
without relaxation of the frozen-in disorder proves that the
disorder is a many-body cooperative effect and not an
accumulation of independent Frenkel defect excitations. XRD
measurement on the quenched samples show also that the F$^-$ ions
are displaced in the <1,1,1> direction toward the four
equivalent positions represented by (1/3, 1/3, 1/3).

DOPING WITH ALIOVALENT CATIONS

Fluorite Structure

    CeO$_2$. CeO$_2$ having the flourite structure and a melting
temperature T$_m$ = 2600°C is a poor ionic conductor at all
temperatures of interest. When doped with Gd^{3+}, Y^{3+}, Ca^{2+},
Mg^{2+}, it becomes a fast O^{2-} anion conductor via V$_o$$^{\cdot\cdot}$ with σ$_{ion}$
~ 0,05 Ω^{-1}cm^{-1} at 700°C and ΔH = 0,70 eV.[3] Research in recent
years was concerned with the nature of interaction between the
dopant ion that has an effective negative charge, e.g. Gd$_{ce}$´,
and the positively charged vacancy V$_o$$^{\cdot\cdot}$. Calculations show[32]
that the interaction increases with the cation effective
charge, as could be anticipated. Thus, the free vacancy
concentration for the same equivalent doping level and
temperature is higher when doped with Gd^{3+} than Ca^{2+}. The
interaction between dopant ion and V$_o$$^{\cdot\cdot}$ is minimized.when the
size mismatch between host and dopant cations is least.[32]

    ^{17}O NMR measurements on the (1-x)CeO$_2$·xY$_2$O$_3$, 0 ≤ x ≤
0.006 yields the temperature dependence of the jump time for
diffusion of V$_o$$^{\cdot\cdot}$, with an activation energy of 0.49 eV for T
< 600K.[33] This activation energy is interpreted as that of
the mobility, ΔH$_m$. AC conductivity measurements on
(1-x)CeO$_2$·xY$_2$O$_3$, x ~ 0.05 - 0.50% yield: ΔH ~ 0.9 eV.[34] The
difference ΔH- H$_m$ being the association energy of Y$_{ce}$ - V$_o$$^{\cdot\cdot}$:
ΔH$_A$ ~ 0.4 eV. AC conductivity measurements[34,35] show that ΔH
has a minimum value ~ 0.73 eV for x ~ 2%. The initial
decrease of ΔH with x has not yet been explained
satisfactorily.

    The associates are expected to be nearest neighbor
(Y´$_{ce}$ V$_o$$^{\cdot\cdot}$) pairs. As the dopant concentration increases, one
would expect to find associates with composition (2Y´$_{ce}$ -
V$_o$$^{\cdot\cdot}$)^x. Anelastic (mechanical) relaxation[36] was measured on
Y^{3+} doped CeO$_2$.[37] The amplitude and phase shift of the
mechanical response of the sample to an applied force
oscillating at a frequency of 8 KHz, was measured as a

function of temperature. The two stronger Debye peaks obtained were fitted by thermally activated reorientation relaxation times. From the number of Debye peaks, their intensity ratio, width, and activation energies of reorientation, a model for defect clusters $(Y_{ce}{}' V_o{}^{\cdot\cdot})^{\cdot}$ and $(2Y_{ce}{}' V_o{}^{\cdot\cdot})^x$ was established (see Fig. 9).

Y_2O_3 and Sc_2O_3 doped CeO_2 exhibit the Debye peak at higher temperatures that correspond to the orientation relaxation of the $Y_{ce}{}' - V_o{}^{\cdot\cdot}$ or $Sc'_{ce} - V_o{}^{\cdot\cdot}$ associates. $(1-x)CeO_2 \cdot xSc_2O_3$ ($x \leq 1\%$ = solubility limit) also reveals a peak at low temperatures that increases with x and decreases if Y_2O_3 is added.[38] This peak is due to $Sc_{ce}{}^{\cdot}$ somewhat displaced from the octahedral site of $Ce^x{}_{ce}$ due to the relatively small radius of Sc, thus enabling a rotational relaxation of Sc in this surroundings. That rotation has a low activation energy $H_r = 0.21$ eV. As Y_2O_3 is added the dissociated $Sc_{ce}{}^{\cdot}$ defect binds the extra $V_o{}^{\cdot\cdot}$ formed, as the binding energy in $(Sc_{ce}{}' - V_o{}^{\cdot\cdot})$ is higher then in $(Y'_{ce} - V_o{}^{\cdot\cdot})$ and the peak due to dissociated $Sc_{ce}{}^{\cdot}$ centers diminishes and finally disappears for $[Y] > [Sc]$.

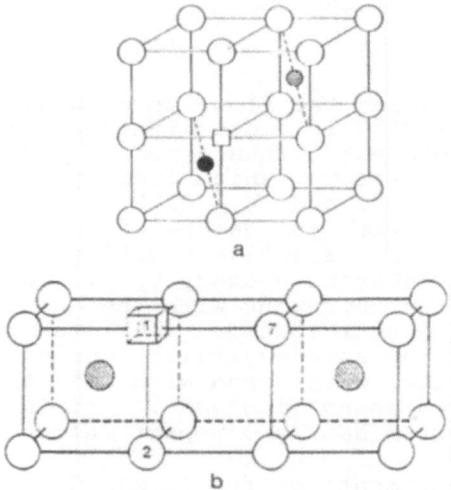

Fig. 9.   a) $Y'_{ce}-V_o{}^{\cdot\cdot}$ associate shown in one half unit cell of CeO_2; b) $Y'_{ce}-V_o{}^{\cdot\cdot}-Y'_{ce}$ associate in CeO_2.   O-Oxygen, O-Y'_{ce}, 1-vacancy, 2, 7 two of 8 equivalent vacancy sites. (From Ref. (37), Figs. 1 and 3).

Neutron diffuse scattering on $0.94CeO_2 \cdot 0.06Y_2O_3$ indicates local lattice displacement. Cations coordinating a vacancy, $V_o{}^{\cdot\cdot}$, move 0.3Å toward the vacancy along <100>.[39]

Doping CeO_2 with a few percent of various rare earth or transition metal cations, maintains the fluorite structure of CeO_2, but not the dimensions. The lattice constant of $0.9CeO_2 \cdot 0.1Gd_2O_3$ is about 0.3% larger than that of CeO_2.[8,40] The increase is due to the vacancies introduced into the lattice.

ZrO_2. In zirconia, the ratio of cation radius to anion radius (0.724) is slightly lower than the theoretical limit for the fluorite structure: 0.732. The fluorite structure is stable only above 2300°C. However, doping with at least ~8 mol% of other cations, e.g. Y^{3+}, Ca^{2+} stabilizes the fluorite structure. These aliovalent ions also introduce oxygen vacancies. The ionic conductivity is then ~0.3 $\Omega^{-1}cm^{-1}$ at 1000°C, and ΔH ~ 1 eV. Stabilized ZrO_2 has been intensively investigated for many years. The new feature attracting attention in recent years is the enhanced mechanical strength, accompanied by only a small change in ionic conductivity,[41] due to partial stabilization of ZrO_2 by using a small amount of dopant.

The EPR method was used to determine the orientation of the associate ($Y'_{Zr}V_o^{\cdot\cdot}$) in a single crystal of ZrO_2 stabilized with 8.7 mol% Y_2O_3.[42] Magnetically active centers were prepared by reduction of the oxide to form a small amount of additional $V_o^{\cdot\cdot}$ and quasi free electrons. At low temperatures the electrons are trapped by the oxygen vacancies to form magentically active F (color) centers. The F center interacts with a neighboring $Y_{Zr}^{'}$. The g factor of the F center depends on the orientation of the associate ($Y'_{Zr}V_o^{\cdot}$)^x relative to the applied magnetic field. The associate was found to be oriented along the <111> direction as shown in Figure 9a.

The same problem was also investigated by a site selective fluorescence method.[43] Eu^{3+} was introduced into Y_2O_3 stabilized ZrO_2. The energy levels of Eu^{3+} depend slightly on the interaction with the surrounding. Using a tunable laser, these finer energy variations can be detected and a site selective spectroscopy is obtained. The local structural properties near Eu^{3+} are inferred from the number, relative intensity, and energetic position of the spectral lines. The conclusions for the fluorite phase are: the dopant ions are randomly distributed. Oxygen vacancies are also distributed at random. Oxygen atoms relax toward a nearby oxygen vacancy, as also observed for doped CeO_2.[39]

Fluorides. The ionic conductivity of ß-PbF_2 below the Faraday transition at T_c was found to increase on doping with BiF_3. The maximum (σ_{ion} ~ 0,03 $\Omega^{-1}cm^{-1}$ at 250°C, $\Delta H = 0.40$ eV) is obtained for $0.75PbF_2 \cdot 0.25BiF_3$.[44] For $T < T_c$, ionic conduction in ß-PbF_2 is through thermally excited fluorine vacancies, $V_F^{'}$ One would therefore expect $Bi_{Pb}^{\cdot}$ to suppress the ionic conductivity by reducing the $V_F^{\cdot}$ concentration. The fact that the opposite occurs is due to the existence of a large concentration of fluorine interstitials as well as an excess of fluorine vacancies, and therefore a substantial increase in the disorder on the anion sublattice. Only when the concentration of BiF_3 exceeds 25 mol% does the doping affect the conductivity and σ_{ion} decreases.

The structure of defects around a higher valency dopant ion can be investigated at low temperatures. Neutron diffraction of $1-xBaF_2 \cdot xUF_4$ with $x = 0.02$, 0.05, and 0.154 show Bragg reflections of the fluorite structure and diffuse quasi-elastic scattering due to defects.[45] $U_{Ba}^{\cdot\cdot}$ introduces two interstitials $F_i^{'}$ and forces two additional nearby F^- ions to relax from their tetrahaderal position (see Fig. 10). The cluster thus formed involves seven species $U_B\cdots +4F_i\cdot + 2V_F\cdot$. It possesses an electric dipole. When the sample is polarized by an electric field, and cooled in the field, the neutron diffuse scattering spectrum is found to change, proving that the clusters are oriented. When the electric field is removed at temperatures low enough to freeze the polarization and the sample is then heated at a constant heating rate (TDSC), peaks appear in the polarization current of the sample. From the temperature dependence of a peak and the heating rate, the activation energy and characteristic jump time for reorientation of a cluster can be calculated. One of the peaks is due to the relaxation of the space charge accumulated under the ion blocking electrodes. The activation energy for ionic conductivity is determined from it. A similar method is that of thermally stimulated polarization (TSP). Here the electric field is applied to the sample at the low temperature at which the dipoles are frozen in. Polarization sets in as the temperature is increased. This also yields peaks in the polarization current. Another method to investigate defect cluster reorientation is by dielectric loss. This is the

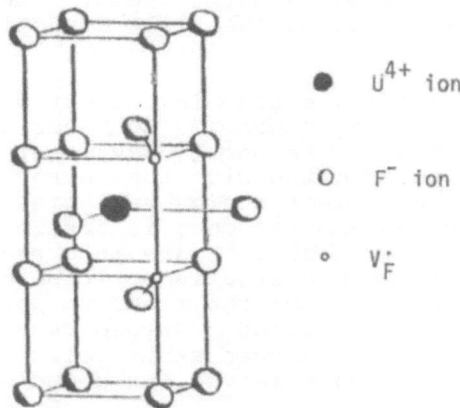

Fig. 10. Defect cluster around U_{Ba} comprising $U_{Ba}\cdots 4F_i\, 2V_F\cdot$ (From Ref. (45), Fig. 6).

electric analogue of anelastic relaxation. As the symmetry of the dipole on a defect cluster in an electric field may be lower than that of a cluster in zero field, the relaxation rates observed by the two methods may be different.[46]

Both TSP and DL methods were used to follow the reorientation of defect clusters introduced into the fluorite BaF_2 by doping with LaF_3[47] and into SrF_2 by doping with CeF_3 and PrF_3[48], DyF_3, and ErF_3[49], and GdF_3.[50]

Neutron diffraction on $Sr_{0.69}La_{0.31}F_{2.31}$ shows that the fluorite structure of SrF_2 is maintained.[51] The defect cluster formed by doping with LaF_3 includes four interstitial fluorine ions at sites represented by (0.42, 0.42, 0.42) surrounding a vacancy at the regular site (0.25, 0.25, 0.25).

There are also F⁻ ions displaced from the (0.25, 0.25, 0.25) position to (0.3, 0.3, 0.3). It is not clear whether this displacement is permanent or a time average due to anharmonic vibration as in βPbF_2 above the Faraday transition temperature.[29]

Tysonite Structure

The anion rich compound LaF_3 is expected to form anion chains, which should allow fast ion conduction of anion vacancies (see Fig. 3). The ionic conductivity of $La_{1-x}Sr_xF_{3-x}$, $0 \leq x \leq 0.14$ was measured by ac impedance spectroscopy, with the sample placed as the dielectric in a capacitor.[52] The maximum conductivity was observed for $x = 3\%$ with $\sigma_{ion} = 10^{-4}$ $\Omega^{-1}cm^{-1}$ at 400K about an order of magnitude larger than that of LaF_3 and comparable to σ_{ion} of $\beta-PbF_2$ at 400K.

The defect associates formed in a similar compound $La_{1-x}Ba_xF_{3-x}$, $x < 0.105$, were investigated by TSDC and DL. For small x values $x < 1.3 \times 10^{-2}$ only simple associates are observed: $(Ba_{La}{}'V_F{}^{\cdot})^x$ in nearest neighbor positions and V_F being confined to the B sublattice.[53]

CHANGE OF STOICHIOMETRY AND MIXED IONIC ELECTRONIC CONDUCTION

Intrinsic Compounds

Ceria is readily reduced at elevated temperatures and low $P(O_2)$ to form new ordered phases $CeO_{1.818}$, $CeO_{1.808}$, $CeO_{1.800}$, $CeO_{1.79}$, $CeO_{1.714}$ at $T < 722K$. The phase diagram of CeO_y obtained by specific heat measurements is shown in Fig. 11.[54,55] The compositions of that series of phases suggests that the oxygen vacancies form a 3D superlattice rather than a 1D or 2D one.[54] HRTEM measurements on CeO_{2-x} seem to support this conclusion.[56]

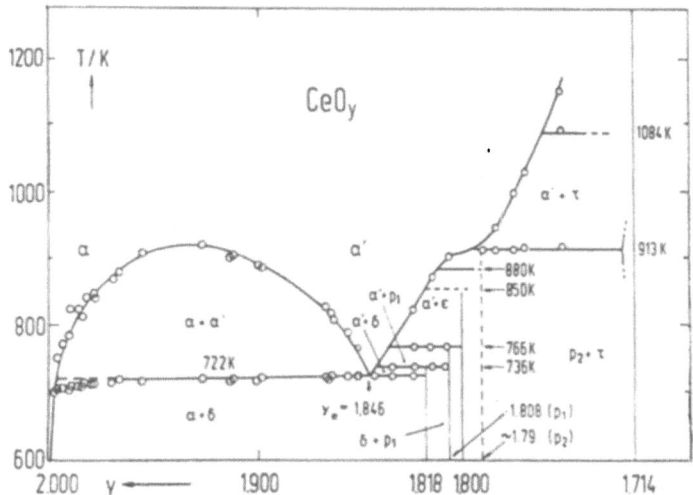

Fig. 11. Phase diagram of CeO_y in the composition range $2 > y > 1.714$ and temperature range $600 < T < 1200K$. (From Ref. 55, Fig. 1).

HRTEM on the similar oxide TbO_{2-x} show that the reduced phases have features in common: a fluorite related structure, and oxygen vacancies ordered in pairs along the <111> direction and centered about a metal atom. In TbO_{2-x} the pairs seem to form a 2d periodic structure.[57] Long period 2D modulations (Magneli structure) have been observed in VO_{2-x} and TiO_{2-x} both with the rutile structure. The CS planes are clearly visible in TEM images.[58] Thick faults are found to be in a twin relation with the VO_2 matrix.

The charge compensating entities of anion vacancies or anion interstitials can be quasi-free electrons or holes, respectively. In reduced ceria, CeO_{2-x}, $V_o^{\cdot\cdot}$ is compensated by $2Ce'_{ce}$. At low temperatures an oxygen vacancy traps an electron and becomes a F center. The concentrations of quasi-free electrons and ionized oxygen vacancies are then of the same order. Due to the much higher mobility of the electrons, CeO_{2-x} is a predominant electronic (n-type) semiconductor. For small deviations from stoichiometry, $x \leq 0.01$, model calculation can relate $P(O_2)$, electron concentration, n, and composition x. Emf measurements of $P(O_2)$ combined with coulometric titration control of x show that the oxygen vacancies are doubly ionized, $V_o^{\cdot\cdot}$.[59] The same conclusion was reached by x vs. $P(O_2)$ measurements, where x was determined from the pressure due to O_2 dissociated from the sample,[60] and by σ_e vs. $P(O_2)$ conductivity measurements.[60,61]

Low Ca impurity concentrations ($\sim$1000 ppm) were reported for the various ceria samples.[59,60,61] It was shown[61] that by counterdoping with Ta_2O_5 at twice that concentration the system behaves as pure ceria where $n = 2[V_o^{\cdot\cdot}]$ and $n \propto P(O_2)^{-1/6}$.

Change in Stoichiometry of Doped Compounds

CeO_2 doped with SrO maintains the cubic fluorite structure of CeO_2 in solid solutions $(1-x)CeO_2 \cdot xSrO$, $x \leq 0.08$. For $0.4 > x > 0.08$ a second phase of $SrCeO_3$ coexists with the fluorite one. $SrCeO_3$ has the perovskite structure. This has little effect on the O^{2-} ionic conductivity for $0.08 \leq x \leq 0.20$.[62] For high $P(O_2)$ the ionic transference number, t_i, is close to unity. σ_{ion} does not depend on $P(O_2)$ (see Fig. 12). As $P(O_2)$ is lowered the sample is reduced. When the degree of reduction is high enough, electronic conduction becomes predominant (see Fig. 12), the conductivity increases and the sample becomes an n-type semiconductor. It should be noticed that by a proper choice of $P(O_2)$ (for given T and x) t_i can be varied from $t_i \sim 1$ to $t_i << 1$.

Uranium in CeO_2 is a donor with an energy level within the CeO_2 conduction band and is therefore ionized to form $U^{\cdot}_{ce}$.[63] For intermediate $P(O_2)$ the main defects in $(1-x)CeO_2 \cdot xUO_2$. $x \leq 5\%$ are therefore quasi-free electrons and $U^{\cdot}_{ce}$. In this region the concentration of electrons, n, being fixed by $[U^{\cdot}_{ce}]$ is independent of $P(O_2)$. For lower $P(O_2)$ enough oxygen vacancies are introduced so that $n = 2[V_o^{\cdot\cdot}] >> [U^{\cdot}_{ce}]$. The ionic current is then carried by $V_o^{\cdot\cdot}$. For higher $P(O_2)$ the concentration of negatively charged oxygen interstitial increases so that $n \ll 2[O_i{}^{\cdot\cdot}] = [U'_{ce}]$. As the concentration of $O_i{}^{\cdot\cdot}$ in this region is a constant, measuring the activation energy of the oxygen diffusion in this region enables the determination of the migration energy, ΔH_m, of $O_i{}^{\cdot\cdot}$.[63] This diffusion measurement was done by measuring the $^{18}O/^{16}O$ profile after exchange using secondary-ion-mass-spectroscopy (SIMS). The enthalpy of migration for O_i is high $\Delta H_m = 1.5$ eV corresponding to the low diffusion coefficient measured.[64]

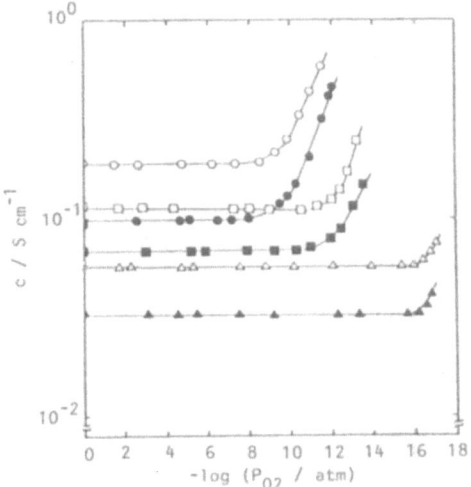

Fig. 12   Electrical conductivities of $(1-x)CeO_2 \cdot xSrO$ vs. logP(O_2). Open symbols o, $\Box$, $\triangle$: x = 0.10; Closed symbols $\bullet$, $\blacksquare$, $\blacktriangle$: x = 0.20. (o, $\bullet$) - 1000°C, ($\Box$, $\blacksquare$) - 900°C and ($\triangle$,$\blacktriangle$) - 800°C. (From Ref. (62), Fig 13).

        Thoria (ThO$_2$) is a fast oxygen ion conductor when doped with fixed lower valent cations. Doping with variable valent Ce (+3, +4) results in a different defect chemistry and electrical properties. Since Ce$^x{}_{Th}$ acts as an acceptor level, it reduces the quasi-free electron concentration. On the other hand, it enhances the formation of positive oxygen vacancies.[65] The result is that at low P(O_2) (P($O_2 \leq 10^{-6}$ atm, T - 1000°C) the reduction of thoria results in an increase in ionic charge carriers $V_o{}^{\cdot\cdot}$ as P(O_2) decreases, while the concentration of electron generated by reduction is small. Thus the ionically conducting region depends on P(O_2).

        For P(O_2) measurements (Cf. Ref. 13, pp. 143, 149) it is desirable that the oxygen ion conductor have only a negligible electronic conductivity. Stabilized ZrO$_2$ is widely used for this purpose. The pressure, P$_0$, at which the ionic and n-type electronic conductivities are equal, was measured for ZrO$_2$ + 11 mol% CaO in the temperature range 1100-1400°C by the Swinkels polarization technique.[66] When the low P(O_2) at one side of the sensor is much lower than P$_0$, the measured EMF reaches a fixed value, which is determined by P$_0$ (cf. Ref. 19, p. 184). At 1100°C, P$_0$ - 10^{-27} atm.[66]

LaMnO$_3$ has the perovskite structure. Doping with a lower valent Sr^{2+} ion introduces oxygen vacancies. Under atmospheric pressure condition the compound is a p-type semiconductor.[67] This can be understood to be due to excess oxygen dissolved in those vacant sites as negative ions, being compensated by positive holes. New interest in oxide perovskite, p-type metallic conductors arises, because certain of these oxides exhibit "high temperature" superconductivity.

CRYSTALS WITH MORE THAN ONE EQUIVALENT SITE PER ANION

δ-Bi$_2$O$_3$ and doped δ-Bi$_2$O$_3$ have received much attention in recent years due to their high O^{2-} anionc conductivity. Replacing some of the Bi^{3+} ions by homovalent lanthanides stabilizes the δ phase to lower temperatures. At sufficiently low temperatures the oxygen vacancies are expected to form a periodically ordered array occupying six out of every eight equivalent tetrahedral sites in the fluorite cubic unit cell. High temperature (774°C) neutron diffraction measuremnt show that the anions are displaced along <111> from their (x,x,x) x = 0.25 position to x + δ with δ = 0.066,[68] i.e., similar to the displacement, mentioned before for βPbF2 above the Faraday transition. Thus, four sites per anion, or 32 sites per unit cell, are available for the six anions. X-ray diffraction shows that as Gd$_2$O$_3$ is added to Bi$_2$O$_3$ the lattice constants decrease. δ vanishes at ~ 30 mol% Gd2O3. It is not surprising therefore that the anionic conductivity decreases with increases in Gd$_2$O$_3$ concentration. For 25-50 mol% Gd$_2$O$_3$ the δ phase was found to be stabilized down to room temperature.

For pure Bi$_2$O$_3$ only the δ-fluorite phase is a "pure" ionic conductor, i.e. t$_i$ ~ 1. X-ray and conductivity measurements have revealed that Bi$_2$O$_3$ with 11-25 mol% CdO is a fast ionic conductor with t$_i$ = 1, but possesses a b.c.c. structure.[69] The latter resembles that of α-AgI, the role of cations and anions exchanged. The Cd and Bi cations statistically occupy the corners and center of the cubic unit cell. The six oxygen anions are distributed over the twelve (per unit cell) tetrahedral sites located four on each face of the cubic cell. This phase is unstable below 640°C.

Tb$_2$O$_{3.5}$ at a concentration of 30-50 mol% in Bi$_2$O$_3$ stabilizes the fluorite structure down to room temperature (quenched sample had the fluorite structure, for composition 10-55 mol% Tb$_2$O$_{3.5}$).[70] A weight increase on cooling was observed due to oxygen uptake into the empty oxygen sites of the δ-Bi$_2$O$_3$ structure with the formation of holes for charge compensation. The ionic transference number thereby decreases with increasing content of Tb$_2$O$_{3.5}$.

ONE AND TWO DIMENSIONAL ANIONIC CONDUCTION

Low dimensional anionic conduction is considered here as the motion of an anion vacancies in a 1D or 2D array of anions. The high conduction in the fluorite structure above the Faraday transition and in the fluorite related δ-Bi_2O_3 anionic conductor is due to the existence of a continuous 3D conduction path for anion vacancies at elevated temperatures.

Anion rich compounds YF_3, LaF_3 doped with lower valent cations draw attention as possible fast anion conductors. One identifies chains of close packed anions with few cation neighbors leaving large windows for anion vacancy propoagation.

Evidence for anion conduction along chains can be obtained from the anisotropy of σ_{ion} of PbClBr. $PbCl_2$ and $PbBr_2$ have the orthorhombic structure. The ion charge carriers are anion vacancies, $V^\cdot_{Cl}$ and $V^\cdot_{Br}$ respectively. X-ray diffraction measurements on single crystals show that there are two inequivalent anion sites (see Fig. 6). For the solid solution PbClBr one kind of site is occupied by Cl^- and the other by the larger Br^- ions.[23,24] These sites form chains and a slight anisotropy is observed in σ_{ion}.

ELECTRONIC BAND STRUCTURE

Ionic conductors usually have wide energy gaps in their electronic energy levels. This is a necessary condition for obtaining anionic transference numbers close to unity. t_i - 1 can be maintained also in many doped materials. This is possible when the defect energy level introduced by the dopant is far from the valence or conduction band to which it can contribute a hole or an electron. On the other hand, the fact that e.g. reduced ceria, CeO_{2-x}, is an n-type electronic conductor shows that the donor level introduced by reduction is close enough to the conduction band to be ionized at -500°C.

The density of states of the valence band of CeO_2 was determined from X-ray photoelectron spectroscopy (XPS), while the conduction band density of states was determined from bremsstrahlung isochromat spectroscopy (BIS).[71] The band gap is 6 eV. The valence band originates from O_{2p} extended states. The conduction band from Ce s states. There are $4f^1$ localized empty states within the gap, close to the conduction band. Therefore electron conduction occurs by hopping within the lower $4f^1$ narrow band and the band gap becomes 3-3.5 eV.

The fundamental absorption edge of $(1-x)CdF_2 \cdot xPbF_2$, $0 \le x \le 1$ with the fluorite structure is - 6.2 eV for CdF_2, but decreases by about 2 eV as small quantities of PbF_2 are added.[72] This decreasse is due to a Pb^{2+} atomic-like (weakly forbidden) $6s^2 \rightarrow 6s\,2p$ transition, peaked at 5.7 eV. The absorption edge for βPbF_2 is - 4.2 eV. These values, for the absorption edges, also give a rough estimate of the energy band gap.

$PbSnF_4$ can be quenched to room temperature maintaining the fluorite structure. The energy gap measured (on the quenched sample) is 3.5 eV.[73]

In the field of fast anion conductors, oxides with the perovskite structure were examined as solid electrolytes and as high temperature electrodes under oxidizing conditions. O^{2-} conductivity at 1000°C, close to that of stabilized ZrO_2, was measured for $CaTi_{0.7}Al_{0.3}O_{2.38}$ (Cf. Ref. 17, p. 306). This was accompanied by p-type electronic conduction (except under reducing conditions), which make these materials interesting mixed ionic electronic conductors with possible applications as electrodes.

In recent months, interest arose in perovskite-type oxides with metallic conductivity which undergo a transition to a zero resistance, superconducting state at relatively high temperatures ~100K. $La_{1.85}Ba_{0.15}CuO_{4-\delta}$[74] was first found to become superconductive below ~35K. $YBa_2Cu_3O_{7-\delta}$ is a superconductor below 93K.[75] Its structure is perovskite related orthorhombic as determined by neutron diffraction.[76] It is shown in Fig. 13. Charge balance should lead to $YBa_2Cu_3O_{6.5}$ having some oxygen vacancies in the structure

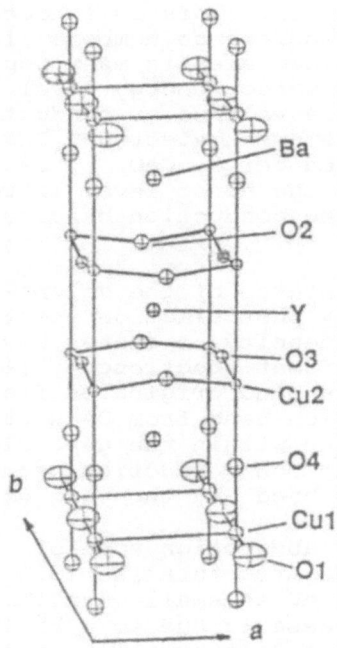

Fig. 13. Perovskite related orthorhombic unit cell of $YBa_2Cu_3O_7$. (From Ref. 76, Fig. 2a.

shown in Fig. 13. Under atmospheric condition, excess oxygen enters the solid, probably into these vacancies, and being negatively charged, are compensated by holes. Under oxidizing conditions, for which the defect concentration is expected to be high, the oxide is found to behave metallic-like. However, at lower oxygen pressures, as the concentration of defects is presumably reduced, the oxide becomes a p-type semiconductor.

Ionic conduction by O^{2-} ions must exist at elevated temperatures as oxygen can be readily exchanged with the atmosphere.

APPLICATIONS

Stabilized ZrO_2 has found wide use in oxygen sensors, fuel cells, water electrolyzers and chemical reaction catalyzers. Stabilized ZrO_2 must be heated. Some fluorine ion conductors exhibit rather good anionic conduction at room temperature. They can therefore be considered e.g. for room temperature batteries. The anionic conductors serve as ion selective electrodes, either for selective sensing or for selectively transporting material. They are used in devices for measuring thermodynamic parameters. Capacitors, timers, and analog integrators can also be prepared from anion conductors in electrochemical devices. Since all these applications have been discussed in detail in the past, we refer the reader to the literature.[77]

An application that evolved in recent years is the use of a room temperature fast anion conductor as an ion source and ion selective electrode to inject anions into another material thereby changing its absorption characteristics (electrochromism). This can be used to control transmission through windows. In the following example, optically transparent thin layers are applied to the window glass in the order: a) SnO_2 - electronic conductor; b) anodic iridium oxide - to be colored by injected F^- anions; c) PbF_2 or $PbSnF_4$ - anion, F^-, selective conductor and anion, F^-, source; and d) gold thin layer - counter electrode. A voltage applied between the Au and SnO_2 electrodes injects or removes F^- ions from the oxide film and changes the oxide film color.[78]

REFERENCES

1. C.R.C. Handbook of Chemistry and Physics, 62nd ed. (1981-82) p. F-175.
2. W.D. Kingery, H.K. Bowen, and D.R. Uhlmann, "Introduction to Ceramics," 2nd ed., pp. 56-61, John Wiley & Sons (1976).
3. L. Börjesson and L.M. Torell, "Raman Scattering Evidence of Rotating SO_4^{2-} in Solid Sulfate Electrolytes," Solid State Ionics 18&19:582-6 (1985).
4. K. Funke and I. Riess, "Debye-Hückel-Type Relaxation Processes in Solid Ionic Conductors," Z. Phys. Chemie NF 140, 217-32 (1984). K. Funke, "Debye-Hückel-Type Relaxation Processes in Solid Ionic Conductors: The Model," Solid State Ionics 18&19:183-90 (1986).
5. L. Slifkin, "Subsurface Effects in Ionic Crystals," Material Science Forum 1: 75-84 (1984).
6. J. Maier, "On the Heterogeneous Doping of Ionic Conductors," Solid State Ionics 18&19:1141-45 (1980).

7. J. Maier, "Space Charge Regions in Solid Two Phase
    Systems and Their Conduction Contribution-II, Contact
    Equilibrium at the Interface of Two Ionic Conductors
    and the Related Conductivity Effect," Ber. Bunsenges.
    Phys. Chem. 89:355-62 (1985).

8. I. Riess, D. Braunshtein, and D.S. Tannhauser, "Density
    and Ionic Conductivity of Sintered
    $(CeO_2)_{0.82}(GdO_{1.5})_{0.18}$," J. Am. Ceram. Soc. 64:479-85
    (1981).

9. W. Schröter and H. Nölting, "Specific Heat of Crystals
    with the Fluorite Structure," J. de Physique
    41:C620-3 (1980).

10. R.W.G. Wyckoff, "Crystal Structures," 2nd ed., Vol. 2,
    Interscience Pub. (1964) pp. 59, 61.

11. P. Shuk and H.H. Möbius, "Überfuhrangszahlen und
    Electrische Leitfahigkeit von Modifikationen des
    Bi_2O_3," Z. Phys. Chemie, Leipzig 266, (1985) pp. 9-16.

12. M.J. Verkerk and A.J. Burggraaf, "High Oxygen Ion
    Conduction in Sintered Oxides of the $Bi_2O_3-Ln_2O_3$
    System," Solid State Ionics 3/4 (1983) pp. 463-67.

13. H. Rickert, "Electrochemistry of Solids, an
    Introduction," Springer-Verlag (1982).

14. P. Hagenmuller and W. van Gool, eds., "Solid
    Electrolytes, General Principles, Characterization,
    Materials, Applications," Academic Press (1978).

15. S. Geller, ed., "Solid Electrolytes," Springer-Verlag
    (1977).

16. O. Toft Sørensen, ed., "Nonstocihiometric Oxides,"
    Academic Press (1981).

17. P. Hagenmuller, ed., "Inorganic Solid Fluorides:
    Chemistry and Physics," Academic Press (1985).

18. R.G. Linford and S. Hackwood, "Physical Techniques for
    the Study of Solid Electrolytes," Chem. Rev. 81:327-64
    (1981).

19. H. Schmalzried, "Solid State Reactions," Verlag Chemie
    and Academic Press (1974).

20. F.A. Kröger, "The Chemistry of Imperfect Crystals," 2nd
    revised edition, Vol. I, II, III, North-Holland Pub.
    Co. (1974).

21. S. Fujitsu, K. Koumoto, and H. Yanagida, "Enhancement of
    Ionic Conudctivity of $SrCl_2$ by Al_2O_3 Dispersion,"
    Solid State Ionics 18&19:1146-49 (1986).

22. J. Corish, B.M.C. Parker, J.M. Quigley, A.R. Allnatt and
    D.C.A. Mulcahy, "Point-Defect Mobility in Thallous
    Chloride Doped with Divalent Cation and Anion
    Impurities," J. Phys. C: Solid State Phys.,
    17:2689-704 (1984).

23. M. Lumbreras, J. Protas, S. Jebbari, G.J. Dirksen, and
    J. Schoonman, "Crystal Growth and Characterization of
    Mixed Lead Halides $PbCl_{2x}Br_{2(1-x)}$," Solid State Ionics
    16:195-200 (1985).

24. M. Lumbreras, J. Protas, S. Jebbari, G.J. Dirksen, and
    J. Schoonman, "Crystal Growth and Characterization of
    Mixed Lead Halides $PbCl_{2x}Br_{2(1-x)}$. I." Solid State
    Ionics, 18&19:1179-83 (1986).

25. V.K. Yanovskii, V.I. Voronkova, Yu.F. Roginskaya, and
    Yu.N. Venevtsev, "Rapid Anion Transfer in Bi_2WO_6
    Crystals," Sov. Phys. Solid State 24:1603-4 (1982).

26. J. Nölting, "Scanning Calorimetry with Adiabatic or Controlled Diabatic Surroundings," Thermochimica Acta 94:1-15 (1985).
27. M. Ouwerkerk, E. Kelder, and J. Schoonman, "Conductivity and Specific Heat of Fluorites $M_{1-x}U_xF_{2+2x}$ (M = Ca, Sr, Ba, and Pb)" Solid State Ionics 9&10:531-36 (1983).
28. M. Ouwerkerk and J. Schoonman, "The Critical Temperature in Fluorite-Type Solid Solutions," Solid State Ionics 12:479-84 (1984).
29. K.Koto, H. Schulz, and R.A. Huggins, "Anion Disorder and Ionic Motion in Lead Fluorite (β-PbF_2)," Solid State Ionics 1 (1980) pp. 355-65.
30. M.H. Dickens, W. Hayes, C. Smith, and M.T. Hutchings, "Anion Disorder in Two Fluorites at High Temperatures Determined by Neutron Diffraction," in "Fast Ion Transport in Solids," Vashista, Munday, Shenoy eds. Elsevier (1979) pp. 225-28.
31. Y. Ito, K. Koto, S. Yoshikado and T. Ohachi, "The Contribution of Anion Disorder to Ionic Conudctivity on Single Crystals of β-PbF_2," Solid State Ionics 15:253-58 (1985).
32. J.A. Kilner and C.D. Waters, "The Effects of Dopant Cation-Oxygen Vacancy Complexes on the Anion Transport Properties of Nonstoichiometric Fluorite Oxides," Solid State Ionics 6 (1982) pp. 253-59.
33. K. Fuda, K. Koshio,. S. Yamauchi, K. Fueki, and Y. Onoda, "^{17}O NMR Study of Y_2O_3 Doped CeO_2," J. Phys. Chem. Solids 45:1253-57 (1984).
34. D.Y. Wang, D.S. Park, J. Griffith, and A.S. Nowick, "Oxygen-Ion Conductivity and Defect Interactions in Yttria-Doped Ceria," Solid State Ionics 2:95-105 (1981).
35. P. Sarkar and P.S. Nicholson, "ac Conductivity and Conductivity Relaxation Studies in the CeO_2-Y_2O_3 System," Solid State Ionics, 21:49-53 (1986).
36. A.S. Nowick and B.S. Berry, "Anelastic Relaxation in Crystalline Solids," Academic Press (1972).
37. M.P. Anderson and A.S. Nowick, "Relaxation Peaks Produced by Defect Complexes in Cerium Dioxide Doped with Trivalent Cations," J. de Physique 42:C5-823-28 (1981).
38. R. Gerhardt-Anderson, F. Zamani-Noor and A.S. Nowick, "Study of Sc_2O_3-Doped Ceria by Anelastic Relaxation," Solid State Ionics 9&10:931-36 (1983).
39. M.P. Anderson, D.E. Cox, K. Halperin, and A.S. Nowick, "Neutron Diffuse Scattering in Y_2O_3-and Sc_2O_3-Doped CeO_2," Solid State Ionics 9&10:953-60 (1983).
40. A. Overs and I. Riess, "Properties of the Solid Electrolyte Gadolinia-Doped Ceria Prepared by Thermal Decomposition of Mixed Cerium-Gadolinium Oxalate," J. Am. Ceram. Soc. 65:606-09 (1982).
41. W. Tinglian, L. Xiaofei, K. Chukun, and W. Weppner, "Conductivity of MgO-Doped ZrO_2," Solid State Ionics 18&19:715-19 (1986).
42. J. Shinar, D.S. Tannhauser, and B.L. Silver, "ESR Study of Color Centers in Yttria Stabilized Zirconia," Solid State Ionics 18&19, 912-15 (1986).

43. J. Dexpert-Ghys, M. Faucher, and P. Caro, "Site Selective Spectroscopy and Structural Analysis of Yttria-Doped Zirconia," J. Solid State Chem. 54:179-92 (1984).

44. J.M. Reau, A. Phandour, S.F. Matar, and P. Hagenmuller, "Optimisation des Facteurs Influencant la Conductivite Anionique dans quelques Fluorunes de Structure Fluorine," J. Solid State Chem. 55:7-13 (1984).

45. M. Ouwerkerk, N.H. Andersen, F.F. Veldkamp, and J. Schoonman, "Neutron Diffraction and TSDC on $Ba_{1-x}U_xF_{2+2x}$ Solid Electrolytes," Solid State Ionics 18&19:916-21 (1986).

46. A.S. Nowick, "The Combining of Dielectric and Anelastic Relaxation Measurements in the Study of Point Defects in Insulating Crystals," J. de Physique 46:C10 507-11 (1985).

47. K.E.D. Wapenaar, H.G. Koekkoek and J. van Turnhout, "Low Temperature Ionic Conductivity and Dielectric Relaxation Phenomena in Fluorite-Type Solid Solutions," Solid State Ionics 7:225-42 (1982).

48. J. Meuldijk, R. van der Meulen, and H.W. den Hartog, "Dielectric-Relaxation Experiments on Cubic Solid Solutions of SrF_3 and CeF_3 or PrF_3," Phys. Rev. B29:2153-59 (1984).

49. J. Meuldijk, G. Kiers, and H.W. den Hartog, "Effect of Clustering on the Space-Charge Relaxation Phenomena in Fluorite Type Solid Solutions $Sr_{1-x}Dy_xF_{2+x}$ and $Sr_{1-x}Er_xF_{2+x}$," Phys. Rev. B28:6022-30 (1983).

50. H.W. den Hartog and J. Meuldijk, "Dipoles in Solid Solutions $Sr_{1-x}Gd_xF_{2+x}$," Phys. Rev. B29:2210-15 (1984).

51. L.A. Muradyan, B.A. Maksimov, B.F. Mamin, N.N. Bydanov, V.A. Sanin, B.P. Sobolév, and V.I. Simonov, "Atomic Structure of the Nonstoichiometric Phases $Sr_{0.69}La_{0.31}F_{2.31}$," Sov. Phys. Crystallogr. 31:145-47 (1986).

52. H. Geiger, G. Schön, and H. Stork, "Ionic Conductivity of Single Crystals of the Nonstoichiometric Tysonite Phase $La_{1-x}Sr_xF_{3-x}$ ($0 \leq x \leq 0.14$)," Solid State Ionics 15:155-58 (1985).

53. A. Roos, M. Buijs, K.E.D. Wapenaar, and J. Schoonman, "Dielectric Relaxation Properties of Tysonite-Type Solid Solutions $La_{1-x}Ba_xF_{3-x}$," J. Phys. Chem. Solids 46:655-64 (1985).

54. M. Ricken, J. Nölting, and I. Riess, "Specific Heat and Phase Diagram of Nonstoichiometric Ceria (CeO2-x)," J. Solid State Chem. 54:89-99 (1984).

55. I. Riess, M. Ricken, and J. Nölting, "On the Specific Heat of Nonstoichiometric Ceria," J. Solid State Chem. 57:314-22 (1985).

56. P. Knappe and L. Eyring, "Preparation and Electron Microscopy of Intermediate Phases in the Interval $Ce_7O_{12}-Ce_{11}O_{20}$," J. Solid State Chem. 58:312-324 (1985).

57. R.T. Tuenge and L. Eyring, "On the Structure of the Intermediate Phases in the Terbium Oxide System," J. Solid State Chem. 41:75-89 (1982).

58. H. Sato, N. Otsuka, H. Kuwamoto, and G.L. Liedl., "Nonstoichiometry and Defects in V_9O_{17}," J. Solid State Chem. 44:212-29 (1982).

59. I. Riess, H. Janczikowski, and J. Nölting, "O_2 Chemical Potential of Nonstoichiometric CeO_{2-x}," J. Appl. Phys. 61:4931-33 (1987).

60. J.W. Dawicke and R.N. Blumenthal, "Oxygen Association Pressure Measurements on Nonstoichiometric Cerium Dioxide," J. Electrochem. Soc. 133 (1986) pp. 904-9.

61. E.K. Chang and R.N. Blumenthal, "The Nonstoichiometric Defect Structure and Transport Properties of CeO_{2-x} in the Near-Stoichiometric Composition Range," J. Solid State Chem. 72:330-37 (1988).

62. H. Yahiro, K. Eguchi, and H. Arai, "Ionic Conduction and Microstructure of the Ceria-Strontia System," Solid State Ionics 21:37-47 (1986).

63. H.L. Tuller and T.S. Stratton, "Defect Structure and Transport in Oxygen Excess Cerium Oxide-Uranium Oxide Solid Solution," in Proceedings 3rd Int. Conf. on Transport in Nonstoichiometric Compounds, Penn. State Univ., University Park, PA, U.S.A., June 10-16, 1984.

64. H.L. Tuller, J.A. Kilner, A.E. McHale, and B.C.H. Steele, "Oxygen Diffusion in Oxygen Excess CeO_2-UO_2 Solid Solutions" in Reactivity in Solids," edited by P. Barrett and L.C. Dufour, Elsevier Science Pub. B.V. (1985) pp. 315-19.

65. H.H. Fujimoto and H.L. Tuller, "Mixed Ionic and Electronic Transport in Thoria Electrolytes," in "Fast Ion Transport in Solids," Vashista, Mundy, Shenoy eds., Elsevier (1979) pp. 649-52.

66. M. Inouye, M. Iwase, and T. Mori, "Mixed Ionic and n-Type Electronic Conduction in Commercial ZrO_2 + 11 mol% CaO Solid Electrolyte," Transactions ISIJ 21:54-5 (1981).

67. M. Kertesz, I. Riess, D.S. Tannhauser, R. Langpape, and F.J. Rohr, "Structure and Electrical Conductivity of $La_{0.86}Sr_{0.16}MnO_3$," J. Solid State Chem. $\underline{42:}$125-29 (1982).

68. K. Koto, H. Mori, and Y. Ito, "Oxygen Disorder in the Fluorite-Type Conductors $(Bi_2O_3)_{1-x}(Gd_2O_3)_x$ by X-ray and EXAFS Analysis," Solid State Ionics 18&19:720-24 (1986).

69. T. Graia, P. Conflant, J.-C. Boivin and D. Thomas, "High Oxygen Ion Conduction in a Bismuth Oxide-Cadmium Oxide Phase: Conductivity and Transport Number Measurements; Structural Investigations," Solid State Ionics 18&19:751-55 (1986).

70. T. Esaka and H. Iwara, "Oxide Ion and Electron Mixed Conduction in the Fluorite-Type Cubic Solid Solution in the System Bi_2O_3-$Tb_2O_{3.5}$," J. Appl. Electrochem. 15:447-51 (1985).

71. E. Wuilloud, B. Delley, W.-D. Schneider and Y. Baer, "Spectroscopic Evidence for Localized and Extended f-symmetry States in CeO_2," Phys. Rev. Lett. 53:202-5 (1984).

72. I. Kosacki and J.M. Langer, "Fundamental Absorption Edge of PbF_2 and $Cd_{1-x}Pb_xF_2$ Crystals," Phys. Rev. 33:5972-73 (1986).

73. G. Couturier, Y. Danto, J. Pistre, J. Salardenne, C. Lacat, J. M. Reau, J. Portier, and S. Vilminot, "The Anionic Conductor $PbSnF_4$: A Study of Thin Film and Ceramics," in Fast Ion Transport in Solids, Vashishta, Munday and Shenoy eds., Elsevier 1979, pp. 687-90.

74. J.G. Bednorz and K.A. Müller, "Possible High T_c Superconductivity in the Ba-La-Cu-O System," Z. Phys. B 64:189-93 (1986).

75. M.K. Wu, J.R. Ashburn, C.J. Torng, P.H. Hor, R.L. Meng, L. Gao, Z.J. Huang, Y.Q. Wang, and C.W. Chu, "Superconductivity at 93K in a New Mixed-Phase Y-Ba-Cu-O Compound System at Ambient Pressure," Phys. Rev. Lett. 58:908-10 (1987).

76. M.A. Beno, L. Soderholm, D.W. Capone II, D.G. Hinks, J.D. Jorgensen, I.K. Schuller, C.U. Serge, KI. Zhang, and J.D. Grace, "Structure of the Single Phase High Temperature Superconductor $YBa_2Cu_3O_7$," Appl. Phys. Lett. 51:57-9 (1987).

77. R.A. Huggins, "Some Non-Battery Applications of Solid Electrolytes and Mixed Conductors," Solid State Ionics 5:15-20 (1981).

78. C.E. Rice and P.M. Bridenbaugh, "Observation of Electrochromism in Solid-State Anodic Iridium Oxide Film Cells Using Fluorine Electrolytes," Appl. Phys. Lett. 38:59-61 (1981).

79. A.E. McHale and H.L. Tuller, "New Tantala-Based Solid Oxide Electrolytes," Solid State Ionics 5:515-18 (1981).

AMORPHOUS FAST ION CONDUCTORS

Harry L. Tuller

Crystal Physics and Optical Electronics Laboratory
Department of Materials Science & Engineering
Massachusetts Institute of Technology
Cambridge, MA   02139   USA

INTRODUCTION

It is now several decades since the subject of fast ion conduction in crystalline materials was first seriously examined. Scientists were and still remain fascinated with the source of the anamalously high ion conductivities and technologists with the possibilities for these materials in high energy density batteries, sensors, displays, etc. Although these materials, as exemplified by α-AgI and Na β-alumina, were crystalline, many found it convenient to discuss transport in these materials as being liquid-like but bound by an immobile periodic lattice which provided easy transport paths.

It is not surprising therefore that glasses, which after all are but frozen liquids, would sooner or later attract the attention of the fast ion conductor (FIC) community. In the last decade, many hundreds of glass compositions have been investigated with the focus of attention on silver and alkali ion conductors. Some interesting work has also been reported on fluorine and more recently on lead ion conductors.

Glassy FIC's do not generally, exhibit markedly higher ionic conductivities than corresponding crystals. They are, however, very attractive because of unique advantages peculiar to the glassy state. Perhaps most important is the ease with which glass can be fabricated into complex shapes. Glass optical fiber systems, thousands of kilometers long, presently criss-cross our oceans, while plate glass windows many square meters in area decorate our modern office buildings. Glass can easily be drawn into thin walled tubes at relatively low temperatures and formed to make impervious seals. Their isotropic character and lack of grain boundaries eliminate major sources of resistive loss and chemical attack particularly important in the aggressive environments of high energy density batteries. Thin walls allow for wider flexibility in the choice of specific glasses. A case in

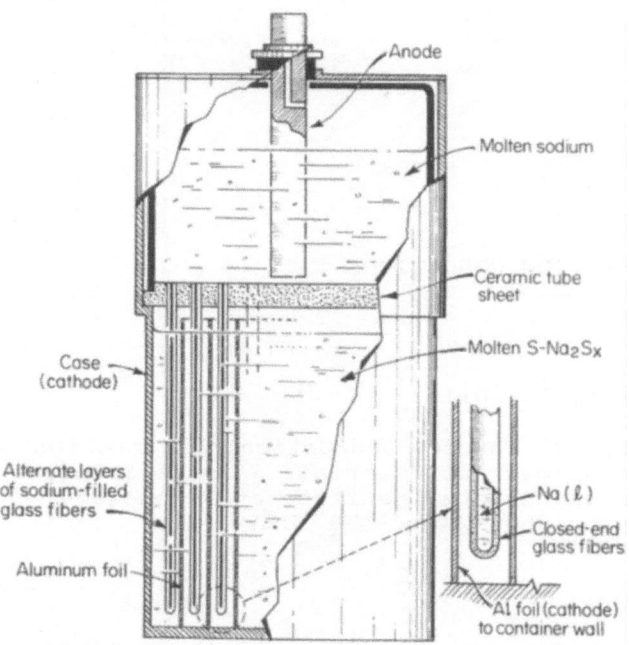

Fig. 1.   Schematic diagram of Dow cell using glass
electrolytes (from Ref. 1).

point is the Dow Chemical Company battery design[1]
illustrated in Figure (1) in which thousands of thin hollow
glass fibers (typically 50μm ID by 80μm OD) are used as the
electrolyte separating the molten sodium anode from the sodium
sulfide cathode thereby providing a relatively low resistance
electrolyte path while using a relatively high resistivity
(10^4 Ω-cm under operating conditions) but chemically stable
glass.

High ionic conductivities alone are insufficient for many
of the applications envisaged for those materials. On par
with conduction, as illustrated above, is the ability of the
electrolytes to withstand corrosive environments. While many
crystalline and glass fast ion conductors exhibit the
requisite level of conduction, they rapidly degrade upon
exposure to agressive environments. Further, since glasses
are not thermodynamically stable, thermal stability becomes
another important paramater. Thus glasses with higher glass
transition temperatures (Tg) are preferred, all other factors
being equal.

Given the need to optimize a number of paramaters
simultaneously, one obviously desires as large a pool of
compositions as possible to choose from. Here again, glasses
provide some unique opportunities. Since glasses are not
restricted to the narrow compositional limits characteristic
of most crystalline compounds, they provide an additional
important level of flexibility in tailoring materials
properties. Figures (2) and (3) for example, illustrate the

iso-Tg and iso-density curves within the broad glass forming regions of the Li ion conducting system $Li_2O-(LiCl)_2-B_2O_3$.[2] Even larger glass forming regions are characteristic of the $Ag_2O-AgI-B_2O_3$ system.[3] Table I summarizes the attractive characteristics of glasses. A number of these will become more apparent as we discuss the transport and structural properties of FIC glasses in more detail.

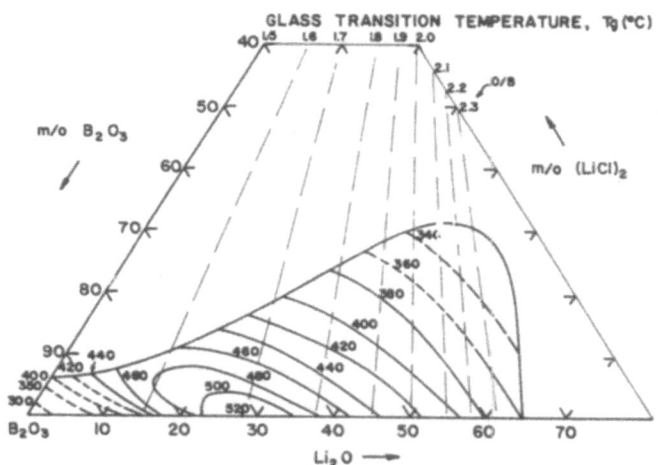

Fig. 2.   Iso-Tg curves shown within glass forming region of system $Li_2O-(LiCl)_2-B_2O_3$. Straight dashed lines correspond to fixed O/B ratios (from Ref. 2).

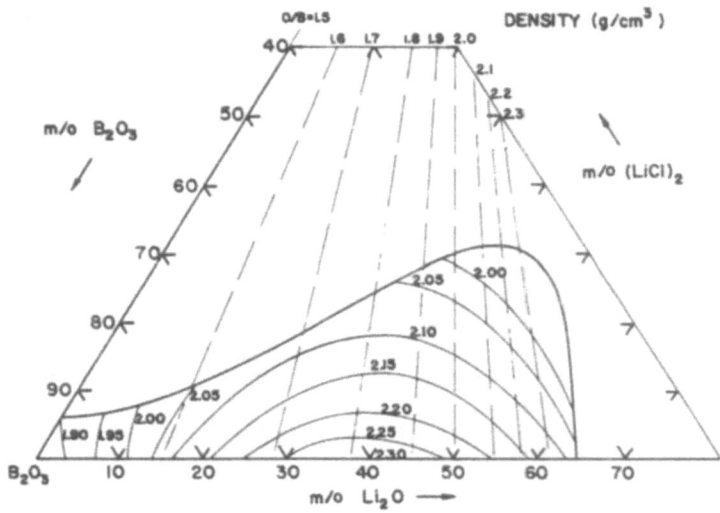

Fig. 3.   Iso-density curves for conditions as in Fig. 2 (from Ref. 2).

Table I.   Characteristics of Glass

Fabrication	1.	Easily formed into complex shapes.
	2.	May often be sealed readily to other materials.
	3.	May be drawn into thin walled structures.
Structural	1.	Isotropic.
	2.	Absence of grain boundaries.
	3.	Open structures.
	4.	Structural flexibility.
Compositional	1.	Wide compositional ranges.
	2.	Accommodate many components.

Fast ion conduction in glasses is now an established fact. Many new systems are being identified yearly. The understanding of the transport mechanisms however, lags substantially behind that in corresponding crystalline conductors. The great strides made in the last fifty years in developing a quantitative description of ionic conduction in crystals has gone hand-in-hand with advances in understanding of crystal structure—both perfect and imperfect. Corresponding correlations between ion transport and structure in glasses have not been emphasized until recently. Clearly the disordered nature of glasses and lack of convenient tools for structure determination (e.g. x-ray and neutron diffraction) have hampered this approach.

Glass FIC´s provide us with a number of challenges. Phenomenologically, can we identify more glass systems which exhibit an optimized mix of properties e.g. high ionic conductivity, chemical and thermal stability, minimal electronic conduction, and ease of fabrication? Can we extend the families of FIC glasses to include other mobile ions as has apparently been demonstrated recently with lead? In other cases we may wish to support mixed ionic and electronic conduction for certain electrode applications.

The challenge from a more fundamental standpoint relates to the clarification of the mechanisms for conduction. In the generalized definition of the electrical conductivity, given by

$$\sigma = \Sigma\sigma_i = \Sigma n_i Z_i q\mu_i \qquad (1)$$

one finds that each partial conductivity σ_i is the convolution of the mobile carrier concentration n_i, the charge $Z_i q$, and the carrier mobility μ_i. Much controversy still remains in the glass literature regarding the relative importance of carrier density and mobility to the magnitude, composition, and temperature dependence of the ionic conductivity.

Notwithstanding these difficulties, much may be learned regarding FIC in general by the study and comparison of crystalline and amorphous conductors. For example, certain common structural features characteristic of FIC´s were generally accepted for many years.[4]

54

These include:

(a) a highly ordered structural array, which may be in the form of tunnels, layers, or three dimensional arrays, generally localized on one of the ion sublattices, and

(b) a highly disordered complementary sublattice in which the number of equivalent sites is greater than the number of available ions to fill them.

The existence of FIC in disordered materials such as glasses contradicts the first of the above structural premises while the latter appears to be characteristic of both FIC crystals and glasses. Long range periodicity is therefore not a prerequisite for support of long range (ionic) transport in glass FIC's. A similar conclusion was arrived at earlier by physicists who studied electronic transport in amorphous semiconductors.

Following the above comments, I will break the following discussion up into two major categories. The first will be a summary of the phenomenological trends as functions of glass composition and the second will concern itself with the interpretation of these trends in terms of transport models. Since structure must play a major role in influencing carrier mobilities, a review of structural trends in glasses with composition will also be reviewed.

FIC GLASSES. PROPERTY TRENDS

In reviewing property trends as a function of composition, it is useful to summarize the nature of the major constituents of FIC glasses. These include

i) network formers (e.g. SiO_2, B_2O_3, P_2S_5)

ii) network modifiers (e.g. Ag_2O, Li_2O, Na_2S)

iii) doping salts (e.g. AgI, LiCl, NaBr)

The network formers typically represent co-valently bonded units (e.g. SiO_4 tetrahedra, BO_3 triangles) which form strongly cross-linked macro-molecular chains and serve as the structural "backbone" of the glass. Figure (4) illustrates those elements which commonly serve as network formers in oxides. Modern x-ray and neutron diffraction techniques support a "random network model" of glasses in which these structural units are interconnected in three dimensional arrays without repeating at regular intervals as in a crystal. This leads to a dispersion of valence angles and bond lengths within and between these units. The manner in which the units are interconnected also depends directly on the anion (e.g. O, S) to network forming ion (e.g., Si, B, P) ratio. In crystalline silicates, for instance, sheets, chains, rings, and three dimensionally interconnected structures of silica tetrahedra are found depending on the O/Si ratio with similar configurations expected in glasses of like composition.

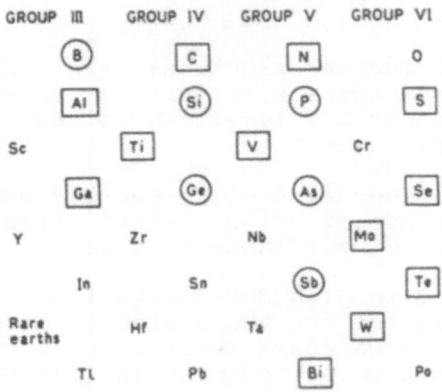

Fig. 4.   Elements of periodic table, the oxides of which are
either glass formers or conditional glass formers
(from A. Paul, Chemistry of Glasses, Chapman and
Hall, London, 1982).

The O/Si ratio in silica glasses for example is readily
controlled by the addition of alkali oxide "modifier" such as
Na_2O which interact strongly with the network formers. While
the alkali ions are incorporated into the interstices of the
network, the excess oxygen is accommodated by the network by
rupturing "bridging" oxygen bonds between adjacent Si atoms
and replacing them with two "non-bridging oxygens" often
designated as "NBO´s." This process may be represented by the
following reaction

$$Si-O-Si + Na_2O \longrightarrow 2Si-O^-:Na^+ \qquad\qquad (2)$$

in which the interstitial Na^+ ion is observed to charge
compensate the unsaturated non-bridging Si-O bond. The
network modifiers, as their name implies, serve to induce
structural changes in the glass network which are reflected in
many physical properties including melting point, glass
transition temperature T_g, density, refractive index and
others (see Figures 2 and 3). At the same time they introduce
ionic bonds into a structure which was previously virtually
covalently bonded. This leads to notable increases in ionic
conductivity as the modifier ion level in the glass is
increased. The manner in which other network formers (e.g.
B_2O_3, P_2O_5) are modified is discussed later.

The last of the major constituents are the doping salts
which, unlike the network modifiers, appear to enter the glass
with little direct effect on the network. Here both the
cation and anion are accommodated into the glass
interstitially with the cations contributing substantially to
the ionic conductivity. Whether these salts form clusters or
are homogeneously distributed still remains a matter of
controversy.

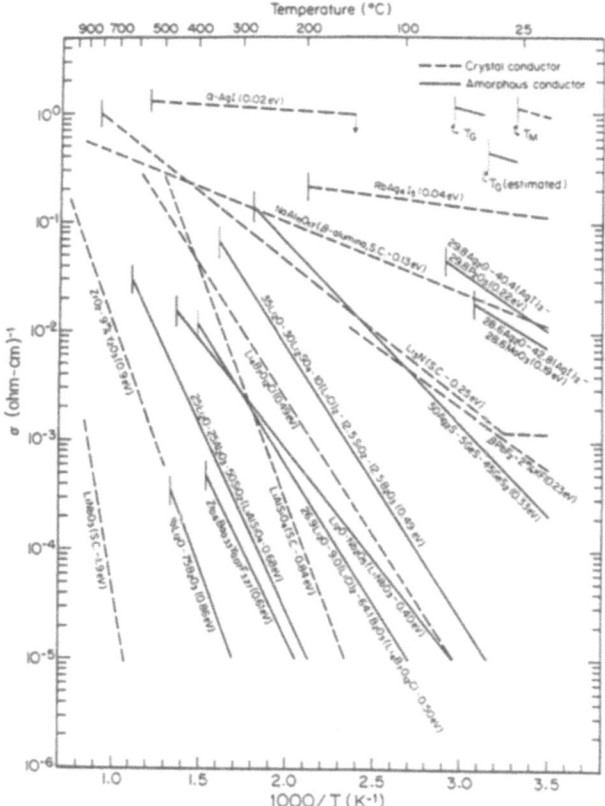

Fig. 5.   The ionic conductivities of a number of fast Ag, Li, and F conducting glasses as a function of reciprocal temperature are compared with better known crystalline fast ion conductors (from Ref. 4).

Several of the first questions which come to mind when considering FIC glass include:

i) How conductive do FIC glasses get?

ii) Does the long range disorder associated with glass formation aid or hinder FIC?

To answer these questions let us begin with Figure (5) which is now becoming a bit dated but nevertheless remains very instructive. First, it is clear that glasses can be highly conductive as exemplified by the silver phosphate and silver molybdate glasses which exibit silver ion conductivities on the order of 10^{-2} S/cm at room temperature. Other glass FIC´s illustrated in this figure, which are not as conductive, include Li and F ion conductors. Second, several comparisons may be made between materials of like composition but of different structure, i.e. crystalline versus amorphous. In the case of $Li_4B_7O_{12}Cl$ in which the crystal is itself FIC, the crystalline and amorphous phases are very similar with the

57

crystal being perhaps twice as conductive as the glass but with near identical activation energy. On the other hand, for the case of LiNbO$_3$ where the crystal is an insulator with activation energy of 1.9eV, the disordered phase is many orders of magnitude more conductive and on the same order of magnitude as the Li$_4$B$_7$O$_{12}$Cl material. Generally, one can expect that the disorder induced by forming a glass, opens up additional avenues for ionic transport in solids by formation of extensive networks of "defects." Long range disorder does not on the other hand appear to markedly disrupt optimized conduction in materials which exhibit it in the crystalline phase.

The majority of studies on FIC in glasses have been performed on Ag and Li ion conductors since they exhibit the highest ionic conductivities and in the latter case the greatest likelihood for practical application. Figures (6) and (7) are log conductivity versus reciprocal temperature plots for silver and lithium ion conductors recently compiled by Souquet and Kone.[5] While crystalline Ag ion conductors (e.g. RbAg$_4$I$_5$) still represent the best Ag FIC's, the reverse is now true for Li ion conductors (e.g. 0.44 LiI, 0.30 Li$_2$S, 0.26 B$_2$S$_3$). Note also that sulfide glasses generally provide considerably higher ionic conductivities than do the

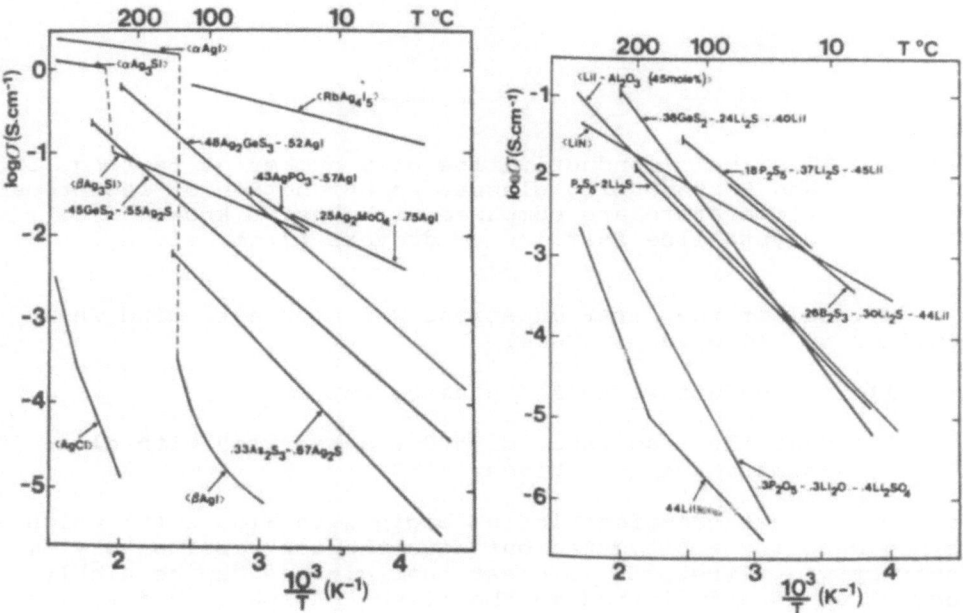

Fig. 6. Comparisons of silver ion conducting crystalline (< >) and glass electrolytes (from Ref. 5).

Fig. 7. Comparisons of lithium ion conducting crystalline (< >) and glass electrolytes (from Ref. 5).

corresponding oxides. It also follows that many glass systems now exhibit requisite electrical properties for ambient electrochemical devices.

As mentioned above, network modifiers provide ionic carriers to networks which are initially devoid of predominantly ionic bonds. It is thus expected that the ionic conductivity would increase systematically with the addition of network modifiers. This is generally the case and is illustrated in several figures below.

The ionic conductivities of oxide glass formers have long been known to increase dramatically with the addition of

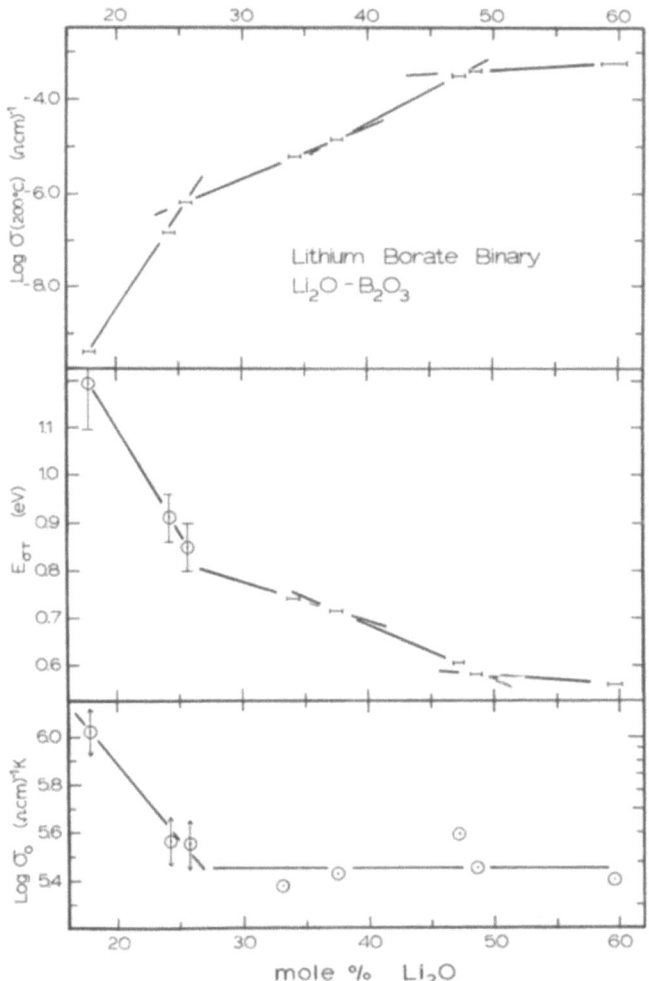

Fig. 8. Log conductivity (200°C), activation energy, and log σ_o are plotted versus mole% Li_2O for binary borate glasses with 18-60 mole% Li_2O (from Ref. 7).

alkali oxide modifiers (Doremus, 1973).[6] This is
illustrated in Figure (8) in which one observes that a two
fold increase in Li_2O in a $Li_2O-B_2O_3$ glass from 18-36mol%
increases the Li ion conductivity at 200°C by over four orders
of magnitude.[7] It is further clear from the figure that the
rapid increase in conductivity which generally takes the form

$$\sigma = \sigma_o/T \exp(-E/kT) \qquad (3)$$

is due, almost in its entirety, to a corresponding decrease of
the activation energy to nearly one half of its initial value.

Similar sharp increases in ionic conductivities at 25°C
with increasing M_2S for glasses in the system $xM_2S-(1-x)GeS_2$
where M = Ag, Li, Na are illustrated in Figure (9). One also
observes the general trend of increasing conductivity in a
given system in the order Na, Li, and Ag. In Figure (10), one
observes the network modifier effect in both oxides and
sulfide glasses.[8] Here one sees, most dramatically, the
enhancing effect of network sulfur over oxygen on conduction.
Given the much higher initial conductivity, it is not
surprising to see the weaker dependence of conductivity on
modifier content for the sulfide glass.

We next examine the remarkable effects that doping salts
have on conductivity. Figure (11) shows the influence of AgI
additives on a series of Ag ion conducting glass systems, both
sulfides and oxides.[9] In the most dramatic case, the room
temperature conductivity is increased approximately four
orders of magnitude in the $Ag_2O-P_2O_5$ glass as the AgI mol
fraction is increased to roughly fifty percent. The nearly

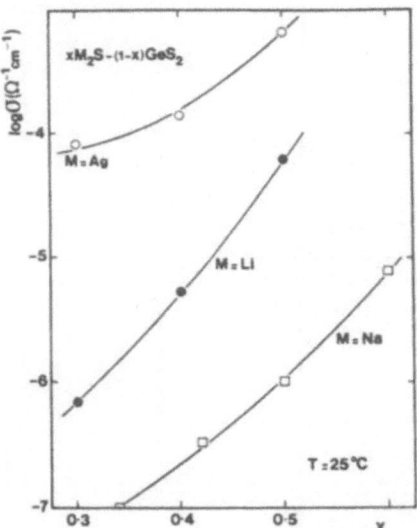

Fig. 9. Log conductivity (25°C)
for glasses in system x M_2S-(1-x)
GeS_2 as a function of M_2S content
(M=Ag,Li,Na) (quoted in ref. 5).

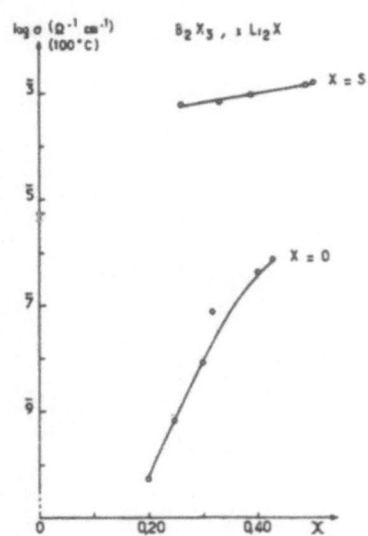

Fig. 10. Log conductivity
(100°C) for glasses Li_2X-
B_2X_3 (X=0,S) vs. molar
fraction Li_2X (Ref. 8).

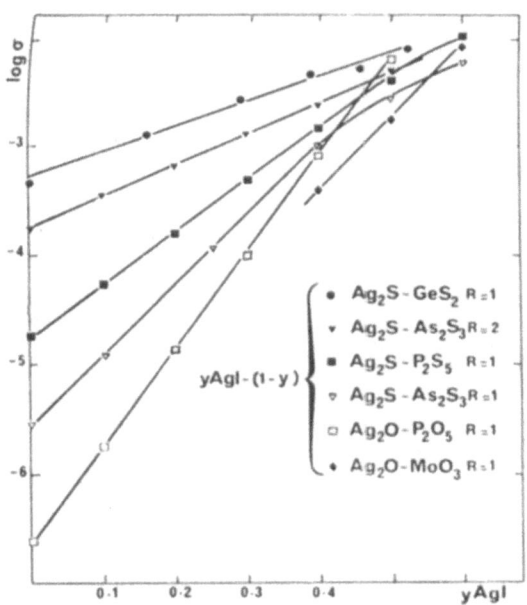

Fig. 11.   Log conductivity for a variety of oxide and sulfide based glasses as a function of AgI content (from Ref. 9).

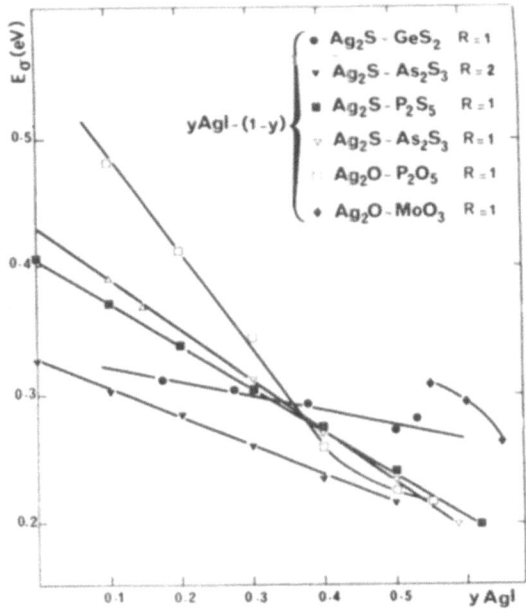

Fig. 12.   Activation energy versus AgI content in various oxide and sulfide based glass (from Ref. 9).

61

linear increase in log conductivity with mole fraction AgI is
particularly striking and is characteristic of the types of
changes induced by doping salts in many glasses. Figure (12)
demonstrates that the linear change in log σ reflects a
corresponding linear decrease of the activation energy, E,
with fraction AgI. The reason for this simple linear
relationship still remains somewhat illusive and controversial
as we shall later see.

The nature of the salt can also play an important role.
In Figure (13), we see the same linear increase in log σ
(25°C) with mole fraction silver halide as before but the
magnitudes and slopes of these curves depend in a sensitive
manner on the nature of the halide.[10] Here, as in many
other cases, one finds that the conductivity increases with
AgX (X = Cl, Br, and I) in the order AgCl, AgBr, and AgI.
Figure (14) demonstrates that the pre-exponential factor σ_0 in
Equation (3) remains nearly the same for all the glasses
confirming that the variation of conductivity with composition
results from the variation of activation energy with
composition.

Both network modifier and dopant salt enhance
conductivity. Can we establish the relative importance of
each in enhancing conductivity when both are added
simultaneously? In attempting to answer this question, Minami
et al[11] plotted log σ (25°C) both versus the total Ag ion
content as in Figure (15) and the Ag ion concentration
contributed by the salt alone as in Figure (16). These

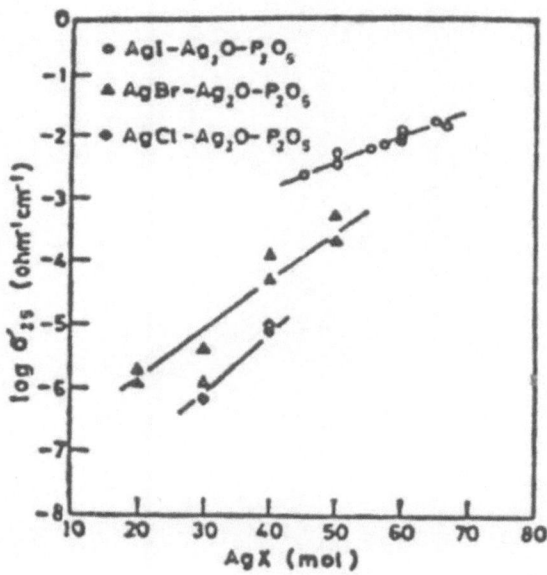

Fig. 13. Log conductivity (25°C) versus AgX content in AgX-
          $Ag_2O-P_2O_5$ glasses where X = I, Br, Cl. (from Ref.
          10).

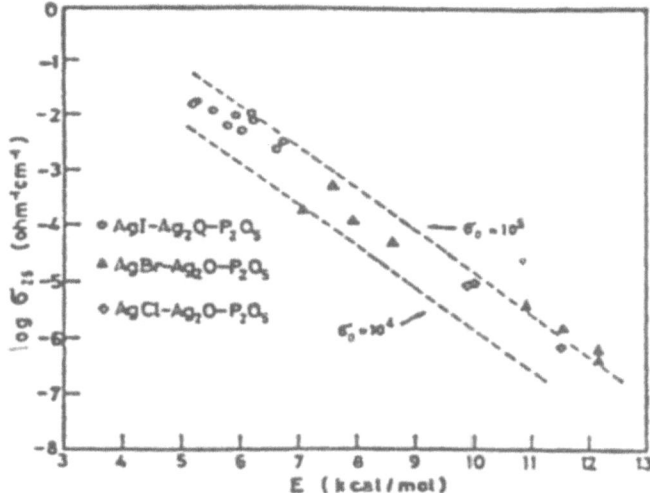

Fig. 14. Relation between (25°C) and activation energy E for glasses indicated. Dotted lines represent values for the pre-exponential $\sigma_o = 10^4$ and 10^5 Scm^{-1}K (from Ref. 10).

results clearly suggest that the total Ag content is not the key criteria but rather that fraction associated with the halides. This observation serves as a key test to proposed transport models.

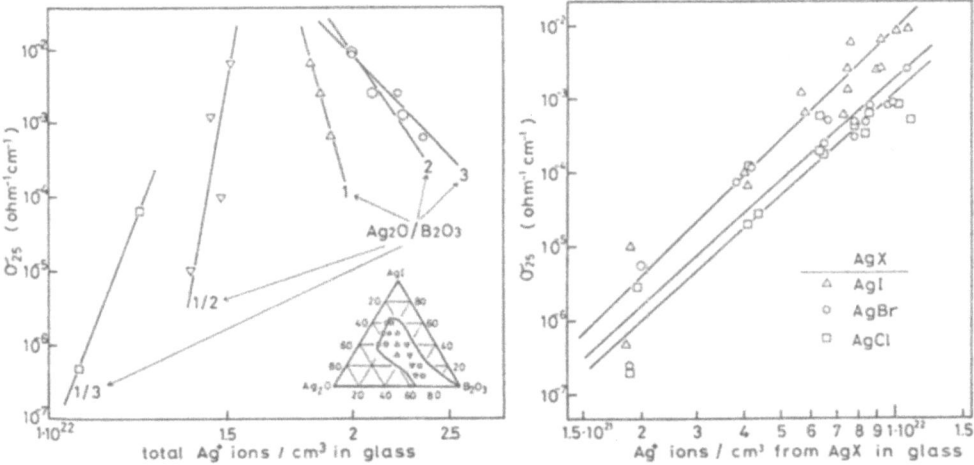

Fig. 15. Conductivities at 25°C as a function of the total concentration of Ag ions in AgI-Ag$_2$O-B$_2$O$_3$ glass. Curves correspond to constant Ag$_2$O/B$_2$O$_3$ ratios (from Ref. 11).

Fig. 16. Conductivities at 25°C as a function of Ag ions introduced into AgX-Ag$_2$O-B$_2$O$_3$ glasses by the AgX components (from Ref. 11).

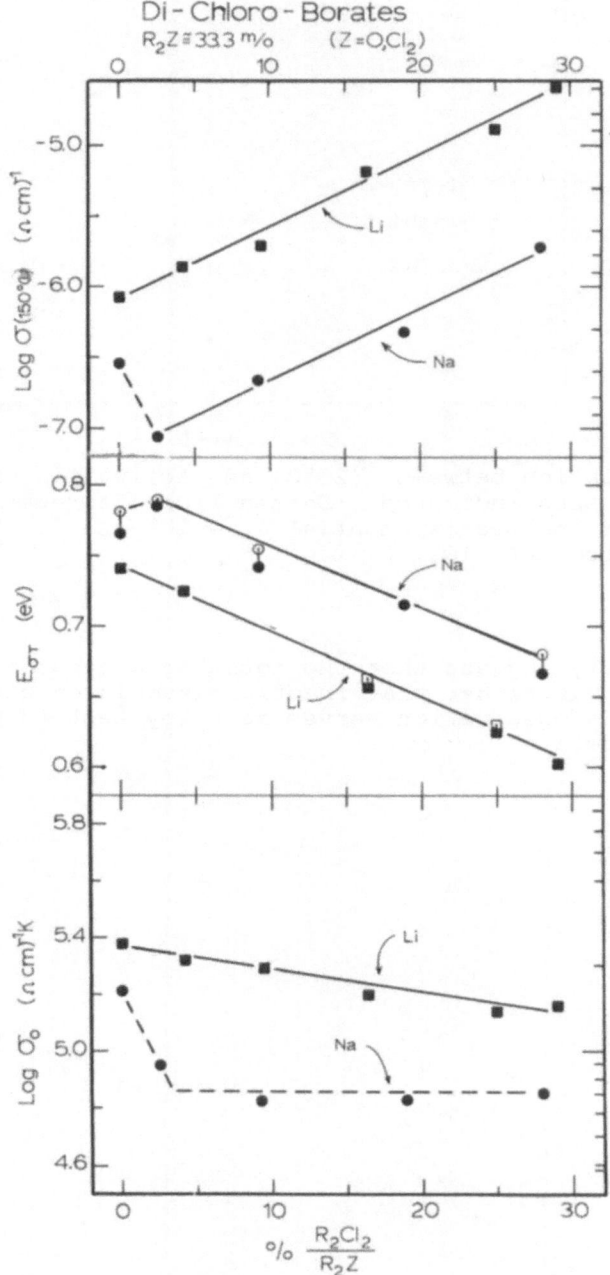

Fig. 17.   Conductivities (150°C), activation energy E, and
pre-exponential are plotted vs. relative Cl content
for a series of glasses $R_2 Z-2B_2 O_3$ (R = Li or Na; Z =
O, Cl_2) (from Ref. 7).

Similar effects of dopant salts are seen in the alkali ion conductors. A case in point is illustrated in Figure (17) in which the total alkali contents of a lithium and a sodium diborate glass are held constant while the fraction alkali halide is increased at the expense of the alkali oxide. Again one notes the characteristic linear increase in log σ and decrease in E respectively as a function of increasing halide fraction. The pre-exponential is again, for the most part, insenstive to halide composition. Sodium, the larger cation, exhibits a lower conductivity and a somewhat larger activation energy than does lithium as observed earlier in the germanium sulfide glasses.

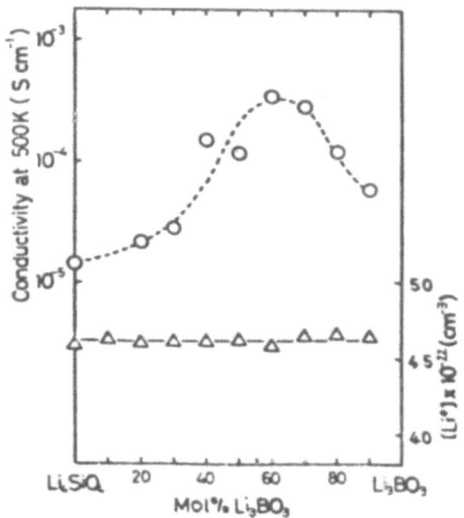

Fig. 18. Composition dependence of conductivity at 500K for Li₄SiO₄-Li₃BO₃ glasses illustrating a mixed anion enhancement. The Li ion concentration in the glasses is indicated by triangles (from Ref. 12).

Although mixed-alkali effects in glasses are well known to result in depressed conductivities, a number of investigators[12,13] have found that mixed anions can result in enhanced conduction as exemplified in Figure (18). The Li ion conductance is seen to peak at intermediate Li_4SiO_4-Li_3BO_3 mixtures.

In completing this section we note that although nearly all FIC glasses investigated so far exhibit an arrhenius temperature dependence below Tg (see Eqn. 3), several exceptions do exist in the silver arsenate and tungstate systems.[14,15] Thermal history appears to play a particularly key role in these glasses. The challenge is to propose transport models which are consistent with these as well as the above observations.

IONIC TRANSPORT MODELS

For a single carrier one obtains an expression for the ionic conductivity given by

$$\sigma = nZq\mu = \sigma_0/T \, \exp(-E/kT) \qquad (4)$$

by combining Eqns. (1) and (3). Examination of the applicable expressions for the carrier density n and mobility μ leads one to understand the significance of both σ_0 and E in the different possible transport models.

Invoking a diffusion model characterized by isolated jumps with random walk one obtains an expression for the diffusion constant

$$D = \alpha d^2 \, \nu_0 \, \exp(\Delta S_m/k) \, \exp-(\Delta H_m/kT) \qquad (5)$$

in which α is a geometrical factor, d the jump distance and ν_0 the vibration frequency of the ion within a potential well which requires the free energy $\Delta G_m = \Delta H_m - T \, \Delta S_m$ for a successful jump. D may be related to μ via the Nernst-Einstein equation $\mu kT = ZqD$, to obtain

$$\sigma = [(n(Ze)^2 \alpha d^2 \, \nu_0)/(kT)] \exp(\Delta S_m/k) \, \exp(-\Delta H_m/kT) \qquad (6a)$$
or
$$\sigma = n(B/T) \, \exp (-\Delta H_m/kT) \qquad (6b)$$

Ionic conductivities range over 20 orders of magnitude between 10^{-20} S/cm for high purity silica to 1 S/cm for αAgI. The two factors which allow for such large variation in σ are the carrier density n and the migration enthalpy ΔH_m.

Assuming for the moment that n is fixed at a value close to the total population of ions on the relevant sublattice, as is conventionally assumed in crystalline FICs[16] then E in Eqn. (4) becomes equivalent to the migration enthalpy ΔH_m. Observed changes in E with glass composition are then expected to reflect changes in mobility induced by corresponding structural modifications.

As pointed out above, mobile carriers can be introduced into the glass network by addition of modifiers or dopant salts. In either case, the positively charged ion is generally compensated by immobile negatively charge species, e.g. a NBO in silica glass or a halide ion. If the electrostatic forces between these centers are not insignificant, then some subset of the potentially mobile ions may remain trapped to form immobilized "associates." This phenomena is commonly observed in insulating crystals in which defects such as vacancies or interstitials are formed by the addition of altervalent dopants, e.g. Cd^{2+} dopants in NaCl which are compensated by the formation of sodium vacancies. As the temperature is increased, the complexes dissociate and begin to contribute to conduction. One may easily predict the temperature and "dopant" dependence of the free carriers by considering the following dissociation reaction.[17]

$$(M^+ X^-)^x \Longleftrightarrow M^+ + X^- \qquad (7)$$

where M^+ is the mobile ion and X^- the NBO, halide or other center of negative charge while $(M^+X^-)^x$ is the neutral associate. Applying the law of mass action gives

$$[M^+][X^-] / [(M^+X^-)^x] = K_\Lambda(T) \qquad (8)$$

Assuming only one type of associate implies that

$$[M^+] = [X^-] \qquad (9)$$

One therefore obtains an expression for $[M^+]$ given by

$$[M^+]^2 = [(M^+X^-)^x] K_\Lambda(T) \qquad (10)$$

For nearly full association

$$[M^+X^-)^x] >> [M^+] \; ; \; [(M^+X^-)^x] \approx C \qquad (11)$$

where C is the total concentration of dopant.

The equilibrium constant K_Λ is given by[17]

$$K_\Lambda(T) = (1/W) \exp(-\Delta H_\Lambda/kT) \qquad (12)$$

where W is the number of orientations of the associate and ΔH_Λ the enthalapy of association. Substituting Eqns. (11) and (12) into Eqn. (10), we obtain for the mobile fraction

$$n = [M^+] = (C/W)^{1/2} \exp(-\Delta H_\Lambda/2kT) \qquad (13)$$

Finally, substituting Eqn. (13) into Eqn. (6b), one finds that

$$\sigma_o = (C/W)^{1/2} B \qquad (14a)$$

$$E = \Delta H_m + (\Delta H_\Lambda/2) \qquad (14b)$$

In other words, the pre-exponential factor shows a square root dependence on dopant density while the apparent activation energy represents contributions from both carrier migration and complex dissociation.

A number of investigators[18,19] have favored such a model often referred to as the weak electrolyte model for describing FIC in glasses. They believe that only a small subset of the alkali or silver ions are mobile at any time and that the activation energy for conduction E represents, in large part, half the dissociation energy. They explain the decreases in E, observed experimentally upon addition of network modifiers or salt dopants, as reflecting corresponding decreases in E_Λ, the reason for which is not always clear. All would agree that for sufficiently small E_Λ or sufficiently high temperature T, all complexes dissociate. Then

$$n = [M^+] = C \qquad (15)$$

and σ_o and E take on the values+

$$\sigma_o = CB \qquad (16a)$$

$$E = \Delta H_m \qquad (16b)$$

Under these circumstances, all else being equal, the conductivity should increase linearly with dopant or modifier while E should remain constant. As we shall later see, however, dopants and modifiers induce structural changes in the glass which can be expected to modify both B, via changes in attempt frequency and jump distance, and the migration energy ΔH_M.

A number of major limitations are inherent in the above discussed models. Implicit in the models is the assumption of a dilute solution of carriers which hop independently between equivalent sites. Given that the glasses of interest to this discussion are FICs, the dilute solution approximation is not valid which implies, in turn, that the motion of carriers is likely to be correlated as assumed in crystalline FICs. Further, since these are glasses, a distribution of sites, some presumeably near equivalent, is more representative of the disordered nature of these materials. One therefore needs to address the questions of how these factors influence the meaning of the pre-exponential and enthalpy in Eqn. 6. Are there average jump distances, attempt frequencies, and migration enthalpies which are appropriate? Is the migration enthalpy limited by the largest barriers to motion or does it represent some weighted average? No clear answers have yet emerged regarding these issues.

In addition to the difficulties associated with applying these classical models to a given FIC glass composition, additional complications are created when one wishes to examine the influence of composition on properties. In crystals, when one discusses dopant effects on ionic conductivity one is often discussing variations in dopant densities over orders of magnitude but under conditions where the total dopant density ranges from parts per million to perhaps parts per thousand. Under these circumstances one does indeed observe the square root dependence and linear dependence of σ_0 on dopant for the associated and dissociated regimes respectively. In nearly all studies on FICs, the modifier or dopant is typically on the order of 10-40 mole% which is no longer a dopant in the classical sense. Such high levels of additive, as we have already discussed, markedly modify the structure of the host and in going across the composition range, we are no longer looking at a small perturbation but, in the limit, at the properties of series of individual compounds . This feature is, unfortunately, often ignored.

A case in point is the weak electrolyte model proposed by Souquet and co-workers[18,20] in which the conductivity is found to follow a square root dependence on the modifier (e.g. Na_2O) or dopant (e.g. AgI) activity (see Eq. 14a). A key assumption is that the ion mobility remains constant over the wide composition range; at best a fortuitous result given the variations in glass structure. Bruce et al[19] while considering a weak electrolyte model for analyzing their data on mixed alkali glasses in fact concluded that the constant mobility picture was inconsistent with their results. It should further be noted that the predictions regarding the doping dependence of conduction for the associated picture

68

(Eqn. 14) apply only to the pre-exponent (Eqn. 14a). Souquet and co-workers have compared σ rather than σ_0 with the activities of the modifiers and dopants. This confuses the issue since the activation energy, E, is known to drop substantially with additives.

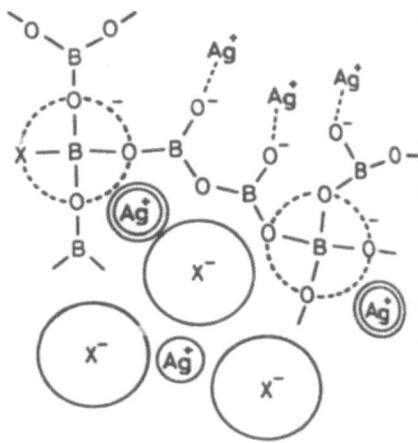

Fig. 19.   Pictorial representation of Ag+ sites in a AgX-Ag$_2$O-B$_2$O$_3$ glass (from Ref. 11).

    A number of models have been tentatively proposed which attempt to describe the near linear decrease of activation energy with composition from a structural point of view. The first, called the "Random Site Model," assumes all ions are potentially mobile and no distinction need be made between populations of "more" and "less" mobile populations. Instead, a wide distribution of sites with varying free energy exist. Glass and Nassau[21], following closely an analysis performed on fluorite FICs[22], demonstrated that such a distribution of sites has a major effect on ion mobility rather than carrier density. Most importantly this model predicts a linear decrease in activation energy with cation modifier, a result of broadening of the distribution of sites with increasing modifier. A simplified pictorial representation of a distribution of sites which a Ag ion is likely to see in a AgX-Ag$_2$O-B$_2$O$_3$ glass prepared by Minami et al[11] is shown in Figure (19).

A somewhat different approach has been taken by the author and his co-workers[2,7]. Here, we assume that structural changes induced by composition variations play a key role in modifying the mobility of carriers primarily via the migration energy. The objective then is to establish whether observed changes in physical properties, which are a reflection of structural changes, in concert with models for the migration energy are consistent with conductivity variations. The formalism used to treat ionic motion in crystals was established in the 1930's[23,24]. This approach has now been substantially updated with the use of computers by allowing the consideration of a significantly larger number of interacting species as well as the relaxation of the rigid crystal lattice assumption. Nevertheless, in the following we use several earlier treatments of this problem since they are useful in illustrating some of the more important principles.

In a relatively recent treatment, Flygare and Huggins[26] calculated the minimum energy paths of cations in the α-AgI structure as a function of cation size. The total energy of the cation was taken as

$$E = e^2 \sum_j (q_i q_j)/(r_{ij}) - e^2 \sum_j (\alpha_j q_i)/(r^4_{ij})$$
$$+ \sum_j \beta_{ij} \exp(r_i + r_j - r_{ij})/\rho \qquad (17)$$

in which (a) the first two terms represent the electrostatic energy for the ith point charge ion in the lattice and (b) the third term is the overlap repulsion term. The sum over j is over all lattice ions with charge q_i, polarizability α_j and radius r_j while r_{ij} is the distance from the mobile ion to the jth lattice ion. Both small and large cations were found to be limited in motion by substantial energy barriers. The small ions due to polarization effects and the large ions due to substantial repulsive effects. For intermediate size ions, these effects largely balance as has been observed experimentally for cation transport in β-alumina.

Anderson and Stuart[27] much earlier proposed a somewhat different approach to calculate the ionic migration energy in glasses which explicitly includes a strain energy component necessary to enlarge a doorway of radius r_D between adjacent sites so as to accommodate an ion of radius r. They estimated this strain energy to be

$$E_s = 4\pi G r_D (r-r_D)^2 \qquad (18)$$

where G is the shear modulus. The electrostatic term which ignores polarization effects is given by

$$\Delta E_b \approx (ZZ_j e^2)/[1/(r + r_j) - 1/(\lambda/2)]/\epsilon \qquad (19)$$

in which r_j is the radius of the nearest neighbor anion, λ the jump distance and ϵ the dielectric constant. This formalism also predicts a minimum migration energy for ions of intermediate size. The attractiveness of this approach is that it allows one to test the validity of the strain formalism by investigating the diffusion of neutral rare gas particles such as He and Ne in glasses. As Shelby[28] and

70

Doremus[6] have shown, this represents an attractive means of probing the interstitial structure of glasses. Such studies are reviewed in the next section.

In summary, although the above models are oversimplified[25] they nevertheless illustrate that the migration energy, made up of an electrostatic and a strain component, is a strong function of the structure which one may view as the composite of the size and polarizability of the mobile ion and the network of ions through which it moves. In our later discussion we attempt to characterize FIC in glasses as a function of both the nature of the mobile carrier and the characteristics of the host network through which it diffuses.

STRUCTURE-COMPOSITION CORRELATIONS

The current consensus on the nature of glasses is best described by the random-network model as mentioned above. Here, the basic network building blocks resemble those in crystallized material but are not arranged in a manner which leads to long range order. In silica glass, the unit is the Si-O tetrahedron. Addition of network modifiers breaks the link between some tetrahedra by forming NBO's. Above the di-silicate composition some tetrahedra bond to only two other tetrahedra and the three dimensional character of the glass begins to diminish. Between disilicate ($R_2O-2SiO_2$) and metasilicate (R_2O-SiO_2) one expects chainlike structures and at even higher modifier concentrations also isolated rings and islands.[6] Although the distribution of the modifier ions in these glasses is uncertain, there is no doubt that the mobile ions find themselves in substantially different environments as the glass compostion is varied.

The development of structural features in alkali borate glasses with alkali additions appears to be best characterized amongst the major oxide glass forming systems. Of particular interest is the evidence for intermediate range order in these glasses.[29] Evidence for existence of polyborate groups including boroxol rings (see Figure 20) comes from a combination of techniques including spectroscopic (Raman, IR, NMR) and diffraction methods.

Alkali oxide additions are initially believed to transform BO_3 units, characteristic of pure borate glass, to BO_4 tetrahedra. In contrast to silicate glasses, the modifier tends to build up the 3 dimensional character and thereby strengthen the network of the borate glass. This transformation is reflected in a concurrent decrease of the $806cm^{-1}$ Raman peak, characteristic of boroxol rings in crystalline borates, and the increase of a new peak at $770cm^{-1}$.[30,31] The latter peak is attributed to six-membered borate rings containing one or two BO_4 tetrahedra. The NMR studies of Bray and O'Keefe[32] have shown that the fraction of four coordinated borons, N_4, in glasses of composition $xR_2O \cdot (1-x)B_2O_3$ (R = Li, Na, K, Rb, Cs) follows the relation

$$N_4 = x(1-x) \tag{20}$$

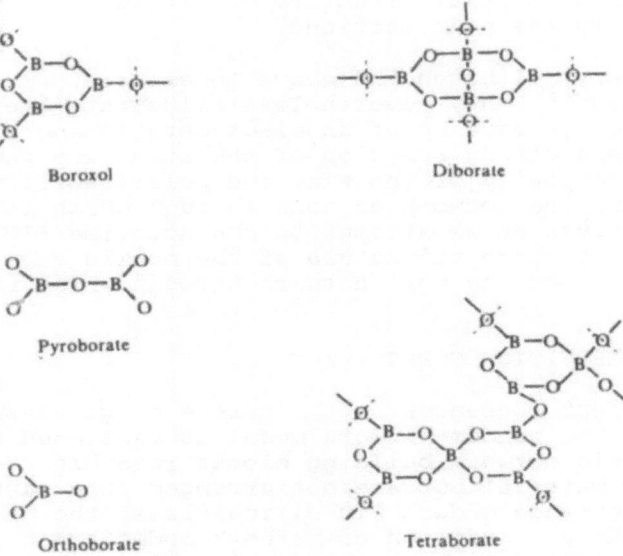

Fig. 20. Typical polyborate groups presumed to exist in borate glasses and compounds. Dashed lines through oxygens indicate that they are bridging (from Ref. 47).

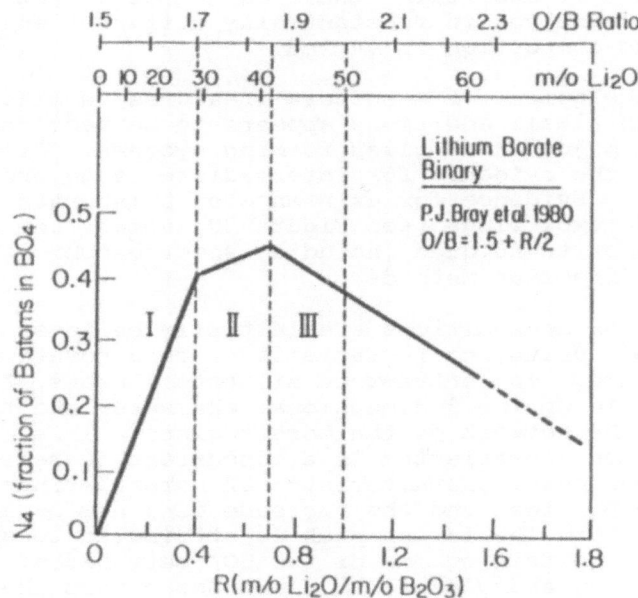

Fig. 21. Plot of the fraction of B atoms in BO_4 tetrahedra, N_4, versus composition of lithium borate glasses (from Ref. 48).

up to approximately x = 0.3 (see Figure 21). Above that point a maximum in N_4 is reached with a subsequent decrease in N_4 coupled to the production of non-bridging oxygens. Krogh-Moe[33] proposed a structural model which views the BO_3 and BO_4 units to be interconnected to form various extended structural units including boroxol, tetraborate and diborate groupings characteristic of crystalline borates (see Figure 20). While boroxol and tetraborate groups are believed to predominate below 20 mol% R_2O, tetraborate and diborate groups prevail between 20 and 30 mol%. Above 35 mol% R_2O, metaborate, pyroborate, and loose BO_3 triangles are reported to exist by Konijnendijk and Stevels[30] in addition to diborate units. More recently, glasses in the ternary system $Li_2O-(LiCl)_2-B_2O_3$ (LCB) were examined by both Raman[31] and NMR[34] techniques. Both techniques found essentially identical spectra for binary and ternary glasses of equal O/B ratios. They thus concluded that LiCl additions produce no major modifications in the vitreous boron-oxygen network and that the Cl ions are accommodated into the network interstices along with the alkali ions. What precise role the Cl plays in modifying these glasses is of great importance since, as we have seen, Cl additions result in marked increases in alkali ion conductivities.

Phosphate glasses, like silicate glasses, are built up of tetrahedra which are bonded to three rather than four other tetrahedra. At high modifier contents the tetrahedra are believed to be linked in chains which become shorter as the R_2O content is increased. In germanate glasses, increased Ge-O linkages occur in that GeO_4 tetrahedra are transformed to GeO_6 octahedra with increasing O/Ge ratio.

In the following, a number of composition dependent physical properties are reviewed which reflect quite clearly the structural changes discussed above which accompany modifier and dopant additions. Changes in these properties will later be correlated to related changes in transport properties.

GLASS TRANSITION TEMPERATURE AND DENSITY

Two experimentally accessible physical measurements, which are a direct reflection of glass structure, are the glass transition temperature, T_g and the density p. Figure 22 shows the results of T_g measurements performed in our laboratory[2] for a wide range of binary (LiCl=O) and ternary glasses in the system $Li_2O-(LiCl)_2-B_2O_3$ as a function of the O/B ratio. T_g for the binary glasses (circles) changes dramatically in concert with predictions of the NMR data; i.e., an initial strengthening of the network by systematic replacement of planar BO_3 units with tetrahedral BO_4 units with increasing O/B followed by a weakening at larger O/B due to increasing numbers of non-bridging oxygens and decreasing N_4.

The ternary glasses allow for an additional degree of freedom, i.e. independent control of O/B and the total alkali content of the glass. Referring again to Figure 22, we find that the binary and ternary glasses exhibit an opposite

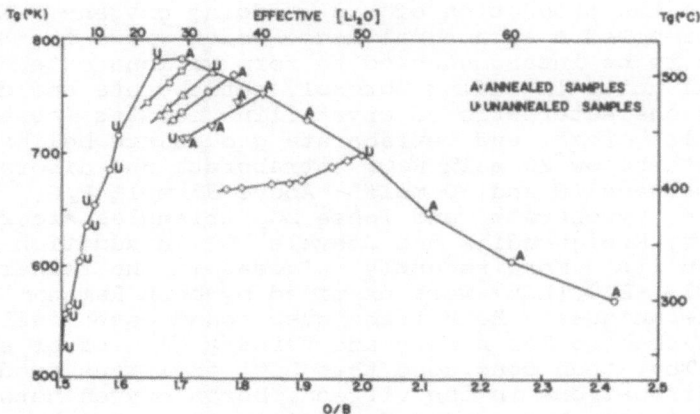

Fig. 22. Glass transition temperatures are plotted versus oxygen-to boron ratio for binary lithium borate (circles) and ternary chloroborate glasses. Data are shown for five series of chloroborate glasses with fixed Li_2Z; 30; 33.3; 36.4; 40; 50 m/o (from Ref. 2).

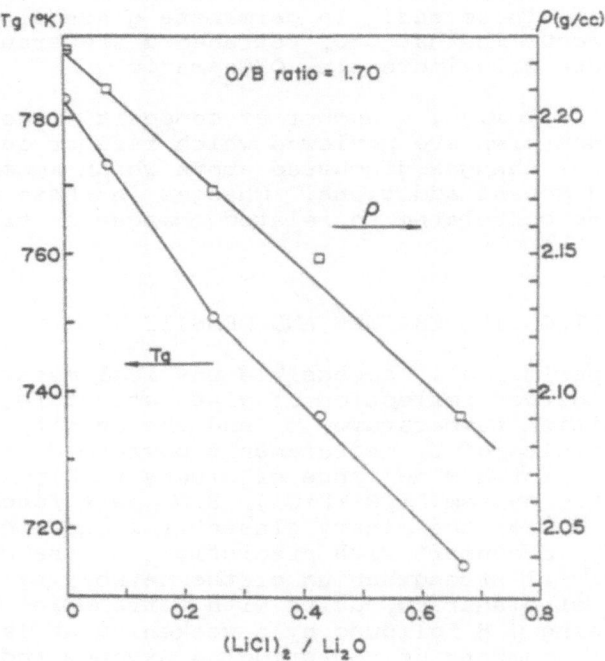

Fig. 23. Glass transition temperatures and densities are plotted versus the ratio $(LiCl)_2/Li_2O$ for a series of glasses with O/B fixed at 1.70 (from Ref. 2).

dependence of T_g on O/B ratio suggesting, in contrast to the conclusions based on Raman and NMR spectroscopy, that Cl does in some way modify the network. A plot of T_g versus $(LiCl)_2/Li_2O$ ratio for glasses with a common O/B ratio = 1.70 shown in Figure 23 demonstrates that increasing levels of Cl serve to systematically weaken the glass network. It is of interest to note that Shelby[35] has recently noted that for a given R_2O/B_2O_3 ratio, T_g decreases with increasing interstitial alkali ion size. Both these results suggest that large interstitial ions, whether cation or anion, serve to weaken the glass structure.

The densities of the same glasses, as in Figure 22, are plotted as a function of O/B in Figure 24. Without exception, the ternary glasses exhibit lower densities than the corresponding binary glasses of equal O/B ratio. This feature is emphasized for O/B = 1.70 in Figure 23. Shelby[36] also reports on the anomalous effects of interstitial ion size on density. He points out, for example, that while molar volumes of R_2O/B_2O_3 glasses tend to decrease with increasing R_2O for the smaller alkali ions (e.g., Li, Na, K) due to the BO_4 formation, due to cesium's large size the molar volumes of $Cs_2O.B_2O_3$ glasses actually increase with increasing Cs_2O over most of the composition range.

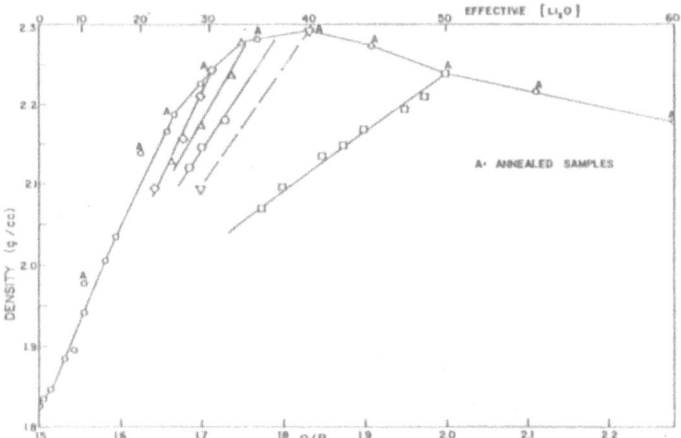

Fig. 24. Density is plotted as function of O/B ratio for a series of binary and ternary lithium borate glasses described in Figure 22 (from Ref. 2).

In Figure 25 we show how the molar volume of a series of alkali diborates and metaborates (i.e., with constant fraction B_2O_3) vary with increasing fraction of alkali in the form of chlorides. Here one readily observes that both excess interstitial chlorine or larger interstitial alkali ions (i.e., Na vs. Li) serve to dilate the glass structure.

The T_g and density results illustrate the limitations of the spectroscopic techniques in detecting changes in structures beyond the basic units i.e., BO_3 and BO_4 units.

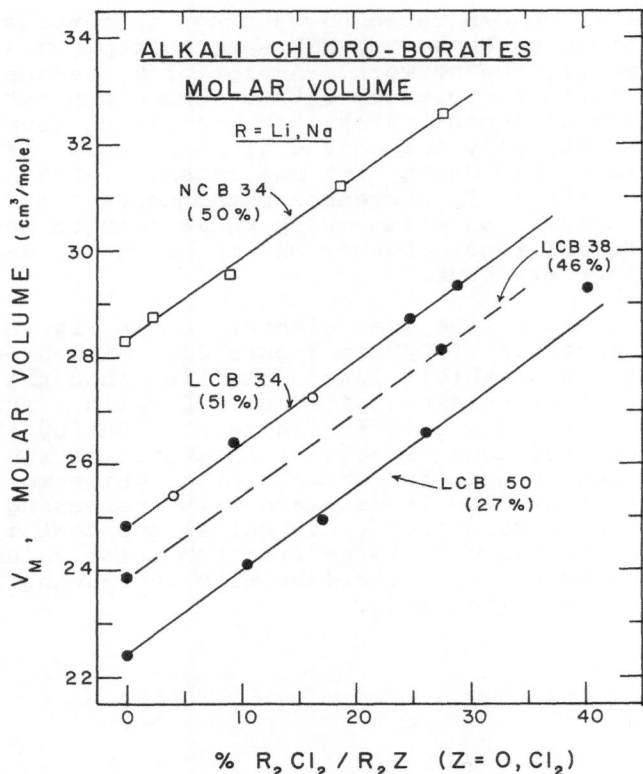

Fig. 25. Molar volume V_m is plotted as function of relative fraction chloride for a number of lithium (LCB) and sodium (NCB) chloroborates with fixed alkali content: $Li_2 Z = 34$, 38, 50 m/o; $Na_2 Z = 34$ m/o ($Z = O$, Cl_2) (from Ref. 2).

Figure 26 shows a tentative structural model that we have proposed[2] which distinguishes the types of structure that binary and ternary glasses of equal O/B ratio might have.

It is interesting to examine the silver borate system given its ability to accommodate large fractons of $Ag_2 O$ and Ag halides. The glass transition temperature dependence on the O/B ratio in $AgBr-Ag_2 O-B_2 O_3$ measured by Minami et al[37] is given in Figure 27. A similar peak in T_g is observed at about 0.3 $Ag_2 O/B_2 O_3$ ratio as in the alkali borates. The data however is available out to a $Ag_2 O/B_2 O_3$ ratio of 3 which shows T_g continuing to fall with increasing modifier suggesting the continued breakup of the network to very small polyborate units at the limits of glass formation. Note at a fixed ratio of 2, T_g drops noticeably with increasing AgBr, similar to the effect noted above in the $Li_2 O-LiCl-B_2 O_3$ case.

Sound velocity measurements by Carini et al[38] in the $(AgI)_x [(Ag_2 O)_y (B_2 O_3)_{1-y}]_{1-x}$ system provide similar

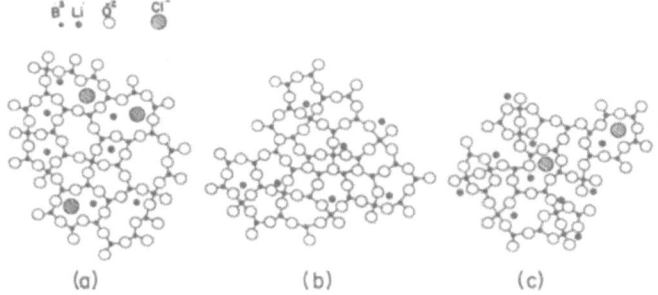

Fig. 26. (a) Two-dimensional schematic of the structure of a
          glass in the system $Li_2O-(LiCl)_2-B_2O_3$ at relatively
          low alkali content. The ternary differs from the
          binary in that additional interstices, not adjaent
          to BO_4 tetrahedra, are occupied by Li^+ and Cl^- ions.
          (b) A schematic of the structure of a glass in the
          binary system but with high alkali content. Note
          that few if any large interstices remain which do
          not contain adjacent tetrahedra. (c) Proposed
          schematic of modifications induced in the above
          structure upon addition of LiCl. The BO_4 tetrahedra
          are now clustered in diborate-like units thereby
          leaving interstices without adjacent tetrahedra for
          occupation by Cl^- ions (from Ref. 2).

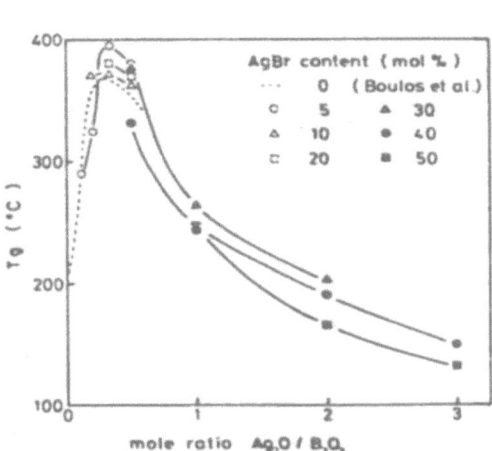

Fig. 27. Composition dependence
of glass transition temperatures
for $AgBr-Ag_2O-B_2O_3$ glasses.
(From Ref. 39).

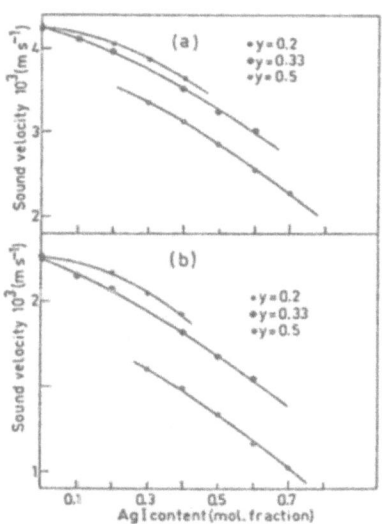

Fig. 28. Sound velocity in
$(AgI)_x[(Ag_2O)_y(B_2O_3)_{1-y}]_{1-x}$
at 295K: a) 5-MHz longitud-
inal waves; b) 5-MHz transverse waves (from Ref. 38).

conclusions. First, the sound velocity at 5MHz peaks at y $\simeq$ 0.25 in the binary system as expected near the peak stiffness of the system. On the other hand, sound velocity drops systematically as observed in Figure 28 as the AgI fraction (x) is increased for constant y or constant O/B ratios. This again supports the conclusion that the halide progressively "softens" the network notwithstanding the fact that it does not "modify" the network directly. Chicodelli et al[39] also note a dilation of the glass upon AgI addition. The rate of dilation appears to depend only weakly on the O/B ratio which they interpret implies AgI is accommodated near BO_4 units.

GAS MIGRATION

Since noble gases such as He, Ne, and Ar exhibit measurable diffusivities in glasses, they serve as ideal probes of the interstitial volume of glasses. Furthermore, since they are uncharged they may in principle be used to probe the "doorways" or "bottlenecks" between mobile ion sites and thereby isolate the strain contribution to the migration enthalpy ΔH_m. A graphic illustration of this principle is illustrated in Figure 29 in which the square root of the activation energy of gas diffusion in vitreous silica is plotted as a function of the gas molecule diameter. The straight line dependence is in agreement with predictions of Eqn. (18) and according to the analysis of Shelby[28] gives an average doorway size in silica of 1.1Å.

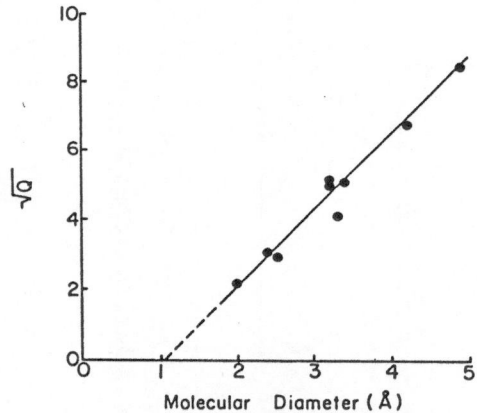

Fig. 29. Relationship between activation energy (Q) for molecular diffusion in vitreous silica and the molecular diameter of the diffusing species (from Ref. 28).

Shelby has examined He migration in a number of glass forming systems as a function of R_2O fraction. As a general rule, the activation energy for diffusion tends to follow the same dependence on R_2O as does the density. For example, in a sodium germanate glass,[40] the He migration enthalpy varies from ~7Kcal/mole(3.6gm/cm³) for a pure germanate glass, peaks at ~18Kcal/mole(4.05gm/cm³) for 18% Na_2O and drops back to ~15Kcal/mole (3.6 gm/cm³) at 33 mol% Na_2O. The values in parenthesis reflect the corresponding densities for those compositions. Note that although the densities are equal for O and 33% Na_2O, the available free volume for migration is decreased in the latter due to Na occupation of the interstices. As in the borates, density decreases with increasing alkali size and along with it a decrease in the He migration energy.

In the alkali borates, except for a small anomaly at low alkali levels, the He diffusivity increases while ΔH_m decreases with increasing alkali ion size. Further ΔH_m tends to increase with increasing R_2O up to 30 mol% R_2O reflecting similar increases in density. Obviously, gas diffusivity results can be extremely useful in probing the interstices of the glass networks and provide important insight into the relationship between structural changes and ionic migration in FIC glasses.

TRANSPORT-STRUCTURE CORRELATIONS

In the previous sections, data was presented which showed that both the structure of various glasses and their transport properties are sensitive functions of composition. In this section we wish to demonstrate that these changes in structure and transport are coupled and can be used to gain insight into the mechanisms controlling the transport process.

Perhaps the clearest correlation between structure and transport exists for the case of inert gas diffusion in borate and germanate glasses. There, in almost every case, one finds the migration enthalpy scales directly with the available free volume of the glass induced either by a change in alkali ion size for a fixed O/B or O/Ge ratio or by a change of O/B or O/Ge for a given alkali modifier. Similarly, for a given glass, the migration energy increases with increasing diameter of the diffusing atom.

Since these atoms are not charged, such experiments confirm that the motion of atoms as small as He are strongly influenced by strain energy considerations in their motion, and that the types of structural changes induced in the glasses by composition variations are sufficiently large to markedly modify these energetics. Similar effects of structure on the transport of charged ions can therefore also be expected.

The influence of excess volume on the energetics of motion of charged particles is illustrated quite explicitly in Figure 30 in which $E_{\sigma T}$ is plotted versus excess volume/mole boron for both the lithium and sodium di-chloroborate systems. Both systems exhibit a near linear decrease of $E_{\sigma T}$ with increasing excess volume, an effect induced by the dilating ability of the interstitial chlorine. A feature which is particularly noteworthy is the offset between the two curves.

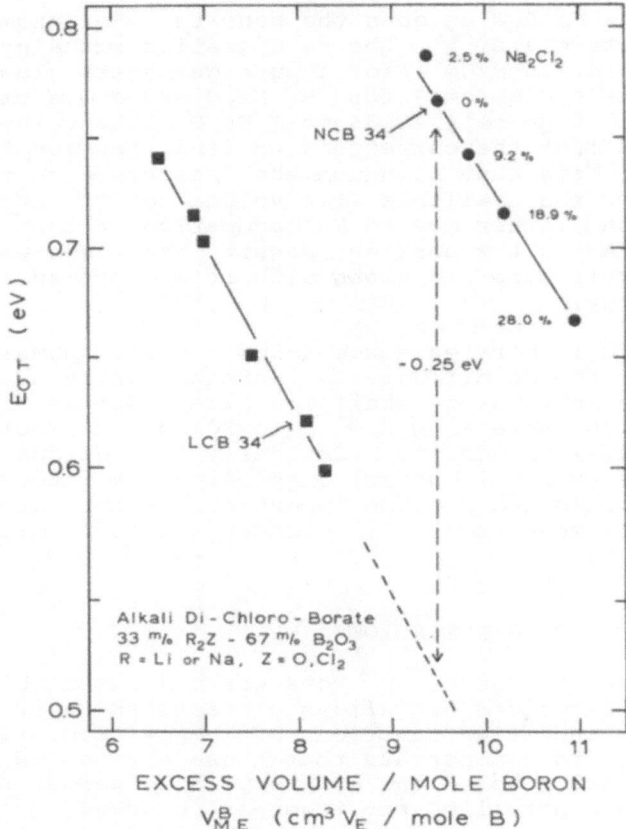

Fig. 30. Activation energy for conduction as function of
          excess volume per mole boron. Note larger
          activation energy for Na conduction compared to Li
          conduction for equal excess volume (from Ref. 7).

The Na conducting glass exhibits a considerably great $E_{\sigma T}$
(~0.225eV) for equivalent excess volumes as might be expected
for the larger ion if strain effects were important in the
migration process. It also becomes clear why lithium and
sodium di-chloro-borates with nominally equivalent
compositions exhibit nearly equivalent $E_{\sigma T}$'s. As it happens,
the larger interstitial Na apparently induces larger excess
volumes in the glass structure than does Li (see earlier
discussion) suggesting that the larger carrier size is in part
compensated by larger doorways between sites.

     An approximation for the Na migration enthalpy ΔH_m in
these glasses using Shelby's[35] He diffusion data can be made
after noting that $r_{He} - r_{Na} \sim 1$ Å. Shelby found that ΔH_m (He)
increases from 0.22 to 0.39 eV while the molar volume V_M
decreases by over 10% (38 to 34 cm³/mole) in sodium borates as
Na_2O increases from 0-10 mol% Na_2O. Above 10 mol% no data for

ΔH_m (He) are available, since the He diffusivity dropped below detection limits, yet V_M continues to decrease markedly up to the diborate composition by another 17%. A reasonable extrapolation of ΔH_m (He) and hence ΔH_m (Na) for the diborate composition yields a value close to $E_{\sigma T}$. This indicates that the measured activation energy for conduction in this composition range is dominated by a strain term in the mobility.

Structural changes were also observed to occur in the binary alkali borates with composition, so it is appropriate to examine how these changes correlate with transport. The activation energy for conduction is plotted in Figure 31 versus excess volume/mole boron for the binary $xLi_2O(1-x)B_2O_5$ system ($17.8 < x < 59.6$ mol%) and several ternary systems. For the diborate compositions and above both the binary and the ternaries exhibit nearly the same systematic decrease in $E_{\sigma t}$ with increasing excess volume. This type of dependence has already been suggested to be related primarily to the strain component of migration.

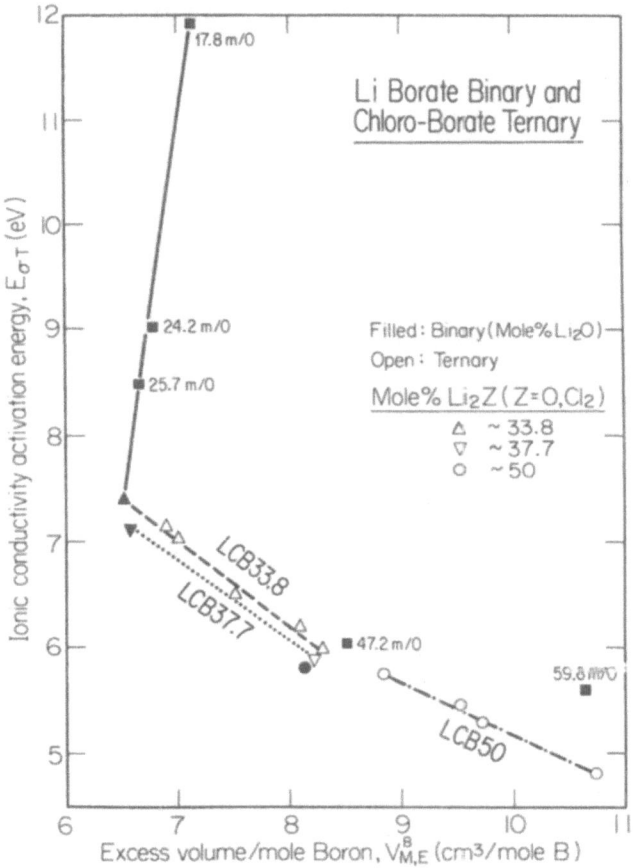

Fig. 31. Activation energy for conduction as function of excess volume per mole boron. Two transport regimes are indicated with the transition occurring at 33 m/o Li_2O (from Ref. 7).

Below 33 mol% Li_2O, however, a nearly vertical increase
in energy is observed although the excess volume increases
slightly. This suggests a change in the conduction mechanism
for the much more highly resistive (i.e., non-FIC) glasses.
As suggested earlier,[41] this dramatic change may be due to
the onset of percolation between neighboring near equivalent
interstitial sites. For lower levels of alkali, the cations
are for the most part associated with centers of charge of
opposite sign in the network, e.g. BO_4 tetrahedra and require
additional energy to dissociate and migrate to an equivalent
site as described in Eq. 14b. A detailed discussion of the
issues related to the onset of percolation may be found in the
thesis by Button.[42]

DISCUSSION

Interest in FIC glasses has grown to the point where
activities in this area now represent a significant fraction
of the total FIC research. Many new FIC glasses continue to
be identified including those which conduct by monovalent ions
(e.g. Li, Na, K, Ag, Cu, and F) and divalent ions (e.g., Pb)
in a variety of networks including many of which are not
traditional glass formers. New processing techniques
including sputtering, evaporation, roller quenching, etc.
allow for a greater variety of systems with differing thermal
histories. The sulphide analogs of borates, silicates, and
phosphates exhibit exceptionally high ambient ionic
conductivities. This together with the characteristic
flexibility in forming complex, thin walled glass structures
has stimulated further interest in these materials as
components of a variety of electrochemical devices.

A number of important phenomenological trends have now
been shown to be fairly general including:

1. Increasing ion conduction in the order K < Na < Li
    < Ag,

2. Increasing ion conduction with modifier
    concentration,

3. enhanced ion conduction with halide salt additions
    in the order Cl < Br < I,

4. Superior ionic conduction in sulphide vs. oxide
    glasses,

5. Correlation of increasing ionic conduction with
    decreasing activation energy rather than increasing
    pre-exponential factor, and

6. Correlation of increasing ionic conduction with
    decreasing density and T_g.

The latter observation focuses on the correlation between
transport and structural trends. As we have emphasized in
this article, major structural modifications accompany
compositional changes in glasses. It is the author's opinion

that such structural changes must be studied and examined in greater detail if we expect to continue to make advances in understanding the phenomenon of FIC in glasses.

Considerable controversy remains regarding transport mechanisms in glasses. Those favoring a weak electrolyte picture view the ions as being for the most part trapped. Conductivity enhancement brought about by temperature or composition variation results in increased dissociation. Others believe that the full ion population is mobile in FIC glasses and that conductivity enhancement is intimately tied to increased ion mobility brought about by modifications in glass structure. The author favors the latter picture while acknowledging that electrostatic interactions remain important particularly for the less conducting glasses. In the following, I summarize some of the key observations supporting this view.

Evidence was presented which demonstrates that the structure of glasses can be readily modified in a near continuous manner by control of composition. It is therefore to be expected that all structure-sensitive properties including ionic conduction will similarly be affected. This is predicted by theoretical models such as those by Flygare and Huggins[26] and Anderson and Stuart[27] and substantiated in crystalline systems by numerous experimental studies including more recent studies on beta aluminas and NASICON related systems.[43]

Structure, relevant to the diffusion of charged particles, represents both the electrostatic environment that the particle must interact with as well as the volumetric aspects which determine the strains necessary to accommodate the particle both at equilibrium sites and at the window saddlepoint between sites. Studies of diffusion of neutral He, an atom of near equivalent size to the Na^+ ion, show that strain related migration enthalpies are sensitive to changes in glass density and easily approach values of 0.5-0.75 eV in sodium borate and sodium germanate glasses respectively.[35,40] Given that Na conduction in borate glasses is characterized by activaton energies of similar magnitude this, on average, allows for a smaller electrostatic contribution to the activation energy.

Dramatic changes in ionic conduction with composition are often attributed to substantial increases in carrier-network dissociation vis-a-vis a weak electrolyte model. This picture, it seems to us, has a number of serious limitations. The first has already been discussed above, i.e. significant strain contributions to transport cannot be ignored. Second, exceptionally large ion mobilities of $\mu \sim 10^{-5} cm^2/v\text{-}sec$ estimated in the silver borate and phosphate glasses by Chiodelli et al[44] assuming 100% participation of Ag ions would become unreasonably large if only a small fraction of the Ag ions were mobile. Third, the observation of $\sigma_{Li} > \sigma_{Na} > \sigma_K$ would be reversed given that larger alkali ions should exhibit a greater degree of dissociation than the smaller ions.

Notwithstanding the above evidence, which supports a FIC model in which the mobile carrier population is large, there is other evidence which suggests that there are sub-populations of ions in a given FIC with different effective conductivities. Thus, as Minami et al[11] demonstrated in Fig. 16, it is the Ag fraction associated with the halide addition rather than the total Ag content which determines the Ag FIC in $AgX-Ag_2O-B_2O_3$ (X = I, Br, Cl) glasses. Along the same lines, a number of investigators have observed that when AgI is added to borate or phosphate glasses, an extrapolation to 100% AgI gives conductivities and activation energies expected of AgI. This would not be expected if the function of the I ion was simply to dilate the glass network.

Corini, Cutroni, Fontana, Mariotto, and Rocca[45] have interpreted their inelastic light scattering data on $(AgI)_x(Ag_2O.nB_2O_3)_{1-x}$ as suggesting that increasing amounts of AgI induce the formation of regions in the glass which "reproduce the structure and vibrational dynamics of crystalline AgI" and whose dimensions increase with the AgI content. Does the borate framework allow for the assembly of such AgI-like regions within its interstices? Such cooperative effects between polymer like frameworks and AgI polyhedra have been proposed in such crystalline systems as $[C_5H_5NH]Ag_5I_6$.[46] Further experimental and theoretical efforts are required to answer these and many other intriguing questions about transport in glasses. The flexibility in manipulating structure and properties in glasses in a near continuous manner should assist greatly in coming to grips with such issues.

We began this article by comparing and contrasting crystalline and amorphous conductors. We can now ask: What can be learned about the process of FIC now that this phenomena has also been observed in glasses? Clearly the first paradigm for the requirement of FIC in crystals, i.e., the need for highly ordered structured arrays, needs to be retired. Instead, it should be replaced by a more general requirement that states (Button et al[41]):

The framework of the immobile structure should be sufficiently open to insure (1) an abundance of physically interconnected and accessible sites and (2) unhindered movement of ions through relatively large windows connecting these sites.

This requirement can be satisfied in both ordered and disordered systems.

Regarding the second paradigm, the requirement of the existence of a "disordered" carrier sublattice; glasses naturally satisfy this condition. This suggests that one should, in principle, expect many more types of FIC glasses than crystals. This seems to be supported by the fact that many crystalline materials which are insulating, e.g., $LiNbO_3$ and $LiBO_2$ become FIC when converted to glasses while FIC crystals such as boracite $Li_4B_7O_{12}Cl$ remain FIC as glasses.[4] Clearly, many opportunities lie ahead for discovering many additional new systems. I expect, in future research, by combining a broad range of spectroscoic and electrochemical techniques together with a new sensitivity to the underlying issues relevant to FIC in glasses, significant progress will be made in this field in the coming decade.

ACKNOWLEDGEMENTS

The substantial assistance provided by F. Fusco in assembling and reviewing material for this chapter is much appreciated as is the continuing and able assistance provided by V. Graw in manuscript and other course-related preparations. I also thank Lawrence Berkeley Laboratories for their support of our research on fast ion conducting glasses under subcontract #4548010.

REFERENCES

1. H.L. Tuller and M.W. Barsoum, "Glass Solid Electrolytes: Past, Present, and Near Future-The Year 2004," J. Non-Crystalline Sol. 73:331 (1985).
2. D.P. Button, R. Tandon, C. King, M.H. Velez, H.L. Tuller, and D.R. Uhlmann, "Insights into the Structure of Alkali Borate Glasses," J. Non-Crystalline Sol. 49:129 (1982).
3. T. Minami, T. Shimizu, and M. Tanaka, "Structure and Ionic Conductivity of Glasses in the Ternary Systems AgX (X=I, Br, or Cl)-Ag_2O-B_2O_3," Sol. St. Ionics 9&10:577 (1983).
4. H.L. Tuller, D.P. Button, and D.R. Uhlmann, "Fast Ion Transport in Oxide Glasses," J. Non-Crystalline Sol. 40:93 (1980).
5. J.L. Souquet and A. Kone, "Glasses as Electrolytes and Electrode Materials in Advanced Batteries," in Materials for Solid State Batteries, eds. B.V.R. Chowdari and S. Radhakrishna, World Sc. Publ. Co. Singapore, 1986.
6. R.H. Doremus, Glass Science, J. Wiley & Sons, New York, 1973.
7. H.L. Tuller and D.P. Button, "The Role of Structure in Fast Ion Conducting Glasses," in Proc. Int. Conf. on Transport-Structure Relations in Fast Ion and Mixed Conductors, eds. F.W. Poulsen, N.H. Anderson, K. Clausen, S. Skaarup, and O.T. Sorensen, Riso Nat. Lab., Denmark, 1985.
8. A. Levasseur, "Lithium Glasses as Electrolyte Materials," op. cit. Ref. 5.
9. E. Robinel, B. Carette, and M. Ribes, "Silver Sulfide Based Glasses: (1) Glass Forming Regimes, Structure, and Ionic Conduction of Glasses in GeS_2-Ag_2S and GeS_2-Ag_2S-AgI Systems," J. Non-Crystalline Sol. 57:49 (1983).
10. T. Minami and M. Tanaka, "Formation Region and High Ionic Conductivity of Glasses in the Systems AgX-Ag_2O-P_2O_5 (X = I, Br, Cl)" Rev. Chim. Miner. 16:283 (1979).
11. T. Minami, T. Shimizu, and M. Tanaka, "Structure and Ionic Conductivity of Glasses in the Ternary Systems AgX (X = I, Br, Cl)-Ag_2O-B_2O_3," Sol. St. Ionics 9&10:577 (1983).

12. T. Minami, "Fast Ion Conducting Glasses," J. <u>Non-Crystalline Sol.</u> 73:273 (1985).

13. A. Magestras and G. Chiodelli, "Silver Borophosphate Glasses: Ion Transport, Thermal Stability, and Electrochemical Behavior," <u>Sol. St. Ionics</u> 9&10:611 (1983).

14. M.D. Ingram, C.A. Vincent, and A.R. Wandless, "Temperature Dependence of Ionic Conductivity in Glass: Non-Arrhenius Behavior in the $Ag_7 I_4 AsO_4$ System," <u>J. Non-Crystalline Sol.</u> 53:73 (1982).

15. A. Magestras and G. Chiodelli, "AC Conductivity and Dielectric Response of the Ionic Conductor $Ag_6 I_4 WO_4$," <u>Electrochimica Acta</u> 26:1241 (1981).

16. R.A. Huggins, "Very Rapid Ionic Transport in Solids," in <u>Diffusion in Solids: Recent Developments</u>, eds. A.S. Nowick and J.J. Burton, Academic Press, New York, 1975.

17. J.A. Kilner and B.C.H. Steele, "Mass Transport in Anion-Deficient Fluorite Oxides," in <u>Nonstoichiometric Oxides</u>, eds. O.T. Sorenson, Academic Press, New York, 1981.

18. D. Ravaine and J.L. Souquet, "A Thermodynamic Approach to Ionic Conductivity in Oxide Glasses, Part I: Correlation of Ionic Conductivity with the Chemical Potential of Alkali Oxide in Oxide Glasses," <u>Phys. Chem. Glasses</u> 18:27 (1977).

19. J.A. Bruce, M.D. Ingram, and M.A. MacKenzie, "Ionic Conductivity in Glass: A New Look at the Weak Electrolyte Theory," <u>Sol. St. Ionics</u> 18&19:410 (1986).

20. A. Kone and J.L. Souquet, <u>Solid State Ionics</u> 18&19:454 (1986).

21. A.M. Glass and K. Nassau, <u>J. Appl. Phys.</u> 51:3756 (1980).

22. K.E.D. Wapenaar and J. Schoonman, <u>J. Electrochem. Soc.</u> 126:667 (1979).

23. W. Jost, "Diffusion and Electrolytic Conduction in Crystals," <u>J. Chem. Phys.</u> 1:466 (1933).

24. N.F. Mott and M.J. Littleton, "Conduction in Polar Crystals. I Electrolytic Conduction in Solid Salts," <u>Trans. Faraday Soc.</u> 34:485 (1936).

25. C.R.A. Catlow, I.D. Faux, and M.J. Norgett, "Shell and Breathing Shell Model Calculations for Defect Formation Energies and Volumes in Magnesium Oxide," <u>J. Phys.</u> C9:419 (1976).

26. W.H. Flygare and R.A. Huggins, "Theory of Ionic Transport in Crystallographic Tunnels," <u>J. Phys. Chem. Sol.</u> 34:1199 (1973).

27. O.L. Anderson and D.A. Stuart, "Calculation of Activation Energy of Ionic Conductivity in Silica Glass by Classical Methods," <u>J. Am. Ceram. Soc.</u> 37:573 (1954).

28. J.E. Shelby, "Molecular Solubility and Diffusion," in <u>Treatise on Materials Science and Technology</u>, Volume 17, eds. M. Tomozawa and R.H. Doremus, Academic Press, New York, 1979.

29. D.L. Griscom, "Borate Glass Structure," in <u>Borate Glasses: Structure, Properties Applications</u>, eds. L.D. Pye, V.D. Frechette, and N.J. Kreidl, Plenum Press, New York, 1978.

30. W.L. Konijnendijk and J.M. Stevels, "The Structure of Borate Glasses Studied by Raman Scattering," _J. Non-Crystalline Sol._ 18:307 (1975).

31. M. Irion, M. Couzi, A. Levasseur, J.M. Reau and J.C. Brethous, "An Infra-red and Raman Study of New Ionic Conductor Lithium Glasses," _J. Sol. St. Chem._ 31:285 (1980).

32. P.J. Bray and J.G. O´Keefe, Nuclear Magnetic Resonance Investigations of the Structure of Alkali Borate Glasses," _Phys. Chem. Glasses_ 4:37 (1963).

33. J. Krogh-Moe, "Interpreation of the Infra-Red Spectra of Boron Oxide and Alkali Borate Glasses," _Phys. Chem. Glasses_ 6:46 (1965).

34. A.E. Geissberger, F. Bucholtz, and P.J. Bray, "^{7}Li, ^{11}B, ^{19}F NMR in Amorphous Alkali Halogen Borate Solid Ion Conductors," _J. Non-Crystalline Sol._ 49:117 (1982).

35. J.E. Shelby, "Thermal Expansion of Alkali Borate Glasses," _J. Am. Ceram. Soc._ 66:225 (1983).

36. J.E. Shelby, "Helium Migration in Alkali Borate Glasses," _J. Appl. Phys._ 44:3880 (1973).

37. T. Minami, Y. Ikeda, and M. Tanaka, "IR Spectra, Tg, and Conductivity of Superionic Conducting Glasses in the Systems AgX-Ag_2O-B_2O_3 (X=I, Br)," _J. Non-Crystalline Sol._ 52:159 (1982).

38. G. Carini, M. Cutroni, M. Federico, G. Galli, and G. Tripodo, "Acoustic Properties and Diffusion in Superionic Glasses," _Phys. Rev._ B30:7219 (1984).

39. G. Chiodelli and A. Magestras, "Short Range Order and Glass Transition in AgI-Ag_2O-B_2O_3 Vitreous Electrolytes," _J. Non-Crystalline Sol._ 51:143 (1982).

40. J.E. Shelby, "Helium Migration in Alkali Germanate Glasses," _J. Appl. Phys._ 50:278 (1979).

41. D.P. Button, L.S. Mason, H.L. Tuller, and D.R. Uhlmann, "Structural Disorder and Enhanced Ion Transport in Amorphous Conductors," _Sol. St. Ionics_ 9&10:585 (1983).

42. D.P. Button, "Structure-Transport Relationships in Amorphous Oxide Systems: Fast Ion Transport in Alkali Borate Glasses," Ph.D. Thesis, 1983, M.I.T. Cambridge, MA, USA.

43. B.J. Wuensch, L.J. Schioler, E. Prince, "Relation Between Structure and Conductivity in the NASICON Solid Solution System," in _Proc. Conf. on High Temperature Solid Oxide Electrolytes, Vol. II--Cation Conductors_, Brookhaven Natl. Lab. BNL-51728 (1983).

44. G. Chiodelli, A. Magestras, M. Villa, and J.L. Bjorkstrom, "Structure and Ion Dynamics in AgI-$Ag_2B_2O_7$ Vitreous Electrolytes," _Mat. Res. Bull._ 17:1 (1982).

45. G. Corini, M. Cutroni, A. Fontana, G. Mariotto, and F. Rocca, "Inelastic Light Scattering in Superionic Glasses $(AgI)_x(Ag_2O \cdot nB_2O_3)_{1-x}$," _Phys. Rev._ B29:3567 (1984).

46. B.J. Wuensch, M.I.T., Private Communications.

47. W.L. Konijnendijk, "Structure of Glasses in the Systems CaO-Na_2O-B_2O_3 and MgO-Na_2O-B_2O_3 Studied by Raman Scattering," _Phys. Chem. Glasses_ 17:205 (1976).

48. P.J. Bray, S.A. Feller, G.E. Jellison, Jr., Y.H. Yun, "B^{10} NMR Studies of the Structure of Borate Glasses," _J. Non-Cryst. Solids_ 38/39:93 (1980).

MULTIPHASE AND POLYCRYSTALLINE FAST ION CONDUCTORS

Joachim Maier

Max-Planck-Institut für Festkörperforschung
D-7000 Stuttgart 80, Heisenbergstraße 1, FRG

INTRODUCTION

Starting from simple (e.g. binary) compounds, the usual strategy of improving the ionic conductivity consists of investigating modified structures (α-AgI, glasses) to form (single phase) multicomponent materials (RbAg$_4$I$_5$, Nasicon, $Zr_{1-x}Ca_xO_{2-x}$). A special case of the latter is the well-known procedure of classical (homogeneous) doping where small amounts of dopants are added (e.g. AgCl(+CdCl$_2$)) in order to influence transport properties rather than to disturb structural properties.

More recently, the investigation of multiphase systems has opened the possibility of increased variability, due to additional boundary effects (composite electrolytes)[1-5]. Many two-phases mixtures of ionic conductors such as halides of Ag$^+$, Li$^+$, Cu$^+$, Tl$^+$, Cu^{2+}, Sr^{2+}, Ba^{2+} together with insulating oxides (mostly Al$_2$O$_3$) have been shown to exhibit much higher conductivities than each of the constituent phases alone (cf. Arrhenius plots in Fig.1[9-11]). Furthermore, in the case of a two ionic conductor materials system, large enhancement effects have been found[6]. It is noteworthy that although several different effects may occur in those mixtures, many analogies and common aspects exist ("Heterogeneous Doping"). The term "Heterogeneous Electrolytes" has been introduced[3] in order to include similar effects at interfaces formed by the grains of conductors of the same composition (grain boundaries) and/or near other higher-dimensional defects (e.g. dislocations). In this paper three types of interfaces will be considered in detail (MX=ionic conductor):
  (i)   MX/A
        (the interface of an ion conductor with a surface active insulator)[2],
  (ii)   MX/MX'
        (the interface of two ionically conductive compounds)[13],
 (iii)   MX/MX
        (the grain boundary interface)[14].
Lastly the question will be discussed:
How will the conductivity properties change in all these cases if the effective thickness of the boundary layer is no longer small compared to the sample thickness (e.g. thin film: one interface is affected by the other one)[15]?

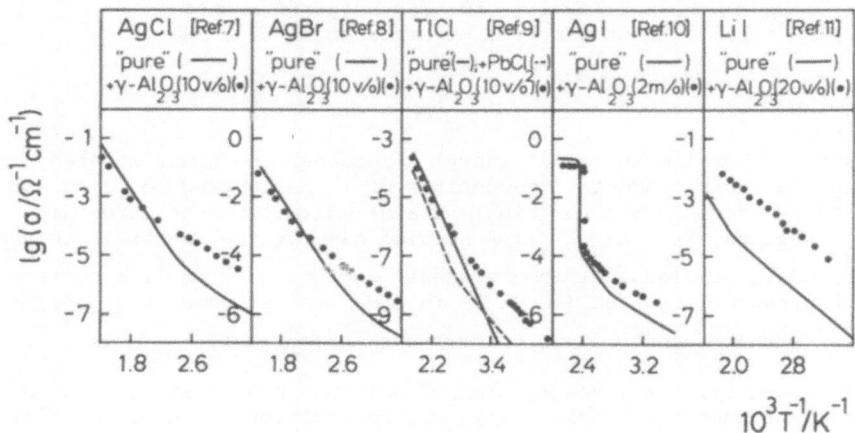

Fig. 1. The effect of γ-Al$_2$O$_3$ ($r_A \cong 0.03$ μm) - admixtures to different
ionic conductors (MX). It is worth noteing that the slopes of
of the Arrhenius-plots in the low T-regime are similar to the
slopes of the low T-regime of "pure" (higher-valent cationic
impurities) or appropriately (homogeneously) doped MX (enrichment of metal vacancies).

Fig.2 gives an overview of what may happen at an interface ionic conductor/insulator. Conduction pathways in the insulator, as well as global reactions or phase modifications that would be detected by X-ray, are not considered. The interface core may provide a fast new migration path. Impurities may be introduced resulting in indirect (doping effects) or direct changes (mobile impurities, e.g. H^+) of the transport properties. Boundary phases that are possibly only stable at the contact due to the modified (Gibbs-) energy situation, may form as a result of a chemical reaction with the oxide or with surface groups or by phase transformation. Moreover interface stress may result in elastic effects, or in plastic effects manifested in the formation of higher-dimensional defects (dislocations). Finally surface interactions have to be taken into account (cf. colloid chemistry). In all cases space charge layers play an essential role, since they are due to the adjustment between surface and bulk defect chemistry.

Although complex situations may and will arise (e.g. slight HCl evolution because of acid surface groups on ("wet") Al_2O_3, hydration of Li^+ by surface groups (chemisorbed (OH) or physi-sorbed H_2O)), the decisive question is: Which conduction pathway determines the observed conductivity? Generally speaking three kinds of pathways can be distinguished: core mechanism (along interface, interaction layer, interphase), space charge mechanism and bulk mechanism. Whereas in the first case more or less independent transport quantities should appear and thus the migration enthalpy and the effective formation enthalpy should be different from the pure bulk (MX), in the latter two cases the mobility and thus the migration enthalpy (weak elastic effects) should be those of pure MX bulk. The main difference between space charge region and bulk (and between possibly doped bulk and pure bulk) is the modified defect concentration (and thus modified effective formation enthalpy).

It has been shown that the enhancement effects mentioned below, are definitely boundary effects (conductivity enhancement is more or less proportional to contact area; in the case of $AgCl:Al_2O_3$ the importance of the physical contact has been highlighted[7]). Nevertheless, it is a striking result (see Fig.3) that the activation enthalpy (E_L) of the excess conductivity ($\Delta\sigma_m$) is in all cases, where a reliable comparison can be made, similar (slightly lower) to the migration enthalpy of metal vacancies in the bulk. This can be easily visualized by comparing typical enhancement curves at low temperatures (see Fig.1) with the extrinsic (enhanced metal vacancy concentration) low T-regime of the "pure" conductors or, in those cases where the extrinsic regime is not reached, with intentionally doped material (cf. $TlCl+PbCl_2$).

If we assume that percolating pathways (see micrographs[4,12,17]) dominate the conductivity for not too low second phase concentrations, we have to discuss three possibilities a) homogeneous bulk doping by impurities whose introduced concentration is roughly proportional to the contact-area; b) space charge regions due to the interface (or interphase) itself or c) indirectly formed around dislocation cores. In view of the activation enthalpy, the only reasonable explanation of "saving" the possibility of a novel mechanism would be that the highly conductive layers are appropriately blocked by MX-material such that the conductivity of the blocked region determines $\Delta\sigma_m$ (see below, Fig.8). This case has been thoroughly discussed elsewhere[3,4] and does not seem to be probable (no third percolation threshold; effective capacitance $\simeq$ bulk capacitance, $E_L \neq$ bulk activation enthalpy).

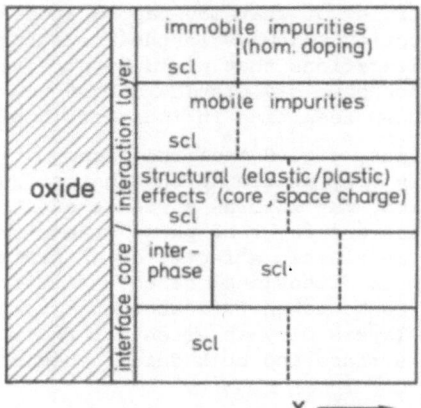

Fig. 2 Possible effects at the interface of an ionic conductor and a second "inert" phase.

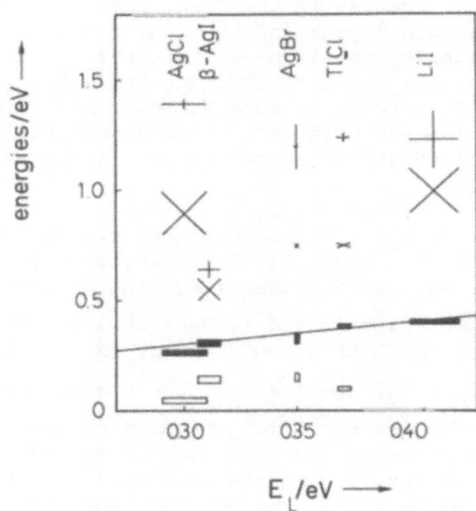

Fig. 3 Activation enthalpies (■: migration enthalpy (h_-) of the cation vacancies (V'_{Ag}, V'_{Li}, V'_{Tl}); □: migration enthalpy (h_+) of the counter defects ($Ag^\bullet_i$, $V^\bullet_{Cl}$); +: formation enthalpy of the bulk defect pair ($\Delta H°$); x: intrinsic activation enthalpy of the ionic conductivity ($\Delta H°/2 + t_+h_+ + t_-h_-$; t: transport number) versus the extrinsic activation enthalpy (E_L) for different sufficiently heterogeneously doped ionic conductors. The line represents the graph $h_-=E_L$. References are given elsewhere[9,12,16]. The size of the symbols represent the error bars.

If homogeneous doping would be decisive, impurities must be introduced by surface groups (acid-base reaction, H^+) or in a diffusion limited way during the preparation (Al^{3+}). Whilst in the case of LiI the possible impurity level seems to be far too low to account for the enhancement effect, in the case of AgCl and AgBr thorough measurements have been done to exclude this possibility[7]. Thus, the effect is independent of the preparation time, even $AlCl_3$ does not cause an appreciable enhancement effect and, most impressively, the effect vanishes if Al_2O_3 is removed. A very strong argument is the distinction between homogeneous and heterogeneous doping via the temperature dependence[4]. Since in the case of AgCl, AgBr (and TlCl), $Ag_i^\bullet$- (and $V_{Cl}^\bullet$-) defects determine the conductivity intrinsically, Cd^{2+}- or Pb^{2+}-doping lead not only to a dominant conductance of the counter defect (V_{Ag}', V_{TL}') at low T, but must also lead to a depression ("knee") in the curve (see $TlCl+PbCl_2$ in Fig.1) in an intermediate T-range (according to the mass action laws) where $\sigma(Ag_i^\bullet)$ or $\sigma(V_{Cl}^\bullet)$ become dominant. This is in sharp contrast to heterogeneous doping where depletion layers are mostly short-circuited. A further argument against homogeneous doping is delivered by the electronic conductivity[18].

As it has also been thoroughly studied and is mentioned below again, there is a perceptible influence of higher-dimensional defects (grain-boundaries, dislocations) on the ionic conductivity immediately after the preparation (particularly after pressing[4,14]). But these effects are deliberately not decisive for the two phase effects discussed here, as argued recently[19]. The influence of thermal treatment[4], in particular the wetting[20] of melted ionic conductors has been studied. The full effect occurs just after the melting process whereas the dislocation effect disappears[4]. Experiments with inactive second phases though treated in the same way, show no effect[7]. It is also probable that these strain effects in the case of separate particles are not conserved as much as for Al_2O_3-fibers which have been used in Ref.19 (where, beyond that, a second phase effect should not appear due to the unfavorable contact aera).

For all these reasons, space charge effects due to surface-surface interactions (in extreme cases due to the formation of an interphase) seem to play a decisive role. As the activation enthalpy suggests, (and as has been proposed earlier[1,2]) the metal vacancy concentration should be enhanced in the above samples due to a stabilization of M^+ at the interface (see Fig.4, first row).

In order to get more information on the described process we performed experiments by changing the surface activity of the second phase by using

a) different oxides (e.g. Al_2O_3, SiO_2)[12]

b) different modifications of a given oxide (α-Al_2O_3, γ-Al_2O_3)[7]

c) particles of a given modification (γ-Al_2O_3) whose surfaces have been treated chemically (($CH_3)_3SiCl$, CH_3Li)[11].

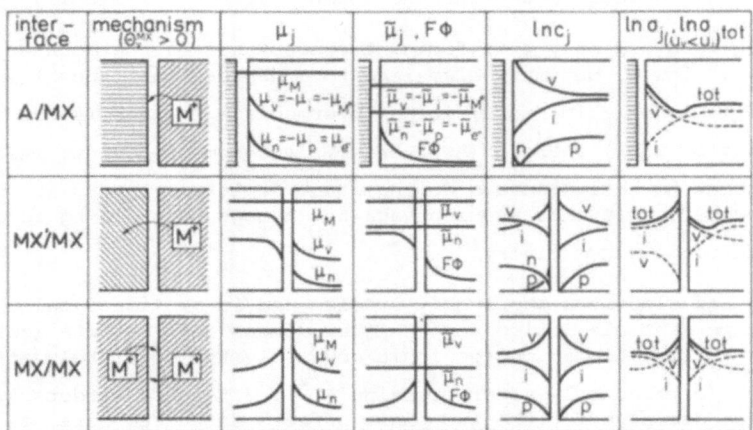

Fig. 4. The mechanisms of defect-redistribution at the interfaces MX/A, MX/MX' and MX/MX (MX; Frenkel-disordered ionic conductor, where $\Theta_V > 0$, i.e. enhanced metal vacancy concentration, is assumed), the profiles (perpendicual to the interface) for the electric potential (Φ), for the (electro-)chemical potentials ($\tilde{\mu},\mu$) of ions, defects and components, as well as for the logarithms of defect concentrations, partial and total conductivities (where u(interstitial defects) $\equiv u_i >$ u (vacancy defect) $\equiv u_V$ is assumed).

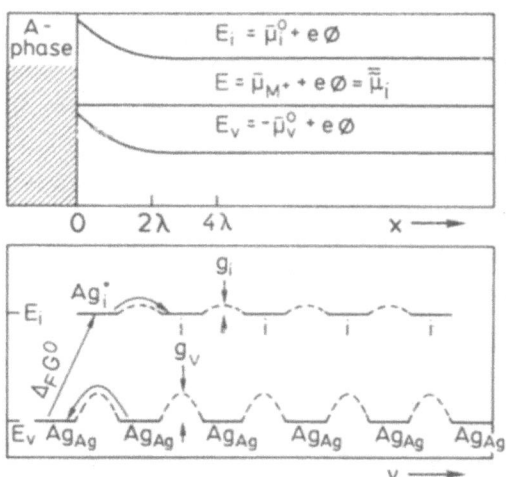

Fig. 5. The thermodynamics of boundary regions near an ideal second phase (no influence on component activities (a_M); a_M has to be fixed by a third phase) can also be represented by constructing the parameters that are analogous to the band model parameters in semiconductor physics.
top: profile perpendicular to the interface (dash relates the quantity to one particle).
bottom: profile (strictly) parallel to the interface (high resolution, negligible outer field, g: hopping free energy).

The results (importance of surface OH-groups; higher basicity of M-OH group enlarges the effect) are fully consistent with the formulation of the defect inducing surface process (mechanism of heterogeneous doping)[17] as:

$$Nu_A + M_M \rightleftharpoons (Nu...M)_A^{\cdot} + V_M'$$ (1)

where the nucleophilic surface groups (Nu) are most probably OH-groups. The detailed interaction of course depends on the surface behaviour. The main problem here is the lack of characterisation of the surfaces involved (see below).

Further evidence for this mechanism in the case of AgCl and AgBr (which describes an internal adsorption of Ag^+) is given by wetting experiments[20]. Parallel to the conductivity results, the decrease in the interaction of AgX with the oxide in the order γ-Al_2O_3, SiO_2, α-Al_2O_3 is reflected in the wetting behaviour of halide melted on oxide crystals (lowering of the contact angle). (See also adsorption experiments on metacaolinite and SiO_2[21]). An analogous behaviour is well-known in colloid chemistry, where the equilibrium with H^+

$$\overline{Al} \text{------} OH + H^+ \rightleftharpoons \overline{Al} \text{------} OH_2^+$$ (2)

is shifted to the left hand-side by replacing γ-Al_2O_3 by α-Al_2O_3 or SiO_2[22]. These significant effects in surface tension are also the reason for the observed "mechanical" stability of the composites at temperatures above the melting point of the halides.

Further support is furnished by recent experiments in which a change of the ionic parallel conductance of thin films has been measured if exposed to NH_3- or $(CN)_2$-gas[20] under appropriate conditions. A nucleophilic behaviour (Eq.1) is probable for these species a priori (free electron pairs, Ag^+-complex-chemistry). It is worth noting that such effects are well-known in the case of semiconductors, where redox-active gases such as H_2 or O_2 can be detected by their effect on the electron concentration in the space charge regions[23].

For these reasons - and also as space charge considerations are the natural extension of bulk defect chemistry to boundary layers - the defect chemical aspects and the conductivity effects involved has been treated quantitatively. Figs.4-6 comprise the thermodynamic procedure and the results of concentration and conductivity effects if structural changes can be neglected and an enhancement of the negative charge carriers is assumed. Details are given elsewhere[3,4,12-15]. The thermodynamic situation in the boundary regions is preferably described by the thermochemical approach (μ, $\tilde{\mu}$, ϕ) used in Fig.4, but can also be represented by energy-level bending as it is customary in semiconductor physics (see E_i, E, E_v in Fig.5).

The result for the parallel extra conductance $\Delta Y''$ (Fig.6) proves that this term can be written as a product of an effective concentration and an effective thickness, as assumed by Jow and Wagner[24]. The result for large enhancement effects is of special meaning in the following. Here the effective thickness is the Debye-length, and the effective concentration is the geometrical mean of bulk concentration (c_∞) and the concentration immediately at the interface (c_o). Since $\sqrt{c_\infty}$ cancels against $1/\sqrt{c_\infty}$ in the Debye-length, the result is independent of impurity effects. (This does not hold for the conductivity in the case that σ of the counter defect is important and also

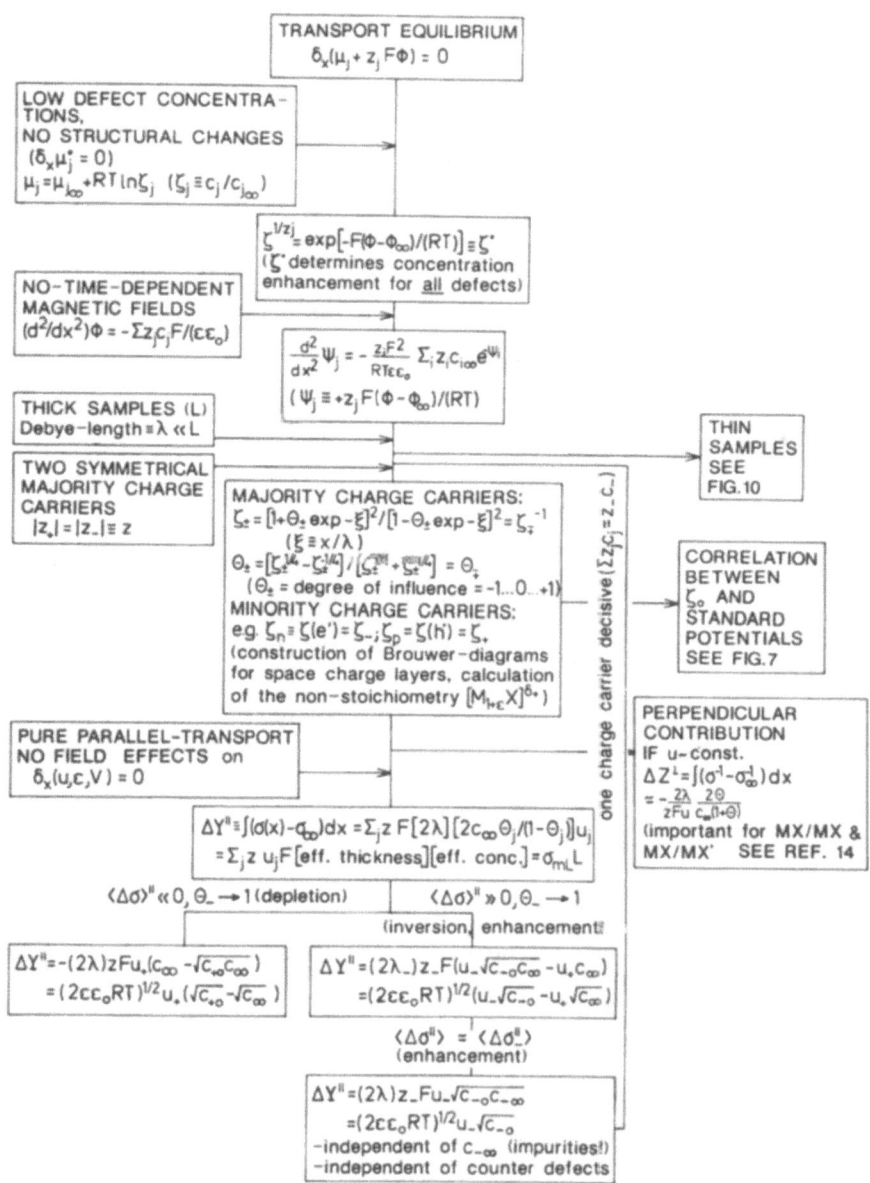

Fig. 6. Defect chemistry and conductivity effects in ideal space charge regions.

EXAMPLE: FRENKEL DISORDER

$$M_M + V_i \rightleftharpoons M_i^{\cdot} + V_M' \qquad (\Delta_F G^{\circ})$$

BULK: local electroneutrality & mass action law;
if no electronic carrier important $\rightarrow$
$c_{i\infty} = K_F / c_{v\infty} = \text{const } [x]$
($\text{const} = \sqrt{K_F}\ (\pm\sqrt{K_F})$ if intrinsic (extrinsic))

SPACE CHARGE REGION:

$$V_M'(\infty) + M_M(x) \rightleftharpoons V_M'(x) + M_M(\infty)$$

transport equilibrium: $\Delta_x \tilde{\mu}_j = 0$ (see Fig.8)
$\rightarrow c_j(x) = f[c_{j\infty}, c_{jo}]$

INTERFACE (Ref.3,4)

$$M_M(0) + V_i \text{ (int)} \rightleftharpoons V_M'(0) + M_i^{\cdot} \text{(int)}$$

a) electrochemical equilibrium: $\Delta_x \langle \tilde{\mu} \rangle = 0$
   (x: reaction progress)
b) continuity of $\varepsilon(\partial/\partial x)\Phi \iff$ global electroneutrality
c) model for potential jump (e.g. charge free zone)
(d) additional intercorrelation with neighbour–phase)

EXAMPLE: MX/MX′ (Ref.13)

$$c_v(0)/c^{\circ} = \left[x(\varepsilon'/\varepsilon)\exp-\Delta_{\alpha\alpha'}G_F^{\circ}/(RT) \right]^{1/2} \doteq (\varepsilon'/\varepsilon)(c_i'(0)/c^{\circ})$$

x accounts for potential jump; $\Delta_{\alpha\alpha'}G_F^{\circ}$ refers to
heterogeneous Frenkel–reaction)

FREE SURFACE   (Ref.3,4)

a) $a_{M^+}(s) = \text{const} \longrightarrow c_{jo} = f[\Delta_j G^{\circ}]$
   $x = \text{const} \longrightarrow$ model of Kliewer and Köhler[27]:
   $c_j \propto \exp\left[-\Delta_j G^{\circ}/(RT)\right]$
b) consideration of surface configurational entropy
   and of limited number of surface sites
   $\longrightarrow$ model of Pöppel and Blakely [28]
c) more accurate see Ref.18

CONTACTED SURFACE IN A KLIEWER–TYPE MODEL (Ref.3,4,12)

$c_{jo} \propto \exp-(\Delta_j G^{\circ} + \Delta_{SI} G^{\circ})/(RT)$
$\Delta_{SI} G^{\circ} = \Delta_R G^{\circ} \{M^+(s) \rightleftharpoons M^+(\text{int})\}$
T–dependence:
$-R(\partial/\partial 1/T)c_{jo} \simeq \Delta_{SI} H^{\circ} + \Delta_j H^{\circ}$ (+correction terms)

Fig. 7. Determination of the parameters c_o and c_{∞}, characterizing the
defect chemistry.

for the electronic part.) Since the approximation for large effects can also be obtained by neglecting concentrations of the counter defects a priori in the basic differential equation, the result is valid irrespective of mobility (if not too large) and of the number and charge of the counter defects (z and c_∞ in λ must be the values of the enhanced defect type).

Fig.7 shows how c_o (and c_∞) may be correlated with energetic and entropic standard values.

Fig.8 outlines the process of calculating the (mean) conductivity (σ_m) of a dispersion which is a complex superposition of different pathways. As our experiments[4,12,17] suggest, continuous pathways can be assumed which are increasingly blocked by insulator-particles at higher second phase concentrations.

Here only the result of large enhancement effects with respect to the negatively charged defect (i.e. $\Theta_- \to 1$, see Fig.8) shall be considered ($\sigma_A \tilde{} 0$):

$$\sigma_m \tilde{} (1-\phi_A)\sigma_\infty + \beta_L \phi_A \Omega_A \sqrt{2\epsilon\epsilon_o RT} \; (u_- \sqrt{c_{-o}} - u_+ \sqrt{c_\infty}) \qquad (3)$$

If the mobility of the counter defect, u_+, is not so high that the second term plays a significant role ($c \ll c_{-o}$), i.e. that a depletion influence can be neglected, we simply have

$$\sigma_m \tilde{} (1-\phi_A)\sigma_\infty + \beta_L \phi_A \Omega_A \sqrt{2\epsilon\epsilon_o RT} \; u_- \sqrt{c_{-o}} \qquad (4)$$

Already this roughly simplified consideration explains the dependences found in the literature.

i) The frequency dependences of the complex impedance can be understood. Highly conductive layers with small capitances short-circuit the bulk impedance: $\hat{\sigma}_m \cong \sigma_m + \sqrt{-1}\ \omega(1-\phi_A)\epsilon_\infty)$; i.e. the semi-circle at higher frequencies is characterised by σ_m but still by ϵ-values close to ϵ_∞.
The low frequency semi-circle may be due to additional transfer resistivities.

ii) The proportionality to the volume fraction (ϕ_A) (see Eq.(3)) has often been observed in the literature ($\delta\beta_L/\delta\phi_A \tilde{} 0$; $\beta_L \sim 0.2\text{-}0.7$) for ideal quasi-parallel switching. Of course, the dependence cannot be observed under all conditions, since σ_{mL} reacts sensitively to the distribution of the two phases. A more detailed consideration is given elsewhere[4,12,17]. For very low concentrations (in particular with large particles) or for a random distribution[25,26], isolated particles exist and the quasi-parallel limit is not reached. But it must be born in mind, that a) with the low particle sizes normally used, continuous pathways may form at concentrations even below 1v/o, and b) random distribution is not realized because of the surface-surface interactions.

Some words shall be added to the apparent discrepancy as far as the superposition procedure used in Ref.24 is concerned. Since electric fields and currents are vectorial quantities, σ_m cannot be obtained by scalar integration over c_j. Such an approach would mean that all regions take part in a parallel switching, i.e. $\beta_L=1$. Moreover in Ref.24, the volume fraction of Al_2O_3 has been neglected in the averaging procedure. Taking these points into account, we end up with the

OVERALL CONDUCTIVITY:

$\sigma_m = \mathcal{P} \{\sigma_\infty, \sigma_{scl}, \sigma_{core}, \sigma_A, \ldots\}$

The percolation operator reacts sensitively on the detailed distribution

HIGH LAYER CONDUCTIVITY (scl,core) $\gg \sigma_A, \sigma_\infty$ (Ref.3,8,12)

$\sigma_m = \Sigma_\alpha \beta_\alpha \varphi_\alpha \sigma_\alpha$ (quasi–parallel–switching)

(α denotes region ∞,L(scl,core,A); β counts measured percolating pathways)

moderate regime (cf. experiments): $(\partial/\partial\varphi_\alpha)\beta_\alpha = 0 = (\partial/\partial T)\beta_\alpha$

blocking regime (high φ_A–values): $(\partial/\partial\varphi_\alpha)\beta_\alpha \neq 0 = (\partial/\partial T)\beta_\alpha$

VERY LOW φ_A–VALUES, RANDOM DISTRIBUTION, FRACTALISED SURFACES, SIMILAR CONDUCTIVITIES:

detailed percolation theory, Monte–Carlo–simulations (Ref.26,29)
effective medium theory (Ref.25)

INTERRUPTED HIGHLY CONDUCTIVE PATHWAYS (Ref.3)

$\sigma_m = \Sigma_\alpha \beta_\alpha \varphi_\alpha [\Sigma_{\alpha'} \beta_{\alpha'} \varphi_{\alpha'} \sigma_{\alpha'}^{-1}]^{-1}$

CUBE–MODEL (cf. polycristalline materials) (Ref.14)

$\sigma_m = (\sigma_\infty \sigma_L^\perp + \beta_L^{\parallel} \varphi_L \sigma_L^{\parallel} \sigma_L^\perp)/(\sigma_L^\perp + \beta_L^\perp \varphi_L \sigma_\infty)$

FREQUENCY DEPENDENCE (Ref.14)

$\sigma_\alpha \rightarrow \hat{\sigma}_\alpha \simeq \sigma_\alpha + \sqrt{-1}\,\omega\,\varepsilon_\alpha$

QUASI–PARALLEL–SWITCHING (Ref.12) ($\sigma_A = 0$)

$\sigma_m = \beta_\infty (1-\varphi_A)\dot{\sigma}_\infty + \beta_L \varphi_{scl} + \beta_L \varphi_{core} \sigma_{core}$

(Ω_A: surface–to–volume ratio of A–particles)

$\sigma_{mL} = \beta_L \varphi_{scl} \langle \Delta\sigma \rangle_{scl} = \beta_L \varphi_A \Omega_A \Delta Y^{\parallel}$

$\qquad = \beta_L \varphi_A \Omega_A \Sigma_j z\, F(2\lambda)(2c_\infty \theta_j/(1-\theta_j))u_j$

$\qquad = \beta_L \varphi_A \Omega_A (2\varepsilon\varepsilon_o RT)^{1/2}(u_-\sqrt{c_{-o}} - u_+\sqrt{c_\infty})$, if $\theta_- \rightarrow 1$

$\qquad = \beta_L \varphi_A \Omega_A (2\varepsilon\varepsilon_o RT)^{1/2}u_-\sqrt{c_{-o}}$, if σ (counter defects) $\simeq 0$

T–DEPENDENCE (Ref.3,4,12) ($\theta_- \rightarrow 1, \sigma_{mL} \rightarrow \sigma_{mL-}$)

$E_L = -R(\partial/\partial 1/T)\sigma_{mL} = h_- + \frac{1}{2}(\Delta_- H^\circ + \Delta_{SI}H^\circ) + \text{corr.} = h_-$

ELECTRONIC CONDUCTIVITY (Ref.18) ($\theta_- \rightarrow 1, \sigma_{mL} \rightarrow \sigma_{mL-}$)

$\sigma_{m,e} = \beta_\infty (1-\varphi_A)\sigma_{e\infty} + \beta_L \varphi_A \Omega_A u_e (2\varepsilon\varepsilon_o RTc_{-o})^{1/2}(c_{e\infty}/c_{-\infty})$

Fig. 8. Conductivity in a two-phase system.

superposition relation used in Fig.8 for quasi-parallel-switching.

iii) The proportionality to the surface-to-volume ratio, Ω_A, or for spherical particles to the reciprocal grain radius, has been found in many examples (but for the above reasons not under all conditions)[4].

iv) As far as the conduction mechanism is concerned, it is very important
that Eq.(4), describes the observed temperature dependence correctly.
Fig.8 shows that for a strong interaction $E_L \tilde{-} h_-$. This result explains
nicely the findings that are visualized in Figs.1,3 and supports the
vacancy mechanism (see Eq.(1)) assumed above. The slight depression at
higher temperature where σ_{mL} is no longer of influence can also be interpreted. This also holds for the range of intermediate temperatures
in the case of TlCl where - as a consequence of $u_+ > u_-$ (inversion
layers at lower temperatures) - depletion layers are formed which are
short-circuited by the bulk and do not perceptibly appear in the σ_m -
graph. Even for large ϕ_A-values (blocking regime) where the superposi-
tion formula $\sigma_m = \Sigma_\alpha \beta_\alpha \phi_\alpha \sigma_\alpha$ only has a formal meaning with respect to the
ϕ_A dependence ($\beta_\alpha = f(\phi_\alpha)$), the temperature dependence is quite accurately described, since the (lowered) β_α-values are still T-insensitive
($\sigma_{mL} \gg \sigma_{mA}$) (see parallel-shifted graphs in Ref.8).

In the following different ionic conductors which have been used in
two-phase mixtures, are discussed separately.

1) AgCl and AgBr

Here extensive studies have been performed in the literature, since
the pure materials are very well investigated[7,8,12,31]. All bulk parameters
for a further analysis of Eq.(4) are known.
It is very interesting, that the σ_m-values can be quantitatively understood
if a maximum effect is assumed, i.e. that the first layer adjacent to the
interface is saturated with defects[7,8,12]. Such large effects are discussed
near grain-boundaries in ceramics. It must not be overlooked in this context
that i) the accuracy of all the parameters manifest in an uncertainty for
c_o of a factor of ~2-3 (cf. $\log \sigma_m$!), ii) such a large concentration exceeds
the assumption of a Boltzmann-distribution and iii) micro-size effects should
be responsible for additional enhancements. The very satisfactory quantitative description of all the experimental curves over the whole T-range for
different ϕ_A- and r_A-values is shown in Fig.9 for the example of AgBr.

2) CuCl

Because of the similarity to AgCl a similar mechanism in the $CuCl/Al_2O_3$
composites is probable. Unfortunately neither the vacancy-mobility nor even
the corresponding activation enthalpy are available for a further test. If
we suppose the validity of the above mechanism for CuCl, $h(V'_{Cu})$ should be
~0.4eV.

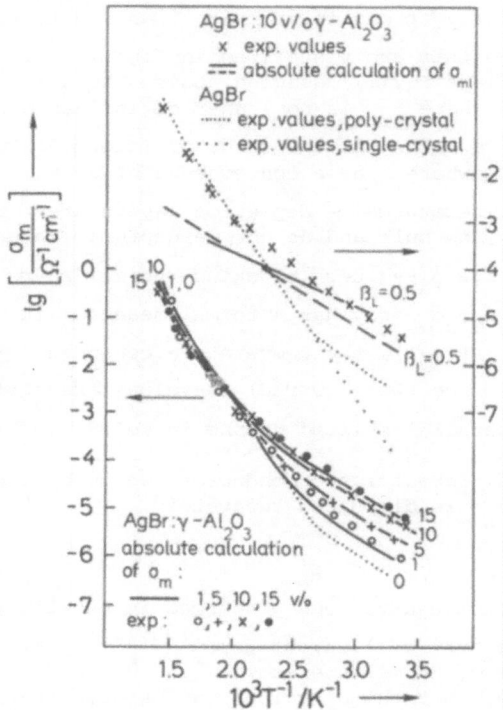

Fig. 9. The conductivity enhancement of the system $AgBr:Al_2O_3$ can be quantitatively described by assuming a saturation effect[8] (top). By taking account of the bulk contribution as well as of a slight T-dependence of the surface concentration, the conductivity curves can be well described over the whole T-range for different volume fractions[8] as well as for different grain sizes[7] (bottom).

3) LiI and LiBr·H$_2$O

Although LiI is the classical example[1], here and also for LiBr·H$_2$O[34], the situation is very unclear. On the one hand the activation enthalpy (see Figs.1,3) strongly supports space charge effects, on the other hand, the u$_v$-data of Jackson and Young[35] seem to yield σ_m-values that are somewhat too low (all the more the data of Haven[31]). But uncertainties in the absolute value[4] or thin film effects (40v/o Al$_2$O$_3$, very narrow spacing) may be responsible for these discrepancies[15] (see below). According to NMR-measurements a large amount of ions in LiI and LiBr·H$_2$O seem to be involved in the boundary effects[37]. In contrast to the other examples, the water-solubility of these compounds represents a special difficulty. Even the precise structure near the interface and thus the mobility and the defect concentrations might change[33,38]. Moreover, these compounds should be highly sensitive to acid-base reactions with the oxide's surface groups. Nevertheless space charge effects must play an important role.

4) AgI

As far as the mobilities (and possibly the structure) are concerned, the situation is similar to LiI[3]. In addition, the exact modifications being present in the different T-intervals are unknown. Recent careful measurements[10] show that the behaviour of the activation enthalpy (see Fig.1) is the same as in the other materials. Early measurements by the group of Wagner Jr. interpreted thermoelectric effects, effects on the temperature of phase transitions as well as excess enthalpies by Ag$^+$-vacancies[10,40].

5) TlCl

Perhaps the most striking evidence is delivered by the two-phase system TlCl:Al$_2$O$_3$[20]. TlCl is intrinsically an anionic conductor. The σ-values above room-temperatures are nearly unaffected by Al$_2$O$_3$ (besides the volume effect) being consistent with a depletion effect. At lower temperatures, however, a clear change towards a Tl$^+$-conduction, induced by the creation of Tl$^+$-vacancies, is observed (inversion layer) in the annealed two-phase mixture. Since no depression at higher temperatures where the annihilation of V'_{Cl} is decisive, is found (in sharp contrast to TlCl+PbCl$_2$, see Fig.1) an impurity effect can be definitely excluded. Just as for AgCl and AgBr all the experiments can be described over the whole T-range for different ϕ_A- and r$_A$-values. The adjusted c$_{vo}$ value is far lower than in the case of the silver halides (smaller interactions?) but the space charge potential is similar. Beyond that, the behaviour of homogeneously plus heterogeneously doped TlCl ((TlCl+PbCl$_2$):Al$_2$O$_3$) is correctly predicted.

CaF$_2$, SrF$_2$, BaF$_2$, SrCl$_2$

Recently some experiments on different earth alkaline halides that are also anion conductors have been reported[41-44]. Unlike TlCl (Schottky-conductor) they exhibit a pure anion-disorder (Frenkel disorder in the anion sublattice). Since a cation conduction should not be observed under any condi-

tions, a nucleophilic effect of the oxide would mean here that the interstitial anion concentration would be enhanced. The effects have been interpreted in this way by some authors[43]. Other authors assume that - by analogy
to our model for the cations - the anions would be adsorbed and thus the vacancy concentration increased[42]. The latter effect is not impossible because
i) oxide's surfaces may behave ambivalent and ii) if so, the mobility of cations would not be sufficiently large to compete with the anions in such an
adsorption process. The relevant mobility values, which are available in the
literature would be sufficiently high to satisfy the measured enhancement in
σ due to space charge effects according to Eq.(4). Up to now, however, there
is no significant hint (e.g. E_L) for space charge effects in the latter materials at all, not to mention for comparably high temperatures of these experiments, possible chemical reactions must be considered.

In any case the variety of relevant processes has to be taken into account. Particularly, it is very important to use thoroughly defined materials
(especially Al_2O_3) to have a chance to identify the correct mechanism.

TWO IONIC CONDUCTORS (MX/MX')

Though the conductivity problem of two phase mixtures of two ionic conductors is at least as interesting as the above case, not many experiments
in this field have been made. Detailed investigations have been performed
on the conductivity behaviour within the miscibility gap of the system AgBr-
β-AgI[6]. Large enhancement effects have been found. Interestingly the $\sigma_m - \phi_A$
characteristic is typical for a random distribution which can be assumed to
occur if the two-phase mixture is prepared by a cooling process[3,13]. If we
extend the above mechanism of re-distribution of ions at the contact zone,
(as shown in Fig.4, second row), we have (as e^- in semiconductors) to assume
that there is a charge transfer from one space charge region to the other.
For large effects the σ-profile (Fig.7) must exhibit an ionic P-N-junction
in the case of the silver halides). Reasonable differences in the Gibbs-
energies for the defects in the different Frenkel-disordered conductors
(see Fig.6) should cause large conductivity effects[13]. The observed activation enthalpies support also the space charge argument. On the other hand,
perceptible excess energies suggest the presence of additional elastic effects[45].

GRAIN BOUNDARIES (MX/MX)(DISLOCATIONS)

In the literature enhanced grain-boundary conduction has been chiefly
analyzed in terms of core conduction (pipe diffusion). In the case of the
above discussed ion conductors, in particular for AgCl, AgBr and TlCl we
have strong evidence for space charge effects[3,4,14,31](cf.also the work of
Kliewer and Köhler[27]). Grain-size dependences and activation enthalpies as
well as the form of the log σ-1/T graph suggest stabilisation of Ag^+ or Tl^+
in the grain's interface (segregation) due to structural pecularities or due
to impurities present. The profile of the thermodynamic parameters, of the
local concentration and the local conductivity is shown in Fig.4 (bottom row)
for such a case.

NH_3 in the grain-boundaries seems to enhance the effects[3], as it does at
surfaces. Impendance measurements on bi- and poly-crystalline Ag-halides
prove that the grain boundaries (and/or dislocations) have a double function

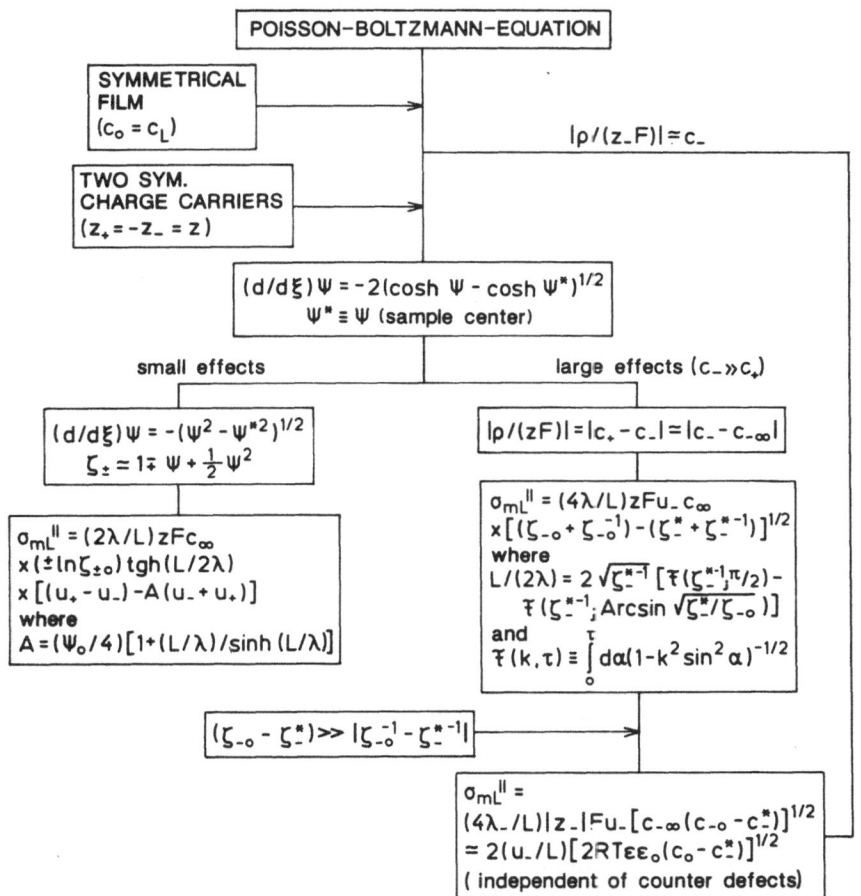

Fig. 10. The calculation of the parallel excess conductivity of symmetrical thin films for the whole thickness range (finite and
semi-infinite boundary conditions) for large (enhancement)
and for small effects, as long as structural changes can be
neglected.

(according to the direction in which the charge carriers move): They act as highly conductive (parallel) paythways as well as additional hindrances for (perpendicular) transport[14]. Analyzing the behaviour by a simple topological model the overall process could be interpreted in a sufficient manner. A lot of phenomena not understood so far may be explained in this way.

MICROSIZE EFFECTS

A fundamental problem affecting thin films, micropolycrystalline samples and two phase mixtures with very narrow spacing of the phases, appears if films or crystals are considered whose thicknesses are no longer large compared to the Debye-length. The defect profile no longer reaches the bulk value for very thin samples in which, as a consequence, the neighbor-phases are of significant importance. Enhancements (or depressions) much larger than that of semi-infinite samples (see Eq.(4)) must be considered[15]. The parallel conductance has been given in an analytical form recently[15]. Fig.10 shows the results for weak and for large effects. In principle, an analysis of the thickness dependence of σ_m may provide far-reaching information. It must not be overlooked, however, that in such microsystems deviations from the idealised assumptions (constancy of μ_j^o, ε, u_j; no elastic effects) can easily arise. Many experiments in thickness-regimes, where the boundary contribution appears as a constant, are reported in the literature (cf. Ref.15). The evaluation of T- or activity-dependences of this contribution must be done in the way outlined in Fig.10. An application of the total solution (thickness dependent space charge contribution) to literature examples has recently been given[15]. The inconsistencies revealed by such a precise procedure may be due to changes in the atomic[13] or in the microstructure (e.g. island-formation). Some work is in progress now to investigate thin layered, multilayered and micropolycrystalline materials.

REFERENCES

1. C. C. Liang, J. Electrochem. Soc 120: 1289 (1973)
2. J. B. Wagner, Jr., Mater. Res. Bull. 15: 1691 (1980)
3. J. Maier, J. Electrochem. Soc. 134: 1524 (1987)
4. J. Maier, Mater. Phys. Chem. 17: 485 (1987)
5. F. W. Poulsen, in: "Transport-structure Relations in Fast Ion and Mixed Conductors", F. W. Poulsen et al., eds., Risø, Roskilde, 1985, pp.67
6. K. Shahi, and J. B. Wagner, Jr., J. Solid State Chem. 42:107 (1982)
7. J. Maier, Solid State Ionics 18/19: 1141 (1986)
8. J. Maier, Mater. Res. Bull. 20: 383 (1985)
9. J. Maier, and B. Reichert, Ber. Bunsenges. Phys. Chem. 90:666 (1986)
10. P. Chowdhary, and J. W. Wagner, Jr., Mater. Lett. 3:78 (1985)
11. J. Maier, Mater. Sci. Monogr. 28A: 415 (1985)
12. J. Maier, J. Phys. Chem. Solids 46: 300 (1985); loc.cit.5, 153 (1985)
13. J. Maier, Ber. Bunsenges. Phys. Chem. 89: 355 (1985)
14. J. Maier, Ber. Bunsenges. Phys. Chem. 90: 26 (1986)
15. J. Maier, Solid State Ionics 23: 59 (1987)
16. J. Corish, and P. W. M. Jacobs, Surf. Def. Prop. Solids 2:160 (1973)
17. J. Maier, Ber. Bunsenges. Phys. Chem. 88: 1057 (1984)
18. J. Maier, to be published
19. N. Dudney, J. Am. Ceram. Soc. 70: 65 (1987)
20. J. Maier, S. Prill, and B. Reichert, to be published
21. E. A. Daniels, and S. M. Rao, Z. Phys. Chem. N. F. 137: 243 (1983)
22. K. Tanabe, in "Catalysis, Science and Technology (Vol.2),

J. R. Anderson, and M. Boudart, eds., Springer, Berlin (1983);
J. A. Schwarz, C. T. Driscolland, A. K. Bhanot,
J. Colloid Interface Sci. 97: 55 (1984)

23. W. Göpel, Progr. Surf. Sci. 20:1 (1986)
24. T. Jow, and J. B. Wagner, Jr., J. Electrochem. Soc. 126:1963 (1979)
25. A. M. Stoneham, E. Wade, and J. A. Kilner, Mater. Res. Bull.
    14: 661 (1979)
26. A. Bunde, W. Dieterich, and E. Roman, Solid State Ionics
    18/19: 147 (1985)
27. K. L. Kliewer, and J. S. Köhler, Phys. Rev. A 140: 1226 (1965)
28. R. B. Poeppel, and J. M. Blakely, Surf. Sci. 15: 507 (1969)
29. H. Böttger, and V. V. Bryksin, "Hopping Conduction in Solids",
    VCH, Berlin (1985)
30. F. W. Poulsen, and P. J. Møller, loc.cit. 5, 159 (1985)
31. A. Khandkar, and J. W. Wagner, Paper 833, Electrochem. Soc.
    Meeting, San Francisco (1983)
32. Y. M. Chiang, A. F. Henriksen, W. D. Kingery, and D. Finello,
    J. Am. Ceram. Soc. 64: 385 (1981)
33. J. Wassermann, T. P. Martin, to be published;
    J. Wassermann, T. P. Martin, and J. Maier to be published.
34. O. Nakamura, and J. B. Goodenough, Solid State Ionics 7:19 (1982)
35. B. I. H. Jackson, and D. A. Young, J. Phys. Chem. Solids 30: 1973
    (1969).
36. Y. Haven, Rec. Trav. Chim. 69: 1471 (1950)
37. T. Asai, and S. Kawai, Solid State Ionics 20: 225 (1986);
    T. Asai, C.-H. Hu, S. Kawai, Mater. Res. Bull. 22: 269 (1987);
    J. L. Bjorkstam, D. Brinkmann, M. Mali, J. Roos, J. B. Phipps,
    and P. M. Skarstad, Solid State Ionics 18/19: 557 (1986);
    R. Dupree, J. R. Howells, A. Hooper, and F. W. Poulsen,
    Solid State Ionics 9/10: 131 (1983)
38. A. Khandkar, and J. B. Wagner, Jr., Solid State Ionics 20:267 (1986)
39. K. Shahi, and J. B. Wagner, Jr., J. Electrochem. Soc. 128:6 (1981)
40. P. Chowdhary, V. B. Tare, and J. B. Wagner, Jr., J. Electrochem.
    Soc. 132, 123 (1985)
41. S. Fujitsu, M. Miyayama, K. Koumotu, H. Yanagida, and Kanazawa,
    J. Mater. Sci. 20: 2103 (1985);
    S. Fujitsu, K. Koumotu, and H. Yanagida, Solid State Ionics
    18/19: 1146 (1986)
42. N. Vaidehi, R. Akila, A. K. Shukla, and K. T. Jacob,
    Mater. Res. Bull. 21: 909 (1986)
43. A. Khandkar, V. B. Tare, and J. B. Wagner, Jr., Rev. Chim. Min.
    23: 274 (1986)
44. T. L. Wen, R. A. Huggins, A. Rabenau, and W. Weppner,
    Rev. Chim. Min. 20: 643 (1983)
45. A. Khandkar, V. B. Tare, A. Navrotsky, and J. B. Wagner, Jr.,
    J. Electrochem. Soc. 131: 2683 (1984)

SPECTROSCOPIC INVESTIGATIONS OF GLASSES

Minko Balkanski

Laboratoire de Physique des Solides de l'Université Pierre et Marie Curie
associé au C.N.R.S.
4, Place Jussieu, 75252 Paris Cedex 05, France

INTRODUCTION

The spectroscopy of glasses is a very broad subject, as can be seen by the different spectroscopy techniques which can be used and by the phenomena which can be investigated. According to the objectives of the investigation different types of spectroscopies may be used which cover an extremely large spectral range, reaching from the X-ray to the Nuclear Magnetic Resonance regions. The investigations may be aimed at the structural or dynamical properties.

When the interest is focussed on structures, the method used gives information either concerning local or long range structure. We shall distinguish between methods revealing the local site symmetry through electronic transitions within one single atom or through the vibrational modes of a given atom. In the first case we are dealing with X-ray and Auger spectroscopy, EPR, NMR and in the second case with IR and Raman spectroscopy.

When the long range structure is of concern we are confronted with methods based on the elastic scattering of X-rays, neutrons or electrons. The allowed structure of the material, considered as an infinite solid, is revealed also by methods related to the electronic energy band structure of the materials or their phonon dispersion relations. The absorption edge observable in the visible and UV range relates to the electron energy band structure whereas the phonon spectra observable in the IR region depend on the phonon dispersion curves.

All the information about glass structure obtained by spectroscopy relies on the relations between well defined structural analysis in perfect crystals and the projected structure of the glass. In accepting the analogies and deduction we shall always keep in mind that one of the great unresolved debates in science concerns the microstructure of glass and its connection to crystalline structure.

The spectroscopy methods are based either on resonance or on scattering phenomena. The elastic scattering of neutrons, electrons and X-rays reveals structure parameters and atomic positions.

The inelastic scattering of photons reveals elementary excitation directly related to the structure of the solid.

The resonance methods can be refered to a very broad frequency range and constitute the different branches of spectroscopy from X-ray to electron spin or nuclear resonances. As we shall discuss the relative merits of the different methods and we shall focus our discussion on two types of glasses GeO_2 and B_2O_3-xLi_2O-yLi_2SO_4 in order to compare essentially the possibilities of the different method, leading to structural models.

In spite of some similarities these two types of glasses are quite different in their structure. GeO_2 or SiO_2 are simple compounds, whereas B_2O_3-xLi_2O-yLi_2SO_4 are solutions in which one can distinguish three components : B_2O_3 is the glass former component, Li_2O is the modifier and Li_2SO_4 the dopant. Spectroscopy methods should allow one to distinguish the role played by each component in the structure of the fast ion conducting glass.

We shall examine the possibilities offered by the different spectroscopic methods beginning at the high energy, high frequency side of the spectrum and progressing towards the lower frequencies. This is equivalent to considering first the electronic transitions of valence electrons, then vibrational excitations and, at the end, electron spin and nuclear spin resonances.

1. METHODS BASED ON SCATTERING PHENOMENA

1.1. Diffraction of X-rays

The response of condensed matter to an external probe reflects either the observable properties of the whole sample or the local atomic or molecular character. Scattering of particles or electromagnetic waves can result in two distinct phenomena both related to the structural properties of the sample.

On the one hand elastic scattering of neutrons, electrons and X-rays reveals structural parameters such as the atomic positions.

On the other hand inelastic scattering of photons reveals elementary excitations related to the site position, site symmetry, of the atoms or molecules constituting the solid, and also the chemical nature of the constituents.

In a perfectly ordered solid a crystalline structure is defined if all interatomic distances and angular disposition of all the atoms in the unit cell are known. The entire lattice can be constructed by performing lattice translations $R = n'_1a_1 + n_2a_2 + n_3a_3$ from an equivalent position in some reference (arbitrary) unit cell. Thus a crystal is composed of a periodic array of atoms and its structure is determined by the diffraction of X-rays by the parallel atomic planes. An incident beam forms diffracted beams when the reflections from parallel planes of atoms interfere constructively.

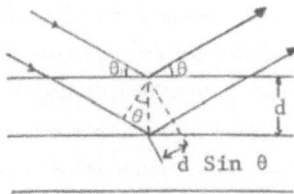

Fig. 1.1

Constructive interference of radiation from successive planes occurs when the path difference for rays reflected from adjacent planes is an integer number n of the wavelength λ. This condition, known as the Bragg law is expressed as

$$2d \sin\theta = n\lambda$$

for a configuration such as that given in figure 1.1 where d is the spacing of parallel atomic planes and θ is the angle of the incident specularly reflected beams with the plane.

Experimentally different methods are used. In the *Laue method*, a single crystal is stationary in a beam of X-ray or neutron radiation of continuous wavelength. The crystal selects and diffracts the discret values of λ for which planes exist of spacing d and incidence angle θ statifying the Bragg law. The diffraction pattern consists of a series of spots. The pattern will show the symmetry of the crystal. In the *rotating-crystal method*, a single crystal is rotated about a fixed axis in a beam of monoenergetic X-rays or neutrons. The variation of angle θ brings different atomic planes into position of reflection.

In the *powder method* the incident monochromatic radiation strikes a finely-powdered specimen contained in a thin-walled capillary tube. Diffracted rays go out from individual crystallites that happen to be oriented with planes making an incident angle θ with the beam satisfying the Bragg equation. Diffracted rays leave the specimen along the generators of cones concentric with the original beam. The generators make an angle of 2θ with the direction of the original beam, where θ is the Bragg angle. The cones intercept the film in a series of concentric rings. The counter recording of the diffracted beam gives a series of sharp lines as a function of the Bragg angle θ.

All these methods lead to the determination of the crystal symmetry and interatomic distances.

The glass is characterised by the lack of long range order ; it is therefore not clear what a glass structure would mean. To understand glass structure one is still in quest of a satisfactory conceptual model of the way crystallization is frustrated in supercooled liquids.

A glass is an X-ray amorphous solid which exhibits the glass transition, the latter being defined as that phenomenon in which a solid amorphous phase exhibits with changing temperature a more or less sudden change in the derivative thermodynamic properties such as heat capacity and expansion coefficient from crystal-like to liquid-like values. The temperature of transition is called the glass transition temperature Tg.

The reason for the difficulty in the investigation of glass structure is that glasses are not in thermal equilibrium and that the atomic structure of glass is not periodic because glasses are not crystalline. In the absence of a periodic structure, such non-thermally-equilibrated material becomes an assembly of interlocking microcosms whose relaxation and organisation has been arrested to varying degrees. A macroscopic glass sample is an universe of regions at different stages of formation.

Because of the nature of the static structures of glasses, which lack long range order, one needs at a given time, a space correlation function to describe the relative position of atoms in a glass. Such a correlation function, while mathematically exact, can only be measured with limited resolution. Thus one usually uses the word structure in disordered condensed phases to describe merely the short range order or the local structure in such media.

Much of the fundamental understanding of the short range order in glasses is derived from diffraction studies and comparison with structural data of the corresponding crystals.

We shall discuss here the spectroscopic techniques, and the theoretical and experimental methods for acquiring more precise data but the main emphasis will be on the information which could be derived from a detailed and careful analysis of the data obtained when approaching the glass-crystalline transition.

1.1.1. Radial Distribution Functions (1)

The diffraction of X-rays, electrons or neutrons by vitreous material yields a few broad halos at low angles rather than the discret lines characteristic of crystalline solids. Structural information for a glass may be obtained from these diffraction patterns by an analysis of the distribution of interatomic distances. The distribution is expressed as a radial distribution function (RDF) which can be computed from experimental scattering data.

The total diffraction intensity from a glass, I_t, is composed of the interatomic interference scattering, I, the coherent atomic scattering, I_c, and the incoherent atomic scattering, I_i, such that

$$I_t(k) = I(k) + I_c(k) + I_i(k) = I(k) + I_b(k)$$

where $k = 4\pi \sin\theta / \lambda$ is the scattering vector, 2θ is the angle between the indicent and the diffracted beam, and λ is the wavelength. An intensity function is obtained by dividing $I(k)$ by Σf^2, the sum of squares of the coherent atomic scattering factors for the unit of composition take out nc, e.g. B_2O_3 corresponds approximately to the scattering intensity from vibrating point atoms i.e.

$$I(k) = I(k) \; [\Sigma f^2 (k)]^{-1}$$
$$= [I_t(k) - I_b(k)] \; [\Sigma f^2 (k)]^{-1}$$

The radial distribution function $D(r)$ is produced by a Fourier sine transform of $kI(k)$

$$D(r) = 4\pi r^2 \rho(r) = 4\pi r^2 \rho_0 + (2/\pi) \int_0^\infty ki(k) \sin(k) \, rdk$$

where $\rho(r)$ is the number of atoms per unit volume at distance r from any arbitrary atom, ρ_0 is the average radial density and $4\pi r^2 \rho(r)$ represents the probability weighted by the product of scattering factors of atoms j and l, divided by Σf^2, of finding atom l in the sample separated by the distance interval $(r, r + dr)$ from the atom j.

The plot of $4\pi r^2 \rho(r)$ versus r will yield a curve oscillating about a parabola representing the average

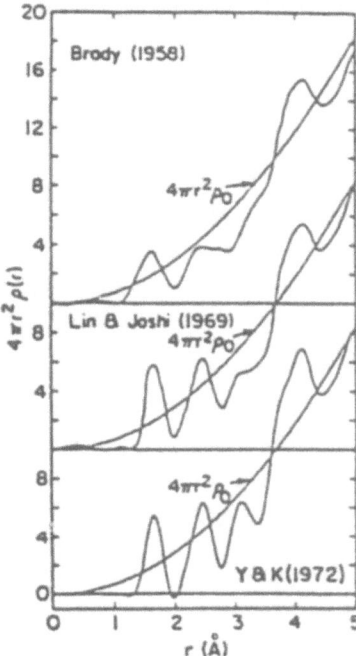

Fig. 1.2   Experimental RDFs for bulk commercial SiO_2, derived from X-ray
diffraction measurements by various authors.

atomic density and will describes the number of atoms in a spherical shell between r and r + dr. The maxima
of this curve corresponds in the average to the most frequently occuring interatomic distances. The number of
atoms N_i in each coordination shall i located at r_i is given by

$$N_i = \int_{r_i - \delta}^{r_i + \delta} 4\pi r^2 \rho (r) dr$$

It is interesting to mention that radial densities δ (r) can be obtained [2] from the extended X-ray absorption
edge fine structure (EXAFS).

Early X-ray diffraction measurements on vitreous B_2O_3 corellated by subsequent nuclear magnetic
resonance studies indicated that the boron atoms are 3-fold coordinated, with a B-O separation of 1.37 Å. Each
oxygen bridges two boron atoms. The major part of the glass, however is made of boroxol rings, B_3O_6,
which are linked together by sharing the corner oxygen atoms with random orientation about the B-O bond

direction at the shared oxygen atoms. Within a ring, the B-O-B angle is 120°.

1.2. Light scattering

Light scattering is a phenomenon which has been studied for a very long time. Since the advent of the laser, light scattering has become one of the most powerful methods for probing the fundamental properties of matter. Compared to the resonance techniques of spectroscopy it has the advantage of combining the two physical processes of absorption and emission, and it is further subject to precise selection rules of energy and momentum conservation.

Already in 1868-1869 Tyndall found that white light scattered at 90° to the incident light by very fine particles was partly polarized and also slightly blue in color. He concluded that both the polarization and the blue color of light from the sky were caused by scattering of sunlight by dust particles in the atmosphere. Lord Rayleigh (1899) gave a quantitative treatment of the phenomenon and showed that the scattered intensity is inversely proportionnal to the fourth power of the incident light wavelength.

$$I_s \propto A(1 / \lambda^4)$$

Rayleigh knew that light is scattered by the gas molecules and that dust particles in the atmosphere are not essential for the blueness and the polarization of light from the sky.

Progress in understanding the frequency spectrum of light scattering was first made by L. Brillouin, who calculated the spectrum of light scattered by the density fluctuations associated with a sound wave. Smekal (1923) predicted the existence of side bands in the scattered spectrum of a system with two quantized energy levels.This was subsequently observed by Raman and Krishnan (1928). Thus the complete spectrum of scattered light has the three main features shown in Figure 1.3.

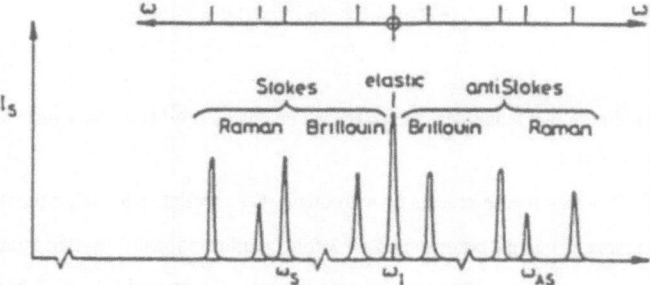

Fig. 1.3 Schematic spectrum of scattered light.

114

For a quantitative understanding of the light scattering phenomenon and its use in the analysis of the fundamental properties of matter, the meeting point of experiments and theory is the scattering cross section.

Measurement of the scattered light for a fixed scattering angle determines a function :

$$(d^2\tau) / (d\Omega \, d\omega_s) = \text{spectral differential cross section.}$$

This is the rate of removal of energy from the incident beam as a result of its scattering in a volume v into a solid-angle element $d\Omega$ with a scattered frequency between ω_s and $\omega_s + d\omega_s$ divided by the product of $d\Omega d\omega_s$. The various quantities are determined inside the scattering medium (sec. figure 1.4).

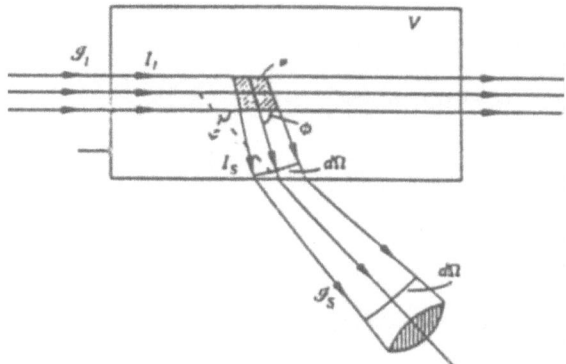

Fig. 1.4   Idealized scattering experiment.

When a laser beam is used in a light scattering experiment its intensity can be regarded as monochromatic with a single angular frequency ω_I. The scattered intensity is distributed across a range of frequencies on both sides of the elastic (Rayleigh) peak, as shown in Figure 1.3. These inelastic contributions are known as Stokes components when the scattered frequencies ω_s are smaller than ω_I and as anti-Stokes components ω_{AS} when the scattered frequencies are larger than ω_I. Each scattered photon in the Stokes component is associated with a gain of energy $h\omega$ by the sample, where

$$\omega = \omega_I - \omega_s$$

The loss of energy $h\omega$ for each scattered photon in the anti-stokes component is

$$\omega = \omega_{AS} - \omega_I$$

The energy gained or lost by the sample corresponds to that of an excited state of the system. In the Stokes process the gain of energy corresponds to the transition from the ground to the excited state : excitation of the system whereas the loss of energy from the system to the radiation field corresponds to the emission of a photon i.e. the de-excitation of the system. Thus a light scattering spectrum measures directly the energy of elementary excitations of the sample. Such excitations can be electronic transitions, vibrational transitions, spin flips : plasmons, phonons, magnons etc...

1.2.1. Classical theory of light scattering by optical phonons

In the elementary theory [3] we shall give here, the radiation field is treated classically and the phonons are treated quantum mechanically.

When an incident light beam with electric field $E(t)$

$$E(t) = E_0 \exp(-i\omega_I t) + E_0{}^* \exp(i\omega_I t) \qquad (1.1.)$$

is applied to on a crystal, an electric dipole is induced :

$$m(t) = \alpha(t) \cdot E(t) \qquad (1.2.)$$

where $\alpha(t)$ is the polarizability tensor (or dielectric susceptibility tensor). $\alpha(t)$ is a function of time because it is modulated by the lattice vibrations. If a particular vibrational mode of frequency ω_q is considered

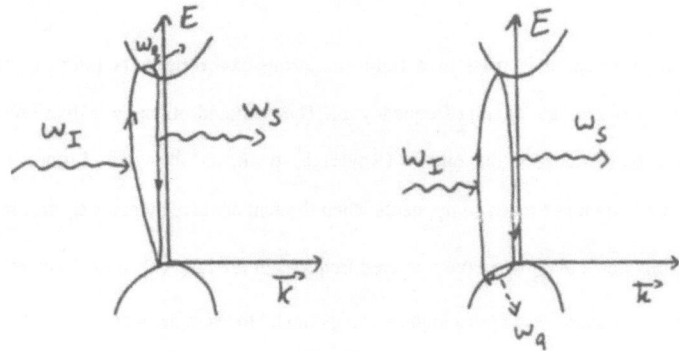

Figure 1.5 shows the elementary processes in a light scattering experiment. Electron-hole pairs are created by the action of an incident light beam ω_I. They are scattered with emission or absorption of phonons ω_q, and then recombine emitting photons of frequency ω_s.

the polarizability tensor may be written as

$$\alpha(t) = \alpha_0 \exp(-i\omega_q t) + cc \qquad (1.3.)$$

and substitution of eq (1.1) and (1.3) into (1.2) yields.

$$m(t) = \alpha_0 E_0 \exp[-i(\omega_I + \omega_q)t] + cc + \alpha_0^* . E_0 \exp[-i(\omega_I - \omega_q)t] + cc \qquad (1.4.)$$

We see that $m(t)$ has contributions which oscillate with the frequencies $\omega_I - \omega_q$ and $\omega_I + \omega_q$. The crystal can therefore emit radiation at these two frequencies, which correspond to the Stokes and anti-Stokes processes.

The intensity of the scattered radiation can be calculated by considering the emission of radiation by the oscillating electric dipole moment

$$m(t) = m_0 \exp(-i\omega_\sigma t) + m_0^* \exp(i\omega_s t) \qquad (1.5.)$$

The scattered energy flux at distance r from the crystal for radiation with unit polarization vector e is given by

$$S_s = [(\omega_s^4)/(2\pi r^2 c^3)][\hat{e}_s . m_0]^2 \qquad (1.6.)$$

The correspondance between the dipole moment amplitude m_0 and the quantum mechanical matrix element between initial and final vibrational states $/v>$ and $/v'>$ is

$$m_0 \quad V < v/\alpha/v' > . E_0 \qquad (1.7.)$$

where V is the volume of the crystal and α is the quantum mechanical electronic polarizability tensor which is a function of the nuclear coordinates. Replacing m_0 in eq. (1.6.) by the right-hand side of eq. (1.7.) yields the result

$$S_s = [(V^2\omega_s^4)/(2\pi r^2 c^3)][< v/\hat{e}_s . \alpha/v'> . E_0]^2 \qquad (1.8.)$$

The energy flux of the incident radiation averaged over a vibrational period is given by

$$S_I = (c/2\pi)[E_0]^2 \qquad (1.9.)$$

The differential scattering cross section can be obtained by considering two different but equivalent expressions for the energy scattered per unit time into solid angle $d\Omega$ when the incident flux is S_I :

$$(d\varepsilon_s)/(dt) = S_s r^2 d\Omega = S_I d\sigma \qquad (1.10.)$$

The differential cross section is obtained by integration of the spectral differential cross section

$$(d\sigma)/(d\Omega) = \int (d^2\sigma)/(d\Omega \, d\omega_s) \, d\omega_s = \text{differential cross section}$$

From (1.10.) we obtain the differential cross section in the form

$$(d\sigma)/(d\Omega) = (r^2 S_s)/(S_I) = (2\pi r^2/c) . (S_s)/[E_0]^2 \qquad (1.11.)$$

Substituting eq. (1.8.) into eq. (1.11.) the differential cross section is given in terms of the quantum mechanical matrix element

$$(d\sigma) / (d\Omega) = V^2 [\omega_s / c]^4 [<v/\hat{e}_s . \alpha . \hat{e}_I/v' >]^2 \qquad (1.12.)$$

where $\hat{e}_I$ is the unit polarization vector of the incident radiation. The matrix elements can be readily calculated, expressing the polarizability tensor α as a function of the nuclear coordinates R_n.

When the matrix element is evaluated for first order Raman scattering by optical phonons the result is

$$(d\sigma) / (d\Omega) = (hV^2\omega_s^4) / 2MN c^4\omega_0 [\hat{e}_s .(\alpha . \xi_0) . \hat{e}_I]^2 \left\{ \frac{n_0 + 1}{n_0} \right\} \qquad (1.13.)$$

Where n_0 is the phonon population factor at thermal equilibrium given by

$$n_0 = [\exp(h\omega_0 / k_B T) - 1]^{-1} \qquad (1.14.)$$

From eqs (1.13.) and (1.14.) we see that the ratio of the intensities of Stokes and anti-Stokes scattering is given by

$$(I_s) / (I_{as}) = (n_0 + 1) / (n_0) = \exp ((h\omega_0) / (k_B T))$$

This expression shows that there will always be asymmetry, the Stokes line being more intense than the anti-Stokes line. This asymmetry increases at low temperature, until finally the anti-Stokes line vanishes at T=0.

The polarizability tensor α reflects the symmetry properties of the crystal. Group theoretical considerations allows one to determine the Raman active modes in the center of the Brillouin zone.

1.2.2. Light scattering spectroscopy in borate glasses

We shall focus our discussion here on the light scattering phenomenon due to vibrational modes. In a borate glass such as B_2O_3-xLi_2O-yLi_2SO_4 for example one can distinguish three parts :

i) the vibration spectrum due to B_2O_3,

ii) the effect of the modifier Li_2O,

iii) the effect of the solute Li_2SO_4.

1.2.2.1. Vibrational spectra of vitreous B_2O_3 (v - B_2O_3)

Vibrational spectroscopy provides one of the most powerful tools in the structural determination of solids. For a perfect crystal, phonon dispersion relations can be calculated using lattice dynamical concepts.

The dispersion curves $\omega(q)$ are characteristic for a solid of given crystalline structure. The lack of translational symmetry in glasses results in a broadening of the dispersion curves $\omega(q)$ with increasing q, corresponding to decreasing "phonon" lifetimes.

There are different methods for calculating the dynamical properties of a disordered lattice. We shall not discuss the theory here and will only consider some of the theoretical results in comparison with experiments.

Early X-rays work [4] indicated that B_2O_3 forms a hexagonal crystal with three formula units per unit cell. The boron is 3-fold coordinated in the crystal in concordance with NMR data [5].

Factor group analysis predicts that there are 14A and 14E optical modes all of which are active both in IR and Raman. The predicted number of modes, however, is far in excess of that observed in both infrared and Raman spectra of v-B_2O_3. There is no direct correspondance between the vibration modes observed in the crystalline state and that in v-B_2O_3.

The Raman spectrum of v-B_2O_3 is strikingly different from the IR spectrum.

Only a few attempts have been made to calculate the vibrational frequencies of B_2O_3 and compare with experiments.

The borate glasses v-B_2O_3 are generally good ionic conductors and electronic insulators. For v-B_2O_3, as for other oxide glasses such as v-SiO_2, v-As_2O_3 and v-$GeSe_2$ there are strong evidence from spectroscopic data, that the structure contains mainly rings whose breathing mode is Raman active and shows up as a sharp peak at 806 cm^{-1}. The existence of such a sharp structure should be surprising for a glass where one should expect to observe only very broad bands. The reason is that the sharp structures are due to highly regular hexagonal rings of bonds connected into a more disordered network. It is supposed that the vibrational mode, at 806 cm^{-1}, of these rings is decoupled from the general network and consequently is strongly localized. A condition for vibrational decoupling of the appropriate ring mode, from the surrounding network has been derived [6].

A detailed analysis [1] has been carried out in the case of v-SiO_2 where the ring assignments are made partly on the basis of nearly exact agreement with the dominant Raman frequencies of isolated cyclosiloxane ring molecules [7]. It is easy to transpose the analysis carried out for v-SiO_2 to v-B_2O_3.

It has been shown [6] that a regular ring which is strongly bonded into the network can have vibrational modes whose motion is weakly coupled so that the other modes appear like localized mode which shows as a sharp line even in an otherwise disordered solid.

Consider the phonon ring shown in Figure 1.6 ; it is connected to the rest of the network by six B-O bonds. In the breathing mode of the ring all three O atoms move in phase along the bisector of ϕ. In general, these O displacements will exert forces on the B atoms, which will communicate their motion to the rest of the network by distortion of the Bethe lattice (BL) bonds. This will normally broaden the breathing mode

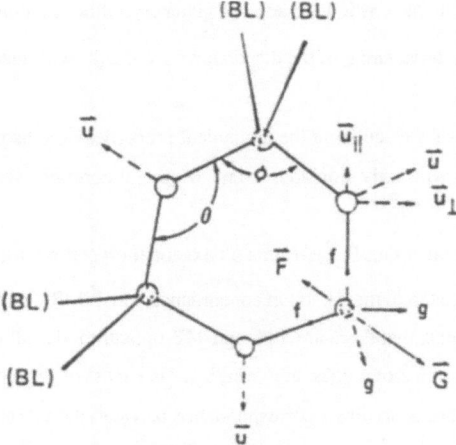

Fig. 1.6   Schematic representation of the breathing mode in a 3-ring. When the net "central force "F on each B atom is canceled by the net "non central" force, each B atom, the mode is vibrationally decoupled from the rest of the network, and there is no B isotope shift for this mode. This occurs because the B atoms are motionless under this condition.

frequency. On the other hand, if there happens to be zero net force on the B atoms they will remain stationary, and the O motion in the ring will not be coupled to the network.

A condition for such decoupling has been derived by Galeener et al. [6] in terms of Born force constants α and β, where α measures the central (or bond-length restoring) force and β "non-central" (or bond orientation restoring) force. The individual central (f) and non-central (g) forces acting on a B atom are shown in Figure 1.6. In the absence of B motion the magnitudes of these forces are given by

$$f = \alpha' u_{||} = \alpha' u \cos \theta/2$$

and

$$g = \beta' u_{\perp} = \beta' u \sin \theta/2$$

where the prime denotes the force constant for a bond within the ring. The magnitude of the oppositely directed total central (F) and non-central (G) forces acting on the B atom are then given by

$$F = 2f \cos(\phi/2)$$

and

$$G = 2g \sin(\phi/2)$$

The absence of B motion requires $F = G$ so   that perfect decoupling occurs when α' / β' has the value.

$$(\alpha' / \beta')_d = \mathrm{Tan}\,(\theta / 2)\,\mathrm{Tan}\,(\phi / 2)$$

120

This decoupling is accomplished by a near cancellation of central and non-central forces within the ring acting on the high-coordination (B) atoms that connect to the network. Therefore the connecting atoms do not move during the ring vibration, and this explains the absence of the mode isotope shift when isotopic substitution is carried out on the connecting (B) atoms. Isotope substitution in borate glasses when ^{18}O is substituted by ^{16}O in the 50 % proportion shows an isotopic shift which produces four bands [8] whose relative intensity is in the ration 1 : 3 : 3 : 1 and the frequencies are $(^{11}B_4\ ^{16}O_3\ ^{18}O_3)$ = 760, 777, 791, 808 cm^{-1}, whereas $(^{11}B_2\ ^{16}O_3)$ = 808 cm^{-1} and $(^{11}B_2\ ^{18}O_3)$ = 760 cm^{-1}. Boron isotopic substitution on the contrary, does not produce any frequency shift.

$$(^{11}B_2\ ^{16}O_3) = (^{10}B_2\ ^{16}O_3) = (^{10}B_2\ ^{11}B_2\ ^{16}O_6) = 808\ cm^{-1}$$

The Raman spectra of v-B_2O_3 with different isotope substitutions are displayed in Figure 1.7

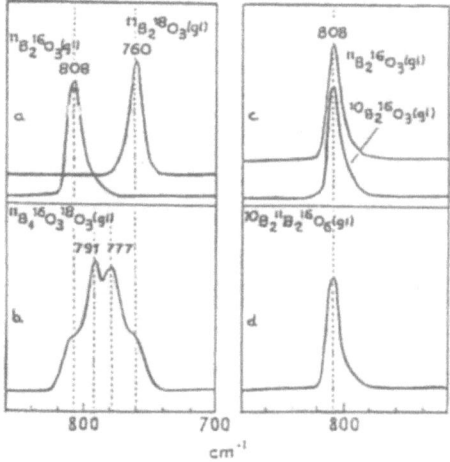

Fig. 1.7   Raman spectra of B_2O_3 in the vicinity of the boroxol ring breathing mode 800 cm^{-1}

a) spectra for $^{11}B_2^{16}O_3$ and $^{11}B_2^{18}O_3$; b) equal amount of ^{16}O and ^{18}O is present in the compound $^{11}B_4^{16}O_3^{18}O_3$; c) isotope substitution of B : spectrum for $^{11}B_2^{16}O_3$ and $^{10}B_2^{16}O_3$; d) spectrum for the compound containing equal amount of ^{11}B and ^{10}B i.e. $^{10}B_2\ ^{11}B_2^{16}O_6$.

From Figure 1.7 it can be seen that substituting ^{16}O with ^{18}O produces a considerable shift of the breathing mode frequency which scales well with the simple law ($\omega = \sqrt{k/m}$) stating that the frequency is inversely proportional to the square root of the mass. The mass being increased from 16 to 18 the mode frequency is decreased from 808 to 760 cm^{-1}. If only half of the ^{16}O are substituted then the vibrational modes of the compound $^{11}B_4\ ^{16}O_3^{18}O_3$ become a combination of the repartition of the isotopes at different sites available. This distribution produces 4 mode frequencies which are clearly observed in 1.7b. 1.7c, but 1.7d shows on the contrary that in spite of full or partial substitution of ^{11}B the frequency of the observed vibrational mode remains constant. This means that boron atoms do not participate in the motion of the groups, which is in good agreement with the theoretical estimations.

1.2.2.2. Raman spectra of B_2O_3 - xM_2O. Effect of the modifier

Numerous investigations on the borate glasses B_2O_3 - xM_2O tend to demonstrate that the concentration of the metal oxide, the modifier of the system, influences the overall structure.

Specifically in the case of borate glasses it has been shown that an excess of oxygen in the form of alkali oxides, like Li_2O for example, tends to increase the number of tetrahedraly bonded boron atoms. Schematically this will be represented by the following reaction.

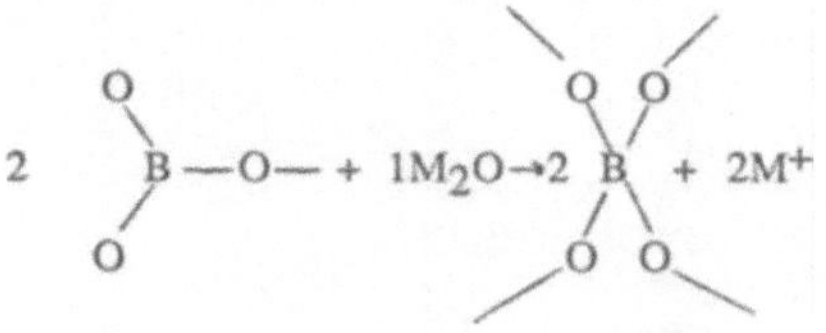

As the oxygen content increases in the glass with increasing concentration of Li_2O, an increasing number of B atoms becomes tetrahedrally coordinated. The maximum number of boron atoms with

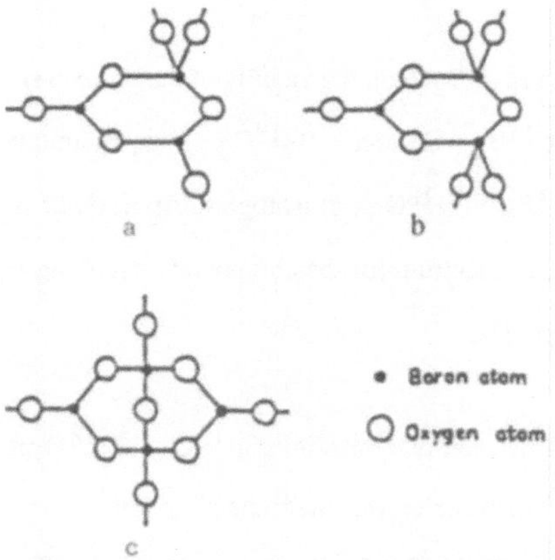

Fig. 1.8   Some of the structural groups appearing in the alkali borates :
a) triborate ; b) ditriborate ; diborate.

coordination number four (BO_4) occurs for a concentration x of Li_2O of the order of 40 % [9]. The addition of the modifier, Li_2O, into the glass B_2O_3, leads to the dissociation of the metal oxide with the formation of a tetrahedral radical and a metallic ion.

A BO_4 can be introduced into a boroxol ring in various ways [10]. The simplest hypothesis is to imagine that the boroxol rings are progressively replaced by triborate groups as shown in Figure 1.8.

The effect of increasing Li_2O concentration on the Raman spectra of B_2O_3 - xLi_2O is shown in Figure 1.9 borrowed from the thesis of M. Kbala [11].

The high frequency region of this spectrum is characterised by the appearance of new bands at 1540 cm^{-1} and 920 cm^{-1} which do not exist in the spectrum of B_2O_3 and slight shift towards higher frequencies of the band at 1250 cm^{-1}. The high frequency band at 1540 cm^{-1} shifts towards lower frequencies down to 1450 cm^{-1} as the Li_2O concentration x increases.

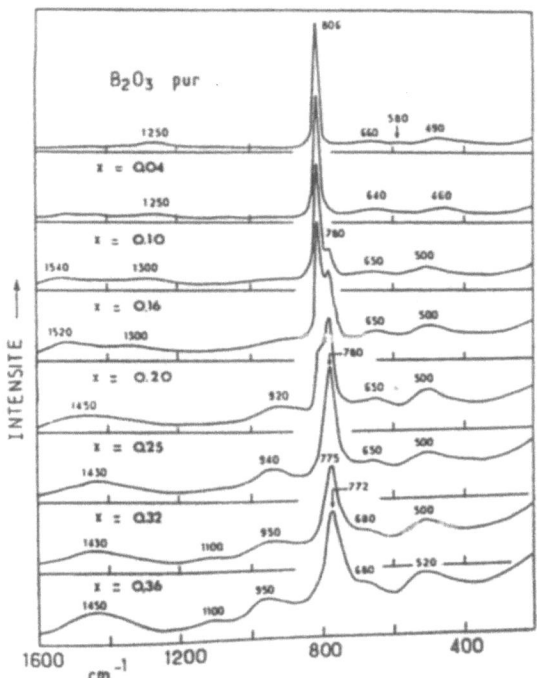

Fig. 1.9   Raman spectra of v- B_2O_3.

The sharp band attributed to the symmetric mode vibration of the boroxol ring is situated at 806 cm^{-1}. The effect of Li_2O concentration x increase on this band is quite interesting. At x = 0.10 a second band appears at 780 cm^{-1} which can be associated with the increase of the number of tetrahedraly coordinated B atoms and formation of new structural groups of the kind shown in Figure 1.8. The increase of the number of tetrahedraly coordinated B corresponds therefore to a softening of the breathing mode. Up to a concentration x = 0.20 the two bands coexist, testifying to a distribution of different configurations. For the concentrations 0.25 only the low frequency peak at 772 cm^{-1} is visible. Further increase of x introduces an

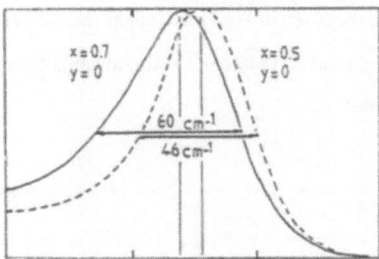

Fig. 1.10 Raman peaks corresponding to the breathing symmetrical
mode of the B$_3$O$_6$ cycles as a function of the Li concentration.

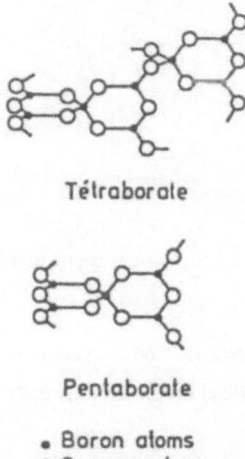

Tétraborate

Pentaborate

• Boron atoms
○ Oxygene atoms

Fig. 1.11 Structural group appearing in the alkali borates :
tetraborates and pentaborates.

asymmetric deformation of this band. A detail of this behaviour is shown in Figure 1.10.

These changes of frequencies and half-width may be due to the appearance in the matrix conformation of more configurations in the B_3O_6 rings such as tetraborate and pentaborate groups shown in figure 1.11.

Increasing the number of different configurations increases the number of different frequencies in a narrow spectral region. Their combination constitutes a broad band and adds to the asymmetric deformation. It is impossible to resolve spectrally each one of the different configurations because they appear in a nearly degenerate range being only slightly different from each other.

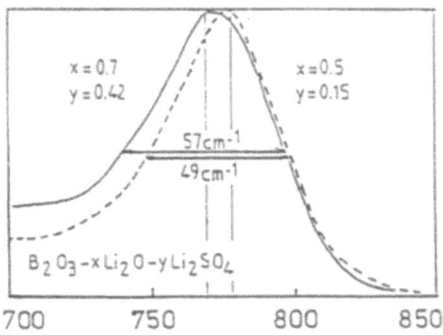

Fig. 1.12 B_2O_3 - $0.5Li_2O$ - $0.15Li_2SO_4$ and B_2O_3 - $0.7Li_2O$ - $0.4Li_2SO_4$

spectra taken at room temperature.

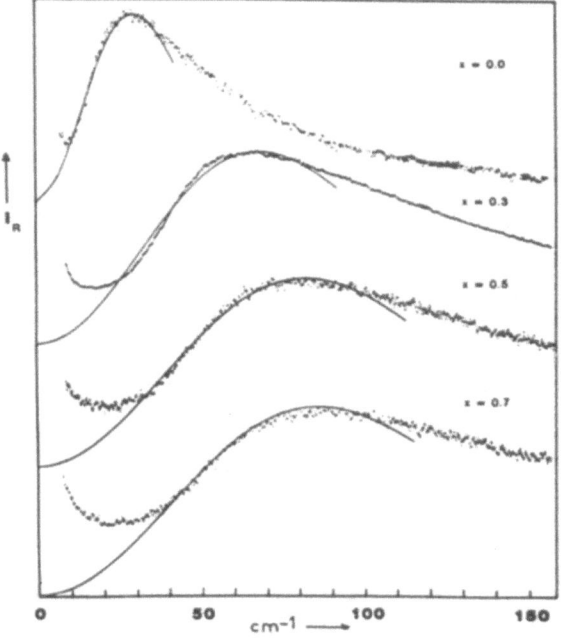

Fig. 1.13   Low frequency Raman spectrum of B_2O_3 - xLi_2O.

The low frequency region would be most interesting for a glass if a quantitative interpretation would be possible. The existence of a sensitive effect is shown in Figure 1.13.

An interesting point is that addition of a salt like Li_2SO_4 in a Li_2O rich borate glass seems to reduce the dispersion in configurations of the glass structure. This effect is seen in Figure 1.12.

1.2.2.3. <u>Raman spectroscopy of B_2O_3-xLi_2O-yLi_2SO_4</u>

B_2O_3 - xLi_2O - yMX are generally considered to be fast ion conductors.

Fast ion conducting borate glasses are solutions of a doping salt, like Li_2SO_4, into a matrix formed by the glass forming material like B_2O_3 and its modifier Li_2O. Thus a glass such as B_2O_3 - xLi_2O - yLi_2SO_4 is a solution where Li_2SO_4 is the solute and B_2O_3 - Li_2O is the solvent.

As early as 1830 Michael Faraday showed that common silicate glass is a solution, not a compound. This point of view has been emphazised only recently and we shall show here how useful is this approach to fast ion conduction, and how strongly it is supported by Raman spectroscopy.

Glasses of the system B_2O_3 - xLi_2O are purely ionic conductors [11] but their conductivity is relatively low as can be seen in Figure 1.14.

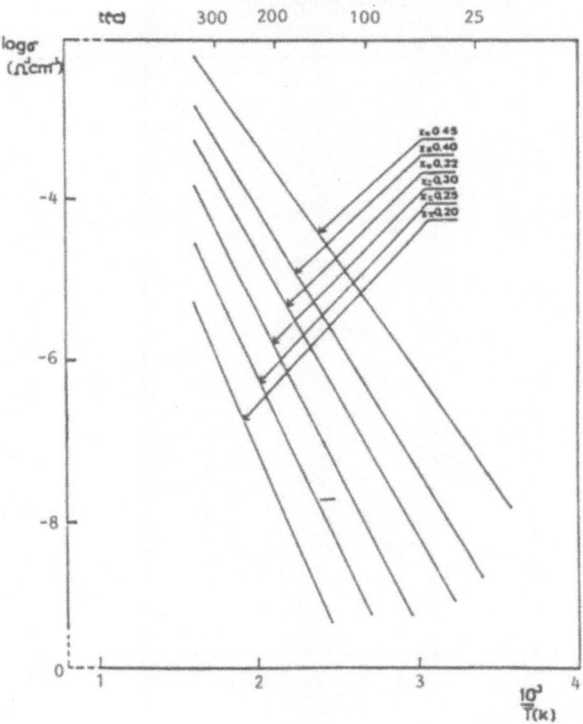

Fig. 1.14 Log of the conductivity of B_2O_3 - xLi_2O plotted as a function of 1/T.

A logarithmic plot of the conductivity [11] of B_2O_3 - xLi_2O - yLi_2SO_4 is a function of 1/T is presented in Figure 1.15.

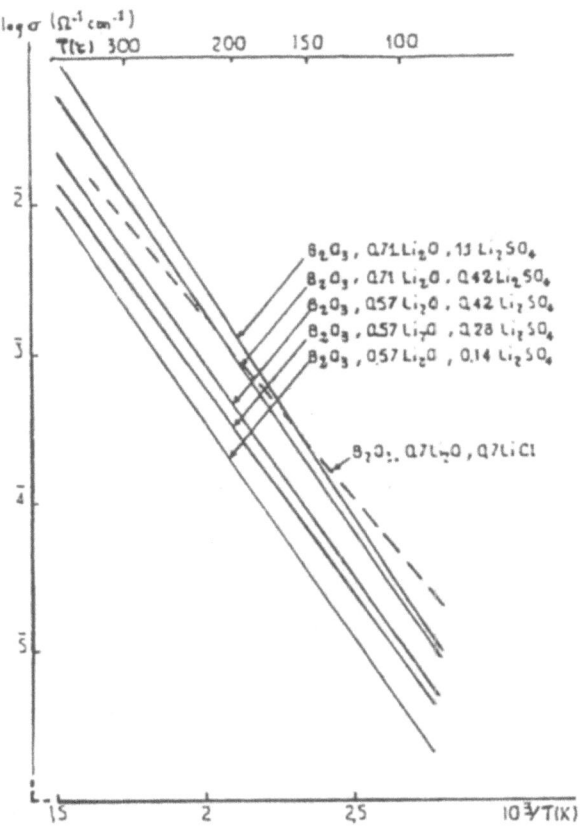

Fig.1.15 Variation plot of the conductivity [11] of B_2O_3 - xLi_2O - yLi_2SO_4
as a function of T.

B_2O_3 - xLi_2O_2 - yLi_2SO_4 is good electronic insulator and shows a reasonable ionic conduction which depends on temperature and Li_2SO_4 concentration [11]. All investigations converge to the idea that the dopant Li_2SO_4 plays an essential role in the ionic conduction. The respective effects of the modifier and the dopant on the ionic conductivity are illustrated on Figure 1.16.

Spectroscopic investigations tending to understand this point will be therefore of a great interest. The contribution of Li_2SO_4 to the Raman spectrum of borate glasses can be readily identified.

The Raman spectrum of B_2O_3 - $0.7Li_2O$ - $0.42Li_2SO_4$ shown in Figure 1.17 is clearly the superposition of the Raman spectrum of the matrix B_2O_3 - $0.7Li_2O$ and that of the solute Li_2SO_4. The peaks such as υ_1, υ_2, υ_3 and υ_4 are those of the characteristic vibrational modes of SO_4^{--}. The sharp peak at 1004 cm^{-1} is that of the breathing mode of the SO_4^{--} ion. The spectrum shown on top of Figure 1.17 is that of the matrix $B_2O_3 - 0.7Li_2O$.

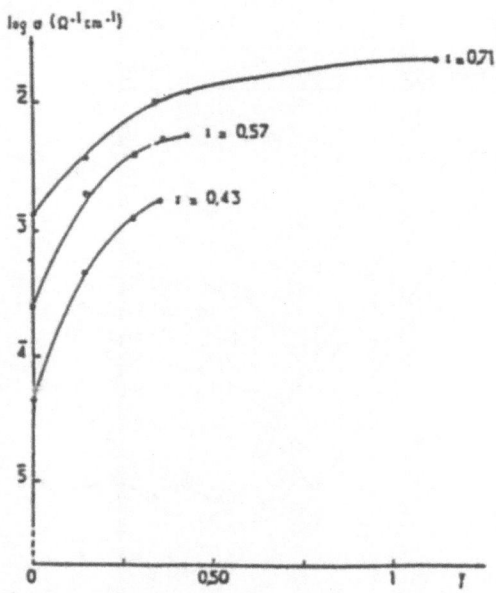

Fig. 1.16   Variation of the conductivity of glasses, with different modifier concentrations x, as a function of the dopant concentration y at 300° C.

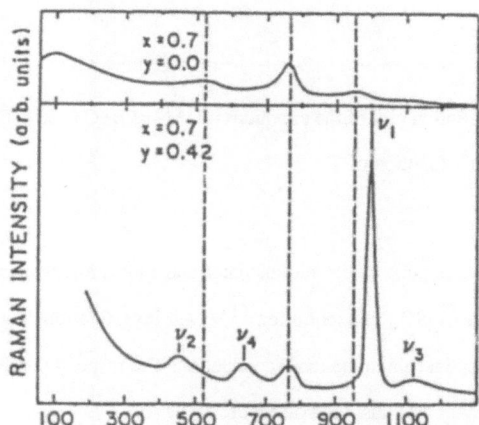

Fig. 1.17   Raman spectra for $B_2O_3 - 0,7Li_2O - y - Li_2SO_4$ for y = 0 and 0,42 at room temperature and laser beam at 514. 5 nm, resolution 2 cm⁻¹.

A schematic representation of the vibrational modes of the ion SO_4^{--}, which belongs to the symmetry group Td, is given in Figure 1.18. All these modes are Raman active.

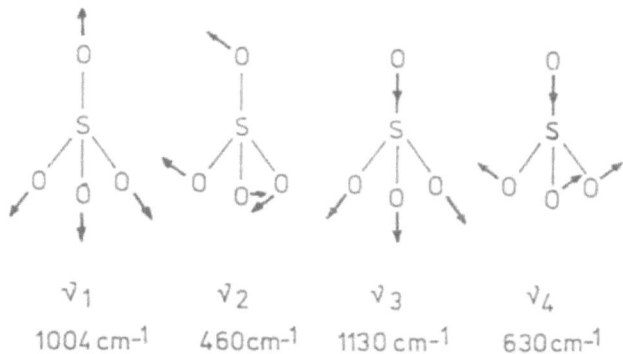

$$\nu_1 \qquad \nu_2 \qquad \nu_3 \qquad \nu_4$$
$$1004\,cm^{-1} \qquad 460\,cm^{-1} \qquad 1130\,cm^{-1} \qquad 630\,cm^{-1}$$

Fig. 1.18 · Schematic representation of the atomic displacement in the intramolecular normal modes of vibration of the ion SO_4^{--}

Systematic investigation of the effect of the Li_2SO_4 concentration y and the substitution of the metal ion Li by K leads to the evidence that the interaction of the SO_4^{--} ion with the Li ion is very weak, practically negligeable. There is no frequency shift or deformation of any of the lines ν_1, ν_2, ν_3 and ν_4 belonging to the SO_4^{--} vibration when the Li_2SO_4 concentration is considerably increased.

Also the addition of Li_2SO_4 up to the concentration, y = 0.42 does not seem to significantly affect the vibrational frequencies of the borate matrix. The matrix is formed by B_3O_6 rings containing one or more B atoms of coordination number 4 whose characteristic frequency is at $780\ cm^{-1}$. In the case of borate glasses with x = 0.5 the rings are probably mainly triborates but the existence of diborates and ditriborates cannot be excluded.

In order to gain further insight in the borate glass structure and its relation with the dopant with regard to fast ion conduction we have considered it interesting to introduce an external parameter susceptible to modify the configuration of the system.

Because all the deductions we make on the spectroscopy of glasses are based on knowledge of the normal modes of vibrations in the corresponding crystal we have adopted the point of view that the way the glass-crystalline transition is approached should be instructive in our search of understanding the glass structure and its relation to fast ion conduction. For this purpose we have conducted detailed studies of temperature and laser annealing.

1.2.2.4. Thermal annealing. Glass-crystalline transition in borate glasses

Progressive thermal annealing [12] of borate glasses leads to the crystalline phase of B_2O_3 - xLi_2O - yLi_2SO_4 through successive configurational transformations of the matrix. The experiments [13] are conducted as follows : the samples are submitted by slow heating to a given annealing temperature T_A and held at that temperature for one hour ; then, are rapidly quenched to room temperature. After this treatment, the Raman spectra are taken at room temperature. We will examine successively the different parts of these spectra.

1.2.2.4a The breathing mode $n_1(A_1)$ of SO_4^{--}

When the annealing temperature increases progressively from 470° C to 550° C the characteristic peak for the SO_4^{--} mode situated initially at 1006 cm^{-1} first decreases in intensity and then disappears. A new

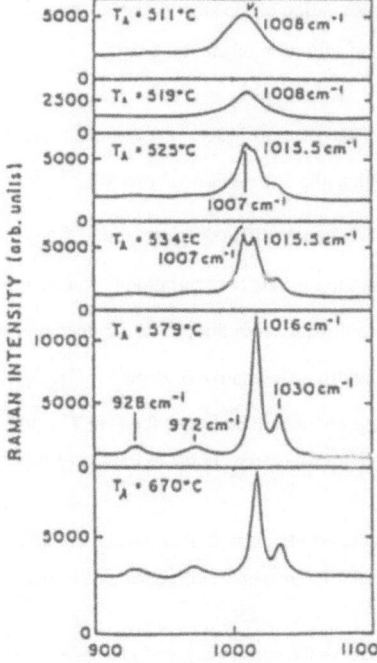

Fig. 1.19 Raman spectra for different annealing temperatures of B_2O_3 - $0.5Li_2O$ - $0.15Li_2SO_4$ centered at the mode ν_1 (A_1) of SO_4^{--}. Spectra are taken at room temperature with a laser beam at 514.5 nm having a power of 250 mW and with an instrumental resolution of 2 cm^{-1}.

peak then appears at 1016 cm^{-1} which has still the frequency of the SO$_4$$^{--}$ breathing mode but now in the crystalline phase of Li$_2$SO$_4$. This transformation can be followed closely in Figure 1.19.

The temperature at which the crystallization of Li$_2$SO$_4$ occurs, signalled by the appearance of the peak at higher frequencies near 1016 to 1020 cm^{-1}, depends on the borate composition. For B$_2$O$_3$ - 0.5Li$_2$O - 0.1Li$_2$SO$_4$ the new peak at 1016 cm^{-1} shows up at 516° C and is fully developed after the disappearance of the peak at 1006 cm^{-1} at 540° C. For B$_2$O$_3$ - 0.5Li$_2$O - 0.15Li$_2$SO$_4$ the peak at 1016 cm^{-1} appears at 525° C and the 1008 cm^{-1} peak disappears completely at 579° C' whereas for B$_2$O$_3$ - 0.7Li$_2$O - 0.42Li$_2$SO$_4$ the second peak is already clearly visible at 475° C and shows a frequency of 1020 cm^{-1}. The first peak initially situated at 1012 cm^{-1} disappears above 511° C. The point to be made here is that the higher the concentration of Li$_2$O and Li$_2$SO$_4$, the lower the annealing temperature at which the Li$_2$SO$_4$ crystallites appear and the stiffer the SO$_4$$^{--}$ ν_1 mode. The stiffening of the breathing mode probably occurs through electrostatic interactions. The higher the ion concentration is, the stronger the repulsive force is among the SO$_4$$^{--}$ ions. This is an effect equivalent to increasing pressure and therefore stiffening the symmetric mode frequencies.

1.2.2.4b The breathing mode of the B$_3$O$_6$ring

In the glasses with the diborate matrix B$_2$O$_3$ - 0.5Li$_2$O above an annealing temperature of 510° C one observes the appearance of a new peak at 725 cm^{-1}, as shown in Figure 1.20.

For annealing temperatures above 560° C the observed spectrum is identical to that of crystalline diborate B$_2$O$_3$ - 0.5Li$_2$O [14]. The crystallization of the borate matrix into the forms of diborate is also accompanied by the appearance of a peak at 1030 cm^{-1} at 525° C as evidenced by spectra shown in Figure 1.19.

One should notice also the appearance in the crystalline phase of a weak peak situated at 772-774 cm^{-1} attributed to the formation of some ditriborates groups [15].

According to the glass matrix composition the crystallization of the matrix may occur at a lower temperature than the dopant salt if x < 0.5 and after the crystallization of the salt if x > 0.5. The respective crystallization temperatures for different glass compositions are listed in table 1, [13].

Apart from the clear definition of the successive phases in the organization of the glass, our investigation shows that each significant step is a threshold phenomenon. The system undergoes a slow evolution before a threshold energy density is reached above which the ordering phenomenon rises rapidly with input power.

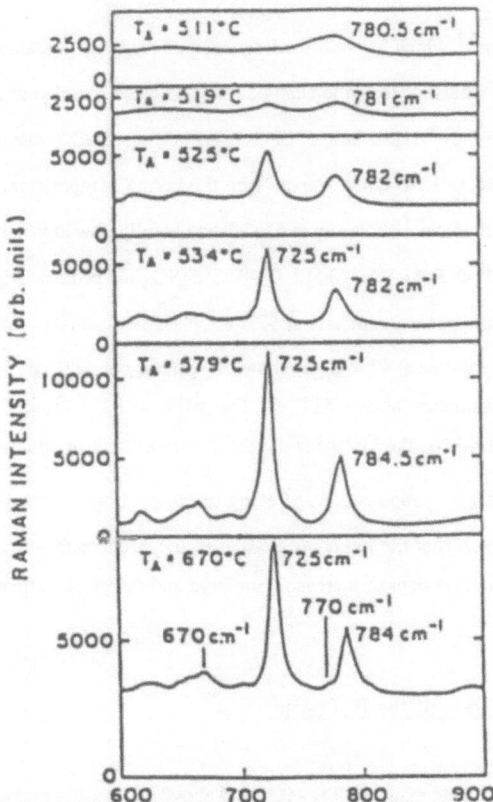

Fig. 1.20 Raman spectra of B_2O_3 - $0.5Li_2O_3$ - yLi_2SO_4 for different annealing temperatures

focussed at the frequency region of the characteristic modes for crystalline diborate.

Spectra taken with a laser beam of 514.5 nm power of 250 mW and the instrumental

resolution of 2 cm^{-1}.

Table 1

B_2O_3 - xLi_2O - yLi_2SO_4		$Tb(^\circ C)$	$Ts(^\circ C)$
X	Y		
0.5	0.05	514	529
0.5	0.1	516	536
0.5	0.15	519	525
0.7	0.2	525	478
0.7	0.2	522	500
0.7	0.6	526	518

1.2.2.5. Laser-induced glass-crystallization kinetics of GeSe$_2$

1. INTRODUCTION

Controlled laser irradiation on GeSe$_2$ glasses produces structural changes which can be analyzed by light scattering which evidences three stages. The first corresponding to low irradiation power is characterized by nucleation and increase of cluster size due to free volume formation. The end of the first stage is heralded by a "precursor effect". The passage to the second stage has a threshold behaviour. At this stage, a metastable state is observed consisting of the coexistence of cluster microcrystallites and crystallites. The increase of input power leads to the "intensity increase effect," whereas the withdrawal of the radiation leads the system to a "dynamical reversal effect". The third state occurs when the system under intense irradiation clamps into the crystalline state.

The experiments which we describe here demonstrate some of the possibilities for investigating structural transitions by light scattering.

Laser irradiation is used to induce structural transformations in the glass and the resulting effects are probed with a light scattering experiment.

Two laser sources : argon and krypton of energies 2.4 and 1.9 eV, well above and well below the band gap respectively, are used.

The progressive transformation evidenced by the Raman spectra revealed the evolution of the system through the following steps :

1. As the energy input in the system grows, the number and the size of clusters randomly nucleated increases. The observed line narrowing of the internal mode A$_1$ of the clusters (200 cm^{-1}) indicates an enlargement of the clusters and constitutes a precursor effect.

2. When the size of the individual cluster becomes sufficiently large, decreasing thereby the average spacing between different clusters, the clusters coalesce randomly in a percolation transition into small crystallites. This is demonstrated by an increase in the intensity of the crystalline normal mode vibration line (210 cm^{-1}). At this stage, the system remains reversible and, without further input of radiation energy, crystallization is prevented by relaxation of the individual clusters towards the glass state. This is a clear demonstration of a dynamical reversal effect.

3. With further increase of the input irradiation power the number of coalescing clusters rapidly increases until the system is definitely clamped into the crystalline state which then remains stable.

Structural transformations in GeSe$_2$ glass and their observation in light scattering.

Light scattering is one of the most powerful tools for studying the structure of disordered solids. This method is of particular interest in the investigation of the transformations between disordered and crystalline phases which can be followed by the correlation between broad bands characteristic of the disordered state and narrow peaks pertinent to the crystal structure. Light scattering reveals the vibrational and rotational modes of the system. The dynamical behaviour is closely related to the immediate environment of the atomic sites. It is therefore a direct probe of the molecular structure and the mutual arrangements of the individual molecules.

Measured Raman spectra of c-GeSe$_2$ have been used [16] to analyze the spectra of g-GeSe$_2$. It has been established that a line at 214 cm^{-1} has a similar behaviour to that of the tetrahedral A$_1$ symmetric breathing mode at 200 cm^{-1} which involves only chalcogen motion. Bridenbaugh et al. (16) have designated the line at 214 cm^{-1} as "the companion A$_1$ Raman line". The model that explains the origin of the companion A$_1$ line is based on the assumption that the structural unit in g-Ge (S, Se)$_2$ is a large cluster which is a fragment of the layer crystal structure still polymerized along the a-axis but is terminated along the b-axis by Se dimers, as shown in Figure 1.21.

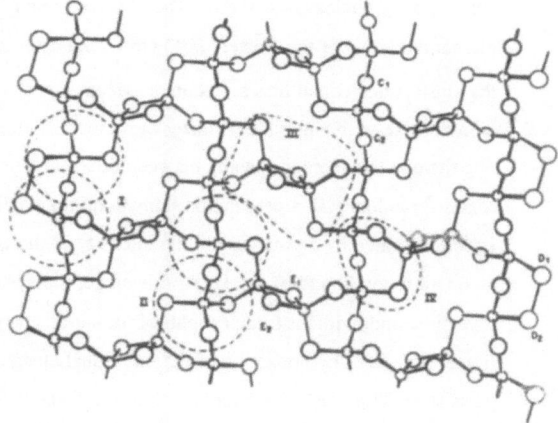

Figure 1.21   Structure of a cluster in Ge(S or Se)$_2$. The companion A$_{1c}$, is associated with
the vibrations of tetrahedra linked by chalcogen dimers, indicated by I here.

Aronovitz et al. [17] and Murase et al. [18] have calculated the vibrational modes of these large clusters and attempted to relate them to the Raman spectra of the crystal and the glass. The identification of all the normal modes is a difficult and in many cases also unrewarding task. It is far more interesting to follow some dominant modes which could bring useful information on the glass-crystalline transformation.

We shall discuss here essentially the effect of a laser power beam on the structural transformations of $GeSe_2$ glass by following the band width and intensity variations of the peaks at 210 and 200 cm^{-1} in connection with the interaction between the clusters and their coalescence.

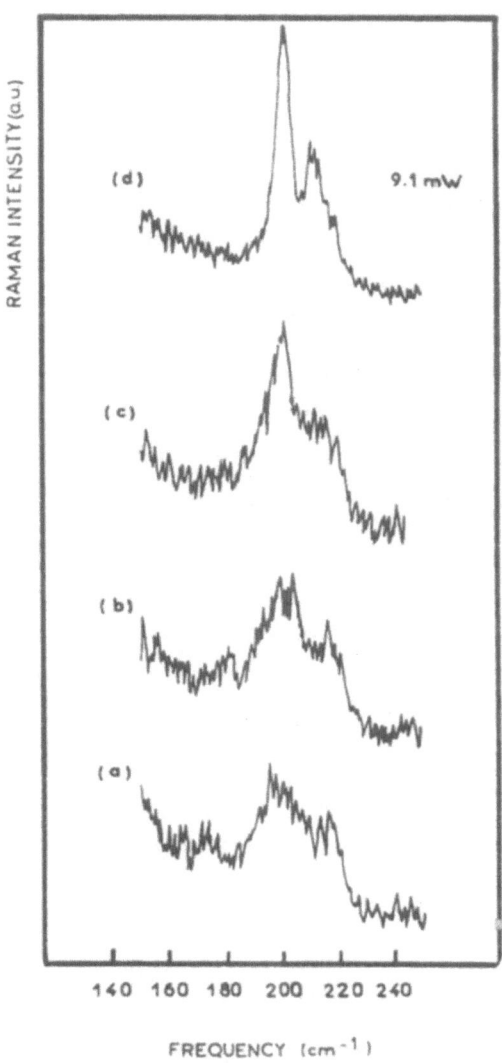

Figure 1.22 Transformation of the A_1 region of the Raman spectrum (a) with probe power 9 mW only, (b) with pump power 18.2 mW, (c) after longer irradiation at 22.8 mW. Power of the probe laser recording beam fixed at 9 mW.

The series of light scattering spectra [19] shown in Figure 1.22 represent different stages of irradiation of g-GeSe$_2$. The samples are irradiated with a laser beam of energy 2.4 eV well above the absorption edge of the material at room temperature. The laser spot on the sample has a diameter of 100 μm. The incident laser energy density is monitored by means of an attenuator.

The reference Raman spectrum (a) is recorded with a probe beam of power P_R = 9 mW which does not induce any structural changes in the glass. This point is confirmed by repeating the recording at sufficiently larger intervals without observing any significant changes in the spectrum. Three bands are observed, centered at 172, 200 and 215 cm^{-1}. We are considering here the spectral region between 150 and 250 cm^{-1}, designated as the A$_1$ cluster Ge-Se vibrational mode in GeSe$_2$. A variety of studies, both by Raman and Mössbauer [20] spectroscopy, have associated the 172 and A$_{1C}$ = 215 cm^{-1} bands with Ge-Ge and Se-Se bonds respectively, although the normal modes of vibration are not necessarily well localized on these atom pairs.[16]

Only in small clusters will the surface bands such as A$_{1C}$ have intensity comparable to that of the inner cluster modes A$_1$. As the cluster size increases, the inner mode density will dominate because the number of the sites inside the cluster with corner-sharing tetrahedra will increase more rapidly than that of outer chains (surface/volume effect). This effect is shown in the spectrum (a) where the intensity of the internal 200 cm^{-1} A$_1$ mode dominates up to the point where the 172 cm^{-1} Ge-Ge mode is not resolved anymore.

The spectrum shown in (b) is recorded after an irradiation with a laser beam of power P_1=18.2 mW. The spectrum (c) is recorded after an irradiation with incident power P_1=22.8 mW. We observe a narrowing and increase in intensity of the inner A$_1$ mode of the cluster which is a demonstration of the tendency for the enlarged clusters to form microcrystallites enlarging the dimensions of the clusters, thus favouring the ratio of the bulk to surface modes. The band narrowing is an indication that the network of corner-sharing tetrahedra is beginning to form a crystal lattice. Further irradiation of the sample enhances even more this tendency as can be seen in the spectrum (d) recorded after a second irradiation with P_1=22.8 mW. This irradiation of a longer duration has led to a more persistent effect. The spectrum recorded with a probe beam of P_R=9.1 mW shows a dramatic narrowing of the peak at 200 cm^{-1} having a half line width Γ_{200}=4 cm^{-1} which is typical for crystalline material. In addition, the normal mode Raman active peak of crystalline [16] c-GeSe$_2$ at 210 cm^{-1} appears at this stage.

Some quantitative changes can also arise from effects of the radiation field. Because the energy of the incident beam lies above the absorption edge, it is strongly absorbed near the surface of the material and does not penetrate more than 0.5 μm. Electronic transitions occur and bonds may break. We consider nevertheless that electronic effects are of secondary importance for the spectral changes under examination which essentially involve structural transitions on the molecular level (clusters of atoms). The transformations occuring in this case are rather a consequence of the softening of some of the intercluster interactions which give more rotational freedom to the system. The thermal energy transferred to the system in the form of cluster rotational excitations leads to the possibility of cluster network reconstruction at the surface of a small cluster thus enlarging the existing clusters and also giving the opportunity for new clusters to nucleate.

transition. It is expressed by the sharp narrowing and the intensity increase of the 200 cm^{-1} peak due to the inner glass cluster mode which just precedes the transformation of clusters into microcrystallites. This is just at the stage where the first sign of crystallization appears with the start of a rise in the intensity of the crystal peak at 210 cm^{-1}. We are now in a nonequilibrium state. The system left on its own will slowly self-anneal and return to the glass state, but on further irradiation, it will tend towards the crystalline state. The precursor effect indicates crystallization if the conditions that brought the precursor are maintained.

1. Intensity Increase Effect

When the radiation power is increased, two types of effects are observed. One is the intensity increase and eventual narrowing of certain bands and the second is the disappearance of certain Raman bands and the appearance of new Raman lines. Both of these effects signal transformation of the glass under laser irradiation. Their evolution is gradual and this makes it possible to study intermediate nonequilibrium states.

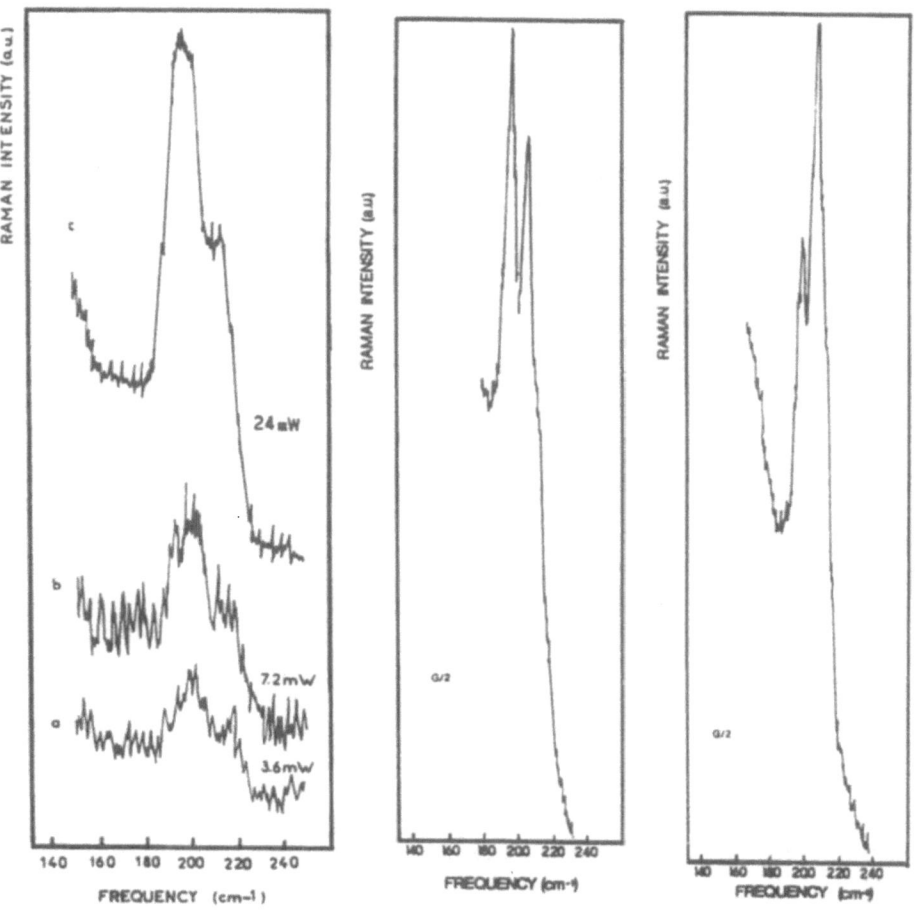

Figure 1.23   Transformation of the A$_1$ spectrum with more closely spaced power levels in the critical region just above P$_1$ = 20 mW. Again the recording power is fixed at 9 mW. G is the gain factor.

137

An almost complete picture of these transformations is given in Figure 1.23. With relatively low radiation power $P_1=3.6$ mW, we observe in the spectrum (a) the main features already described for the glass material. At this laser intensity, no transformation occurs in the glass. When the radiation power is increased and a beam of $P_1=7.2$ mW is used, the recorded spectrum (b) shows an increase in the light scattering strength without modification in the shape of the spectrum.

The spectrum (c) is recorded with an incident laser power of $P_1=21.6$ mW. A significant narrowing of the band at 200 cm^{-1} is clearly observed here and one can almost discern a shoulder at 210 cm^{-1}.

Further increase in the power of the incident beam leads suddenly to substantial changes in the spectrum under irradiation with a beam of $P_1=24$ mW, the recorded spectrum (d) shows clearly two peaks at 200 and 210 cm^{-1}. While the Raman peak at 200 cm^{-1} due to the inner cluster mode A_1 still persists, one sees clearly a new peak at 210 cm^{-1}. This simultaneous observation demonstrates the coexistence of two phases ; on one side large, glass clusters attested by the persistence of the peak at 200 cm^{-1} and on the other side, microcrystallites large enough to show the peak corresponding to the normal mode in the crystal lattice of c-GeSe$_2$, obeying the k=0 selection rule.

As shown in spectrum (e), irradiation under a power beam $P_1=26.4$ mW produces steady growth of the crystalline phase but the peak at 200 cm^{-1} still persists. The picture that one may get from these spectral data is that increasing the incident laser power leads to the formation of crystallites which coexist with the clusters in the glass phase. The clusters transform gradually. Some of them may grow by addition of newly oriented molecules on their surfaces, others may fuse together to form larger clusters or microcrystallites which are still embedded in the glass. The distance scales for these different formations may be estimated as follows. Suppose that the clusters are about 50 Å in diameter, [21] then the critical Gibbs diameter for nucleation of microcrystallites would be about 150 Å for GeSe$_2$, [22] and formation of crystallites will occur when the diameter becomes larger than 200 Å. Above this size, polycrystalline regions separate and macroscopically distinct from the glass regions may form.

In the beginning of this sequence, only a few microcrystallites have formed, let us say a fraction of 0.1 ($f_c \approx$ 0.1) which would be too few to give enough intensity to the band at 210 cm^{-1}. Now there is an important difference between the crystal and the glass as regards density. The density of the crystal is $\rho_c=1.1 \ \rho_g$. When f_c is of the order 0.1, in effect about less than 1% of the surrounding glass becomes "free volume". Many phase transitions involve density changes of a few percent. Thus we expect that when the free volume increases above 1% the 50 Å clusters may be much freer to rotate than they were before. As a result, a number of the smaller clusters will fuse together to form larger clusters, with an average size that increases rapidly.

The evolution of the light scattering spectra suggests that we have to consider three states with regard to the structure of the material. The first is the equilibrium glass state where the contribution to the light scattering is essentially from localized molecular types of vibrations. The third is the crystal state where light scattering occurs through normal lattice vibrational modes characteristic of the infinite crystal with translational invariance. The second state is an intermediate state where clusters (< 100 Å in diameter) embedded in the glass matrix coexist with metastable three-dimensional microscrystalites with a diameter of the order of 150 Å as well as crystallites large enough to be stable.

138

2. Threshold for Microcrystallite and Crystallite Formation

Another demonstration of the continuous evolution of the system as the incident laser energy density increases is given in Figure 1.24. With energy density at a sufficiently low level $P_1=13.2$ mW, when no significant changes in the system have occured, the spectrum is that of g-GeSe$_2$ shown in (a). When the incident energy density is increased, the evolution of the spectrum is shown consecutively in (b), (c), and (d) for incident beams $P^b_1=35$ mW, $P^c_1=39.6$ mW, and $P^d_1=48$ mW, respectively. The intensity of the crystalline peak at 210 cm^{-1} does not reach its final value immediately after the power of the laser beam is incident on the sample : t_1 is the instant at which a laser beam of a given power is incident and t_2 is the "stabilization time", i.e., the time after which the Raman spectrum has reached a nearly constant value and remains stable. Figure 1.24a shows the variation of the Raman intensity of the peak at 200 cm^{-1}

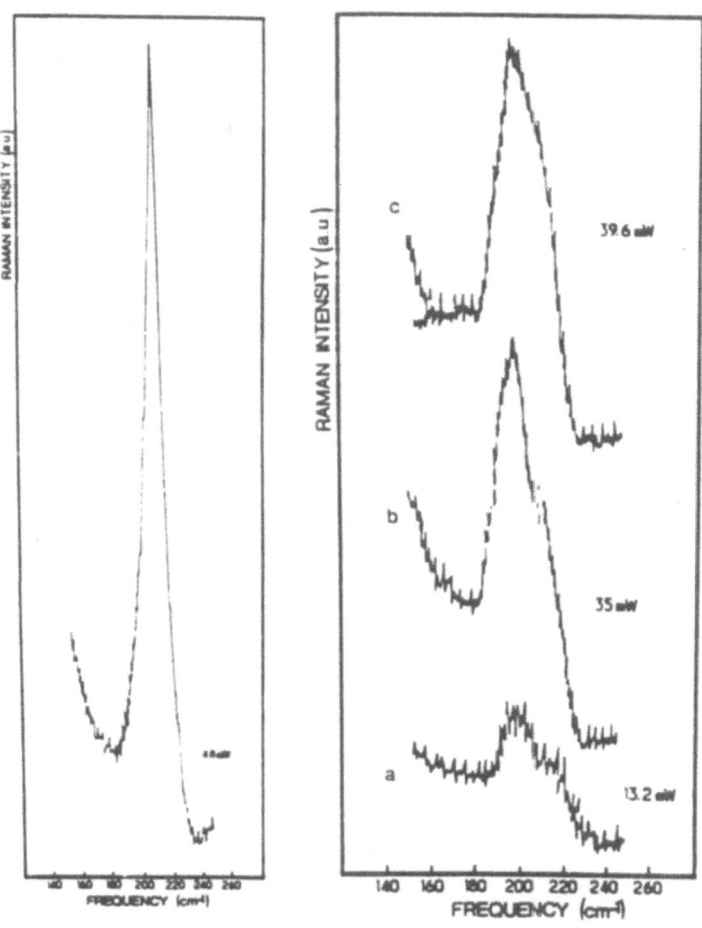

Figure 1.24   Similar to Figure 1.22 but with shorter exposure times,
so that the critical power level is higher.

characteristic of the internal tetrahedral A_1 vibrational mode of the cluster. For relatively low power of the incident laser beam, the Raman intensity changes very slowly with increasing power, whereas the intensity rises abruptly beyond a certain power level. Above a threshold power, the density and the size increase of the clusters develops very rapidly. Figure1.23b shows the same threshold behaviour for the crystallite formation. In this figure, a dramatic increase of the intensity of the crystalline peak at 210 cm^{-1} is shown above a critical incident laser power. Below the threshold power, the crystalline peak is scarcely discernible, but as soon as a power of 25 mW is reached, the intensity of the 210 cm^{-1} peak rises sharply.

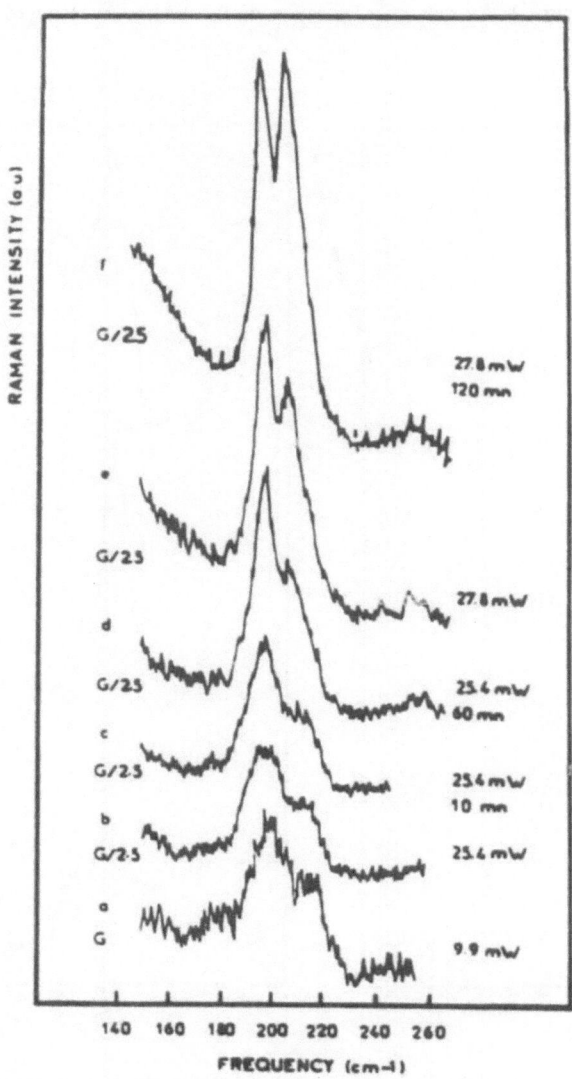

Figure 1.25 The effects of varying exposure times near threshold.

Below threshold, no changes in the structure of the Raman spectrum are observed. The overall intensity of the spectrum increases slightly with the increase of incident power but the shape of the bands remains unchanged. As soon as threshold power is reached band narrowing occurs and the line intensity increases very rapidly with incident power. In the case of the crystalline peak at 210 cm^{-1}, the effect is even more dramatic. Below threshold (E^c_m=25 mW), the peak at 210 cm^{-1} is not observed but as soon as threshold is reached a peak at 210 cm^{-1} appears in the spectrum and its intensity rises far more rapidly than that of the peak at 200 cm^{-1} as power increases. Ultimately, the peak at 210 cm^{-1} dominates with such a great intensity that no peak at 200 cm^{-1} is perceptible.

The most significant effect in the laser induced phase transition is the observation of the coexistence or the continuous transformation of the three phases : glass, microcrystallites, and crystal. This observation is illustrated by the spectra given in Figure 1.25. The spectrum shown on the bottom of this figure is the reference spectrum of the glass stage. This spectrum is taken under irradiation at a power of P_1=9.9 mW. This spectrum recorded for irradiation below the threshold power remains unchanged if the radiation is maintained at this level even for long times. The spectrum is recorded as soon as the incident intensity is increased to a power of 25.4 mW and is shown in (b). No drastic changes are observed in comparison with the spectrum (a) except for a general increased intensity. Ten minutes of irradiation with the same intensity (P_1=25.4 mW) induces only a slight narrowing of the band due to the inner cluster mode at 200 cm^{-1} but the surface cluster mode near 215 cm^{-1} still persists as can be seen in spectrum (c).

At this stage for relatively low energy density very near the threshold energy and sufficient time of irradiation a certain number of clusters are formed and some of them are sufficiently enlarged that some narrowing of the band at 200 cm^{-1} is perceptible. After 60 minutes of irradiation with the same power, the picture changes completely (see spectrum d). The band at 215 cm^{-1} disappears and a new peak arises at 210 cm^{-1} which is the frequency of the normal mode of vibration of c-GeSe$_2$ and the peak at 200 cm^{-1} is now significantly narrowed. This spectrum demonstrates that for long enough time with an irradiation just above the threshold energy, large crystallites are formed under irradiation but clusters although enlarged still persist embedded in the glass. The recording (d) represents a sort of precrystallization state. The irradiation power is just at the threshold energy. The band at 200 cm^{-1}, attesting the existence of incipient microcrystallites and which is now a narrow peak remains pratically constant with time. The crystalline peak at 210 cm^{-1} has just appeared demonstrating the existence of crystallites which remain in equilibrium with the microcrystallite density so long as the radiation power remains just at the threshold energy. Further increase of the incident radiation energy density even by a very small amount induces immediately a drastic change in the picture. The spectrum (e) is recorded with a radiation power of P_1 = 27.8 mW. The two peaks reflecting the microcrystalline state (200 cm^{-1}) and the crystalline state (210 cm^{-1}) are now in competition.

Continuous irradiation at the same power (P_1 = 27.8 mW) for 120 minutes gives rise to the spectrum shown in Figure 1.25f which reflects again a "metastable" situation. Pratically equal amounts of microcrystallites having small dimensions d < 150 Å and crystallites that are of a larger ordered region with d >150 Å remain in metastable equilibrium as long as the radiation power is maintained at the same level. An increase of the radiation power will lead to a sudden rise of the peak at 210 cm^{-1} to an extent such that the 200 cm^{-1} band will be hardly perceptible. The system will transform into the crystalline state which, if the irradiation power is sufficient, will be a thermally irreversible state. In contrast, if the radiation is withdrawn the system tends to return to its initial state. The microcrystallites are dissolved into the glass state.The kinetics observed here are still far removed from equilibrium in the sense that the evolution of the system towards the ordered state requires a continuous energy input at an increasing rate. Even when the transformation is trigerred, if the energy supply is stopped, the system will revert to the glassy state. The intermediate state can be steadily maintained against thermal revitrification only under constant supply of external energy. The kinetic evolution is not describable in terms of near-equilibrium relaxation processes, and we observe metastable states under well-defined external conditions. If these conditions are withdrawn, the system evolves towards its disordered state.

2. METHODS BASED ON RESONANCE PHENOMENA

U.V. spectroscopy is one of the most extensively used methods for the determination of the electron structure of solids. Absorption of U.V. radiation by a solid corresponds to the promotion of an electron from its ground state to an excited state. The excited states can be either localized orbitals which should lead to a discret spectrum or continuum electronic states, energy bands, which give rise to an absorption edge and a continuum of absorption. The absorption coefficient in the U.V. region is generally very high 10^4-10^6 cm^{-1}. All atoms participate in the process. The absorption measurements require then very thin films. For this reason one generally uses reflectivity measurements.

The two measurable quantities are then the reflectivity and transmissivity. They are defined [3] as the fraction of the incident electromagnetic energy that are reflected and transmitted by the crystal, respectively. The transmissivity is determined in part by a quantity known as the absorption coefficient. The latter is defined as the inverse of the distance an electromagnetic wave must travel in a crystal in order for its intensity to decrease by a factor 1/e of its original intensity.

Since the intensity is proportional to the square of the magnitude of the electric field, we can readily obtain an expression for the absorption coefficient, $\alpha(\omega)$, by considering the electric field of the form.

$$E(r,t) = E_0 \exp [i\omega \ [(N(\omega) / c) \ n \ . \ r \ - \ t)] \tag{2.1.}$$

where n is a unit vector in the direction of q and the refractive index $N(\omega)$ is given by

$$N(\omega) = [\varepsilon(\omega)]^{1/2} \tag{2.2.}$$

The phase velocity of the wave is $c / N(\omega)$. $\varepsilon(\omega)$ is the dielectric function which is in general a complex quantity of the form.

$$\varepsilon(\omega) = \varepsilon'(\omega) + i\varepsilon''(\omega) \tag{2.3.}$$

Hence $N(\omega)$ is also complex

$$N(\omega) = n(\omega) + iK(\omega) \tag{2.4.}$$

where $K(\omega)$, the imaginary part of the refractive index, is the so-called extinction coefficient. Substituting eq.(2.3.) and (2.4.) into the square of eq. (2.2.), we obtain the following relations

$$\varepsilon'(\omega) = n^2 - K^2 \tag{2.5a}$$

$$\varepsilon''(\omega) = 2 \ nK \tag{2.5b}$$

By taking the square of the magnitude of eq. (2.1.) and using eq. (2.4.) we obtain the expression for the absorption coefficient :

$$\alpha(\omega) = (2\omega K(\omega)) / (c) \tag{2.6.}$$

Alternatively $\alpha(\omega)$ can be expressed in terms of $\varepsilon''(\omega)$ with the aid of eq. (2.5b) :

$$\alpha(\omega) = (\omega \, \varepsilon''(\omega)) / (cn(\omega)) \tag{2.7.}$$

Frequently $\varepsilon''(\omega)$ posesses structure in a frequency region where $n(\omega)$ is slowly varying. The structure in $\varepsilon''(\omega)$ is then revealed directly by $\alpha(\omega)$.

The reflectivity at the surface of a semi-infinite solid can be calculated by considering incident and reflected waves in the vacuum and a transmitted wave in the solid. At normal incidence the amplitude of the electric field in the reflected wave is related to the incident wave by the equation

$$E_{ref} = r(\omega) \, E_{inc} \tag{2.8.}$$

where $r(\omega)$ is given by

$$r(\omega) = \varphi \, (\omega) \exp \, (i\rho(\omega)) = (n + ik - 1) / (n + ik + 1) \tag{2.9.}$$

$\rho(\omega)$ is the modulus of $r(\omega)$, and $\rho(\omega)$ is the phase difference between the electric field of the reflected and incident waves.

The experimentally measurable reflectivity, $R(\omega)$, relates the incident and reflected intensities.

$$I_{ref} = R(\omega) \, I_{inc} \tag{2.10.}$$

where

$$R(\omega) = [r(\omega)]^2 = \rho^2(\omega) = [(n - 1)^2 + K^2] / [(n + 1)^2 + K^2] \tag{2.11.}$$

In order to determine both n and K we need a second piece of experimental information. The two pieces of information can be obtained by measuring the reflectivity at two different angles of incidence or with two different polarizations at normal incident or by measuring both the reflectivity and transmitivity at normal incidence.

For a slab of thickness t bounded on both sides by vacuum the reflectivity at normal incidence is :

$$R(\omega) = \frac{2\rho^2(\omega) \, [\cosh \, (2K\eta) + \cos 2\eta\eta]}{\exp(2k\eta) + \rho^4(\omega) \exp(-2K\eta) + 2\rho^2(\omega) \cos[2(\rho + \eta\eta)]} \tag{2.12}$$

where ρ is given by eq. (2.11), $\eta = (2\pi / \lambda_o) \, t$, λ_o is the wavelength of the light in vacuum, and the phase angle ρ is specified by

$$\tan\varphi = 2K / (n^2 - 1 + K^2) \tag{2.13}$$

The transmissivity is

143

$$T(\omega) = \frac{\tau^4(\omega) \exp(-2K\eta)}{\exp(2K\eta) + \rho^4(\omega) \exp(-2K\eta) + 2\rho^z(\omega) \cos[2(\rho + n\eta)]} \qquad (2.14)$$

where

$$\tau^4(\omega) = \frac{16(n^2 + K^2)}{[(n+1)^2 + K^2]^2} \qquad (2.15)$$

The pair of equations (2.12.) and (2.14.) can be solved numerically for n and K in terms of $R(\omega)$ and $T(\omega)$. As they appear in eq. (2.5a) and (2.5b) n and K are not independent : they are related through the real and imaginary parts of the dielectric function $\varepsilon(\omega)$ by the Kramers-Kronigs dispersion relations.

Strong variations in the absorption coefficient $\alpha(\omega)$ in the visible may concern localized states atoms or molecules UV regions of the spectrum are on due to electronic transitions from the ground state to excited states. Such transitions in a solid or extented states valence and conduction bands. When the transitions are due to a localized states, they give rise generally to discrete peaks in the absorption spectrum, whereas when they concern transitions between valence and conduction bands they give rise to an absorption edge followed by a continuum.

2.1.1. Optical absorption in borate glasses

The absorption edge of borate glass, v-B_2O_3 for instance is in the region of 6.6 [24] to 7.2 eV [25]. The addition of alkali or alkali earth oxides shifts the absorption edge to longer wavelengths as a consequence of the formation of non-bridging oxygen to which the excitable electrons are less strongly bound.

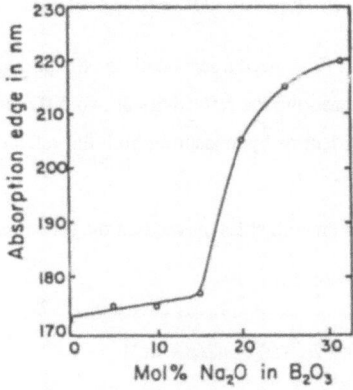

Fig. 2.1 Absorption edge of sodium borate glasses as a function of Na_2O concentration.

In the binary alkali borates a marked change in the cut-off is observed at 15% mol % of Na_2O. The jump is attributed to the appearance of non bridging oxygens at composition ≥ 16 mol % Na_2O.

The wavelength of the absorption edge λ_o follows the order $K > Na > Li$. This observation should be reconciled with the evidence that sodium borate glasses are actually two-phase composites consisting of a B_2O_3 rich phase containing some 7% mol Na_2O and a second vitreous phase containing some 24% Na_2O. One should in fact expect to observe two marked absorption edges which is not the case.

As we shall not be concerned here with colored glasses the visible spectroscopy is not going to be a topic of extensive development. We shall only mention here that the absorption in doped borate glasses is due to localized transitions. An example is the U.V. absorption of Cr(VI) in binary alkali [26] borate glasses. At concentrations below a critical value the absorption spectra of Cr(VI) resemble closely that attributed to the $[HCrO_4]^-$ ion in aqueous solution. With a further increase of metal oxide the spectra approach rapidly that of the $[CrO_4]^{2-}$ ion in aqueous solution.

2.1.2. Reflectance spectra in silicate glasses

    a) Vitreous SiO_2

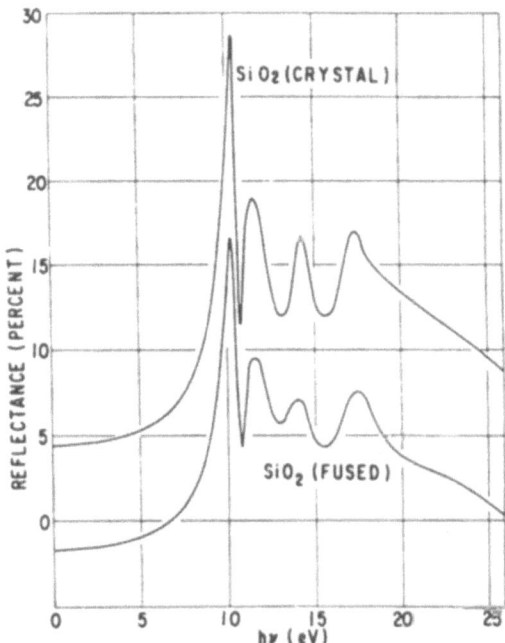

Fig. 2.2 Reflectance spectra of α-quartz and vitreous silica. The values for fused quartz have been lowered by 5% for clarity.

The reflectance spectra of vitreous and crystalline quartz in the region of 1-26 eV [27] shown in Figure 2.2 are remarkably similar, indicating the dominant importance of short range order on the U.V. transitions in SiO_2.

Kramers-Kronig analysis of the reflectance data yields four absorption peaks 10.2, 11.7, 14.3 and 17.2, eV, all of which show shifts of less than 0.1 eV from the corresponding reflectance maxima. The absorption edge obtained by transmission measurements occurs at lower energies between 7.7 to 8.2 eV.

The sharp peak at 10.2 eV in the reflectance spectra is generally identified with a Wannier exciton created by the breaking of a single Si-O bond. The broad peak at 11.7 eV is attributed to band to band transitions. Different calculations aiming at the electronic structure of these materials have been performed but there is not yet universal agreement on the electronic transitions involved in SiO_2.

b) Binary silicate glasses·

The reflectance spectra of alkali binary silicate glasses [28], shown in Figure 2.3, exhibit significant differences with that of SiO_2. These spectra are characterized by three broad reflectance bands at 8.5 eV (145 nm), 9.3 eV (130 nm) and 11.7 eV (105 nm). The reflectance spectra for different types of modifying cations are very similar. The appearance of electronic transitions at lower energies is attributed to the creation of non

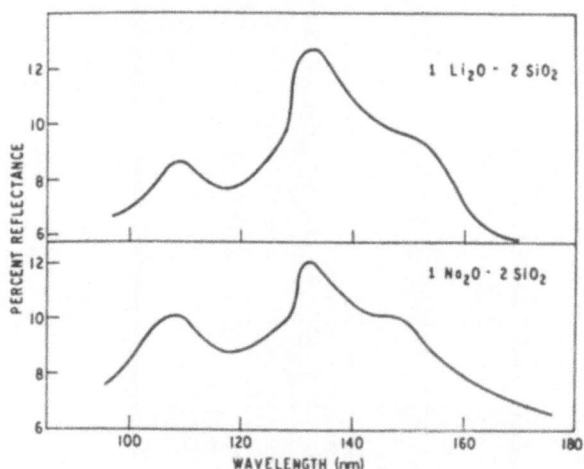

Fig. 2.3   Reflectance spectra of two simple alkali binary silicate glasses.

bridging oxygens by the introduction of network modifiers into the SiO_2 lattice. Some authors do not hesitate to compare the 9.3 eV peak with the 10.2 eV peak in pure SiO_2 and infer the existence of Wannier excitons in glasses;

The significant feature is that the intensity of similar peaks existing in glass and in crystalline matrix is always greater in the crystal.

2.2. Infrared absorption due to optical phonons

In an optical mode of vibration the atoms vibrate against one another, and if they are positively and negatively charged, produce an electric dipole moment. The resulting polarization can be written in the form

$$P = N e_T (u_1 - u_2) / V + (1/4\pi) (\varepsilon_\infty - 1) E \qquad (2.16)$$

where u_1 and u_2 are the displacements of the two atoms and E is the applied electric field.

In order to obtain the dielectric function it is necessary to calculate the displacement $u = u_1 - u_2$ as a function of E. The equation describing the driven motion can be written in the form

$$u + \gamma u + \omega^2_{TO} u = (e_T / M) E \qquad (2.17)$$

where M is the reduced mass of the positive and negative ions, e_T is the so-called transverse effective charge, γ is a phenomenological damping constant and ω_{TO} is the frequency of the transverse optical mode. To solve (2.17) let us assume that

$$u = u_0 \exp(- i\omega t) \qquad (2.18a)$$

$$E = E_0 \exp(-i\omega t) \qquad (2.18b)$$

Substituting into eq. (2.17) we find that

$$u = [(e_T/M) E / (\omega^2_{TO} - \omega^2 - i\omega \Gamma)] \qquad (2.19)$$

Combining eqs. (2.16) and (2.19) and comparing with the equation which relates the polarization vector and the electric displacement vector D

$$D = E + 4\pi P$$
$$= (I + 4\pi\chi) . E = \varepsilon . E \qquad (2.20)$$

where ε is the dielectric tensor, we obtain the following result for $\varepsilon(\omega)$

$$\varepsilon(\omega) = \varepsilon_\infty + [(4\pi e_T^2/Mv_0) / (\omega_{TO}^2 - \omega^2 - i\omega \Gamma) \qquad (2.21)$$

where v_0 is the volume of a unit cell. Setting $\omega = 0$ in eq. 2.21 yields the static dielectric constant.

$$\varepsilon_0 = \varepsilon(O) = \varepsilon_\infty + [4\pi e_T^2 TO/Mv_0 / \omega^2] \qquad (2.22)$$

Eliminating the quantity $4\pi e_T^2/Mv_0$ from eqs. (2.21) and (2.22) gives

$$\varepsilon(\omega) = \varepsilon_\infty + \frac{\omega_{TO}^2 (\varepsilon_0 - \varepsilon_\infty)}{\omega_{TO}^2 - \omega^2 - i\omega\Gamma} \qquad (2.23)$$

The real and imaginary parts of $\varepsilon(\omega)$ are

$$\varepsilon'(\omega) = \varepsilon_0 + \frac{\omega_{TO}^2 (\omega_{TO}^2 - \omega^2)(\varepsilon_0 - \varepsilon_\infty)}{(\omega_{TO}^2 - \omega^2)^2 + \omega^2 \Gamma^2} \qquad (2.24a)$$

$$\varepsilon''(\omega) = \frac{\omega_{TO}^2 \omega \Gamma (\varepsilon_0 - \varepsilon_\infty)}{(\omega_{TO}^2 - \omega^2)^2 + \omega^2 \Gamma^2} \qquad (2.24b)$$

By choosing values for the parameters ε_0, ε_∞, ω_{TO} and Γ one can calculate $\varepsilon'(\omega)$ and $\varepsilon''(\omega)$ as functions of frequency. The real and imaginary parts of the dielectric function are also given in terms of refractive index n and extinction coefficient K

$$\varepsilon'(\omega) = n^2 - K^2 \qquad (2.25a)$$

$$\varepsilon''(\omega) = 2nK \qquad (2.25b)$$

which are related to the experimentally measurable $R(\omega)$

$$R(\omega) = [(n-1)^2 + K^2] / [(n+1)^2 + K^2] \qquad (2.26)$$

Using eqs. (2.25) and (2.26) one can calculate $n(\omega)$, $K(\omega)$ and $R(\omega)$. By adjusting ε_0, ε_∞, ω_{TO} and Γ to give the best fit of the calculated reflectivity curve to an experimental curve, one can determine the value of these four parameters for a particular material.

If eq. (2.23) is substituted in the equation :

$$c^2 q^2 = \omega^2 \varepsilon(\omega) \qquad (2.27)$$

giving the dispersion relation for electromagnetic waves propagating in a dielectric medium, we obtain the dispersion relation for an electromagnetic wave propagating in a polar cubic crystal in the frequency range of optical phonons :

$$(c^2 q^2) / \omega^2 = \varepsilon_\infty + [\omega_{TO}^2 (\varepsilon_0 - \varepsilon_\infty) / (\omega_{TO}^2 - \omega^2)] \qquad (2.28)$$

where the damping is ignored by setting $\Gamma = 0$. This equation may be simplified by introducing the frequency of longitudinal optical phonons of long wavelength ω_{LO}, through the Lyddane-Sachs-Teller relation :

$$\omega_{LO}^2 = \omega_{TO}^2 (\varepsilon_0 / \varepsilon_\infty) \qquad (2.29)$$

One then obtains :

$$[c^2 q^2 / \omega^2] = [\varepsilon_\infty (\omega_{LO}^2 - \omega^2)] / (\omega_{TO}^2 - \omega^2) = \varepsilon(\omega) \qquad (2.30)$$

The eq. (2.30) yields two solutions, ω_+ and ω_-, which are plotted as function of q in Figure 2.4.

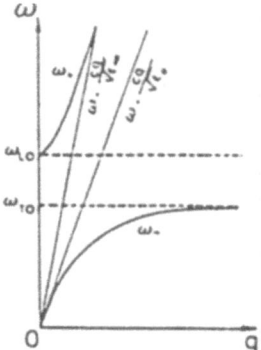

Fig. 2.4   Dispersion relation for the interacting system of
photons and optical phonons.

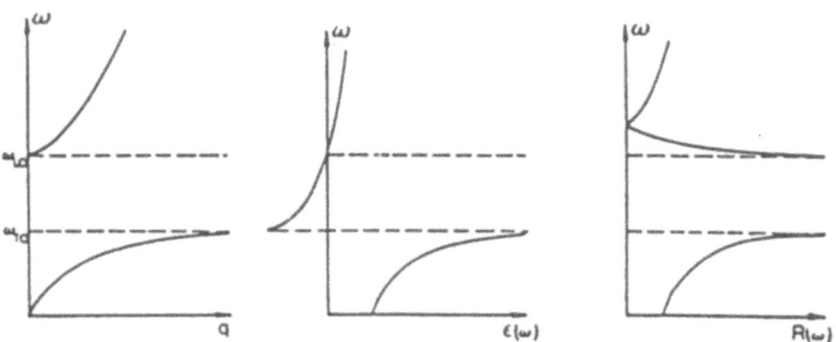

Fig. 2.5   Dielectric function plotted as a function of frequency in the optical phono
region compared to the dispersion relations and the reflectivity function.

We note in this plot that no electromagnetic wave propagates in the crystal in the frequency interval between ω_{TO} and ω_{LO} when damping is neglected. This is the reststrahlen region. The absence of propogation in the region can be understood in terms of the dielectric function which is plotted in Figure 2.5 as a function of frequency.

We see that between ω_{TO} and ω_{LO} the dielectric function is negative and hence the reflective index is imaginary $N(\omega) = [\varepsilon(\omega)]^{1/2}$. Consequently the electromagnetic wave

$$E(r,t) = E_G \exp [i\omega [(N(\omega)/c] \quad q \cdot r - t]] \tag{2.31}$$

incident on the crystal will decay exponentially into the crystal and will not propagate. Therefore the crystal is 100% reflecting in is region. Experimentally the measured reflectivity is not 100% but it is within the order of magnitude. In an experimental reflectivity spectrum one notes that ω_{LO} is the frequency at which the dielectric function goes to zero and ω_{TO} is the frequency at which it goes to infinity as may be seen in Figure 2.5 b and c and equation (2.30).

2.2.1. Infrared spectra of alkali silicate glasses

Infrared spectroscopy reveals the same physical phenomena, the vibrational modes of the system, as the

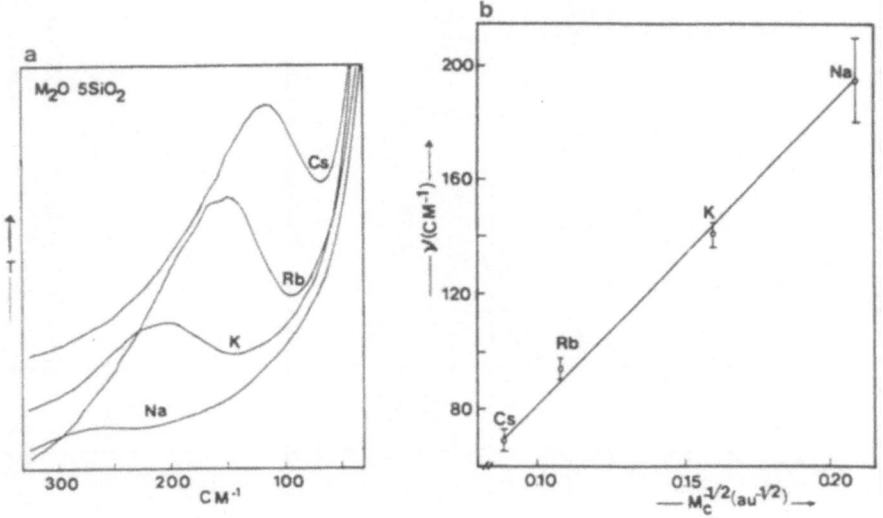

Fig. 2.6a   The far-infrared spectra of single alkali pentasilicate glasses : $M_2O - 5SiO_2$; where M-Na, K, Rb, Cs, as marked on the figure.

Fig. 2.6b   A plot of the cation-motion frequencies of the alkali pentasilicate glasses, $M_2O - 5SiO_2$ versus $M_c^{-1/2}$, where M_c is the mass of the cation.

Raman spectroscopy. In many cases these two techniques are complementary, the selection rules being different under given experimental circumstances. A more complete picture of the structure of the material modes investigation may be gained if both techniques are used.

The vibrational modes in M_2O - $5SiO_2$ involving the vibrations of the metallic ion M have been systematically investigated [29]. Figure 2.6 summarizes results of the series where M = Na, K, Rb, Cs.

The fact that the frequency shift with metal ion mass follows the simple rule $\omega = \sqrt{K/m}$ indicates that the mass substitution occurs without force constant changes. That means that the matrix remains unchanged in the different alkali metal compounds.

The band width also changes systematically with cation decreasing in the order Na > K > Rb > Cs. The spectrum of phase-separated Li_2O - $5SiO_2$, not shown here, gives a very broad featureless absorption between 300-500 cm^{-1}.

The bandwidth can be explained by the fact that in glasses there are different sites which can be occupied by the metal ions. As the vibrational frequency of the cation depends on its mass and specific site force constants, the observed broad bands are the envelope of bands due to individual cation-site interaction. The decrease of band width in the order Na > K > Rb > Cs indicates that the smaller the cation the wider the range of interaction between the individual sites and the broader the distribution of these sites and wider is the variety of their configuration. The larger the ion, more reduced is the variety of sites it occupies, the more localized it is.

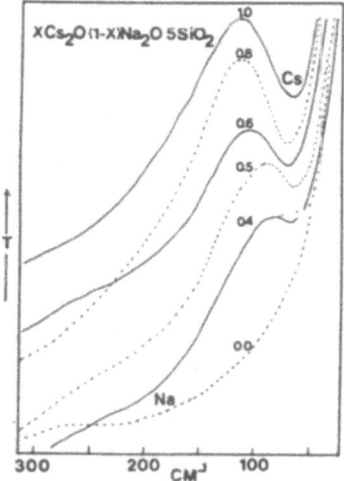

Fig. 2.7   The far infrared spectra of mixed alkali pentasilicate glasses :

xCs_2O (1 - x) Na_2O . $5SiO_2$.

The infrared spectra for the ternary system x Cs_2O - (1 - x) Na_2O - $5SiO_2$ are given in Figure 2.7.

In these spectra the cation-motion bands due to Na and Cs are observed. As x increases from O to 1 their relative intensity changes. Thus the broad featureless Na band gradually decreases in intensity while the Cs-motion band grows in intensity and shows a direct dependence on Cs concentration. The authors claim that each cation motion band remains essentially unshifted in frequency for different x. This is seen more clearly for Cs where the effect of the part of the background due to the Na band is removed.

Raman spectra for the same sequence of glasses show analogous features which tend to demonstrate that the glass network is not modified when different cations are introduced. Each ion has its specific occupations.

2.3. Electron spin resonance

For an electronic system, in the presence of a magnetic field H, the Zeeman interaction between the electronic magnetic dipole μ_J, and the field is

$$U = \mu_J \cdot H = g_J \beta HM \tag{2.32}$$

when M is the component of the electronic angular momentum J along the field acting on the atom, β is the Bohr magneton $\beta = (eh / 2mc)$, g_J is the g-factor when both orbital and spin momenta are present so that the resulting angular momentum is associated with the number $J = L + S$.

When an alternating electromagnetic field of frequency υ is applied at right angles to H, magnetic dipole transitions are induced according to the selection rule $\Delta M = \pm 1$ and an energy quantum

$$h\nu = g_J \beta H \tag{2.33}$$

is absorbed so that the spin of the electron is flipped from a direction parallel to the magnetic field to that of

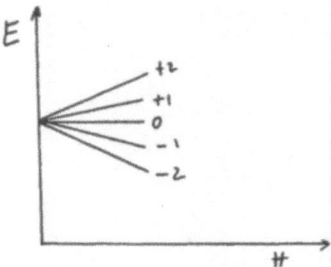

Fig. 2.8   Zeeman levels of a paramagnetic center with $J = 2$ plotted as a function
of the magnetic field H.

the anti-parallel direction. For the reverse process induced emission will be observed. If the system is in thermal equilibrium more spins are in the parallel state because it is of lower energy so that there is a net absorption power. The separation $g_J \beta H$ between the Zeeman levels increases linearly with the magnetic fields as shown in Figure 2.8.

We shall see further that electron paramagnetic resonance and nuclear magnetic resonance have many aspects in common.

For $g = 2$ and $H = 3300$ Gauss, the resonance frequency is 10 GHz which falls within the so-called X-band waveguide where the commercially available spectrometers are.

For a free electron, the g-factor is the ratio of the spin magnetic moment in Bohr magnetons to the spin angular momentum in units of h and its value is $g = 2$.

When both orbital and spin momenta are present in an atom, the value of g becomes a function of the coupling between these momenta.

The transitions between the Zeeman energy levels of the orbital ground state, observed in ESR spectroscopy, depend in a complicated manner on the particular paramagnetic center, the symmetry and strength of the crystal field, the spin-orbit coupling and other effects such as the hyperfine interaction between electrons and nucleus.

The observed resonance data can be described by the use of a spin Hamiltonian in a rather simple way. The spin Hamiltonian yields a concise expression of the coupling of the spin to its surroundings i.e., the extent to which the spin is not free. A great deal about local structures can be learned from the spin Hamiltonian which can be determined uniquely from knowledge of the atomic structure about the paramagnetic center.

Two classes of paramagnetic centers are generally encountered in the study of ESR in monocrystalline solids : i) substitutional impurities such as transition metal or rare-earth ions and ii) radiation induced ESR centers.

In the first transition series where the paramagnetic 3d electrons are the outmost electrons and, therefore, exposed to the electrostatic field of the ligands, the interaction between the electrons and the ligand field becomes so great that the orbital motion is quenched. This simply means that the orbital motion is "locked into" the field of the ligands, and is not able to contribute to the magnetism. The electron spin, on the other hand, with its magnetic moment, does not couple directly with the electrostatic field, and remains free to orientate in an external magnetic field. Thus, the magnetic behaviour of the 3d ions is essentially that of spin-only magnetism. In the case of rare-earth ions, however, the unfilled electron shell is an inner 4f shell, and the interaction with the ligand electrostatic field is much less. The paramagnetism of these ions can be approximated to that of the free ions, having contributions from both spin and orbital motions.

2.3.1. Spin Hamiltonian

The transitions between the Zeeman levels of the orbital ground state observed in Electron Paramagnetic Resonance are generally described by a "spin Hamiltonian" in a rather simple way.

The most general form of the spin Hamiltonian contains a large number of terms, representing the Zeeman interaction of the magnetic electrons with an external field, level splitting due to indirect effects of the crystal field (often referred to as five structures), hyperfine structure due to the presence of nuclear magnetic dipole and electric quadripole moments in the central or ligand ions and the Zeeman interaction of the nuclear moment with the external field. As an example of a spin Hamiltonian let us consider :

$$H = \beta\, g_{ij}\, H_i\, S_j' + D_{ij}\, S_j'\, S_j' + A_{ij}\, S_i'\, I_j + Q_{ij}'\, I_i\, I_j - \gamma\, \beta_I\, H_i\, I_j - R_{ij}\, H_i\, I_j \qquad (2.34)$$

which contains terms in H_i up to order one and terms in S_i and I_j up to order two. The first term describes the Zeeman splitting of the lowest energy levels, the second term describes the zero magnetic field splitting produced by ligand field or spin-spin interactions and D_{ij} is a measure of the splitting of the ground state in a non-cubic field. The remaining four terms describe hyperfine interactions involving a nucleus of spin I.

The first of these is the magnetic hyperfine interaction and contains a nuclear spin coupling parameter A_{ij} ; the second is the quadrupole interaction and the last two are the direct interactions between the nucleus and the magnetic field.

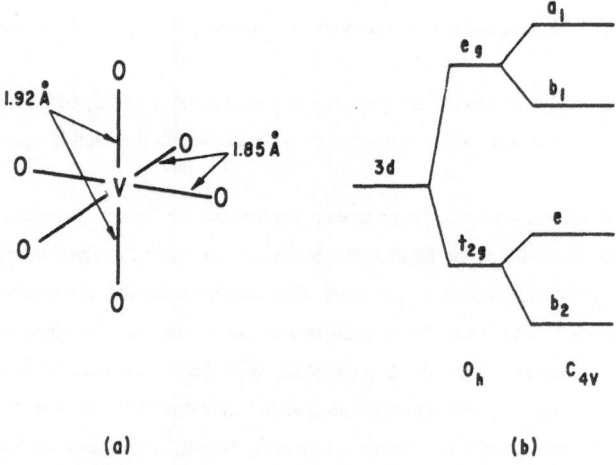

(a)   (b)

Fig. 2.9   (a) Local structure of V^{4+} in tetragonal GeO_2 ; (b) Energy level scheme for

V^{4+} in TiO_2 or GeO_2 in the crystal field approximation.

An example of energy level diagrams to be taken into account for the transition observed in EPR is given in Figure 2.9.

In this figure is shown the energy level scheme for V^{4+} in GeO_2 in the crystal field approximation. Although V^{4+} is usually 6-fold coordinated, its local symmetry is generally not octohedral, but axially distorted to a lower symmetry such as C_{4v}. This axial distortion in mainly responsible for the observed variations of the g-tensor components in the EPR spectrum. From the deviation of the spectra from that expected for the free ion is deduced the influence of the environment and the structure of neighboring sites.

2.3.2. Radiation induced EPR centers in alkali borate glasses

In low alkali borates, trapped hole centers are found to be stable at room temperature. These yield a "five line plus a shoulder spectrum [30]. The spectrum is shown in figure 2.10.

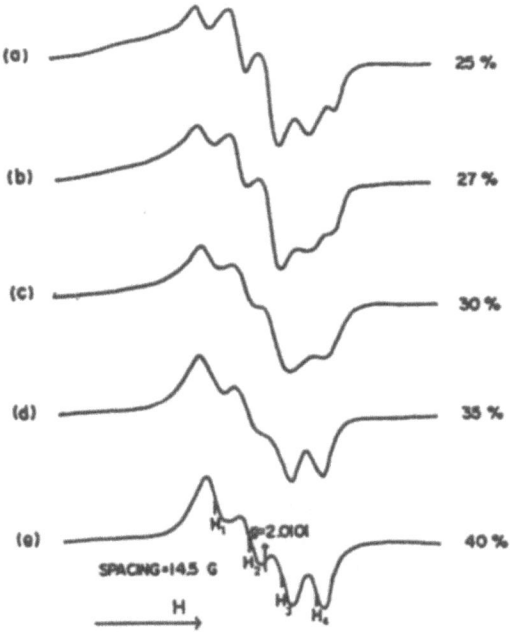

Fig. 2.10 . EPR spectra of a series of γ-ray irradiated (1×10^8 R) sodium borate glasses.
Mol % Na_2O is indicated for each spectrum.

In the model Hamiltonian used to interpret [31] these data the following values of g have been used : $g_1 = 2.002$; $g_2 = 2.010$ and $g_3 = 2.035$. The model for the hole center consistent with these parameters and with the associated ~10 G hyperfine coupling constants is a hole on a bridging oxygen connecting a three coordinated boron with a four coordinated boron.

After heat treatment at 250° C, the five line spectrum disappears and a single resonance with g = 2.0095 remains.

2.4. Nuclear magnetic resonance

For a nucleus that precesses with a magnetic moment μ and an angular momentum h I, the two quantities are parallel ,

$$\mu = \gamma \, h \, \mathbf{I} \tag{2.35}$$

and the gyro magnetic ratio γ is constant. I denote the nuclear angular momentum measured in units of h. In the presence of a magnetic field H, the interaction between the magnetic dipole moment μ and the field is

$$U = \mu \cdot \mathbf{H} \tag{2.36}$$

If the field is applied in the z direction $\mathbf{H} = Hz$, the energy of interaction is

$$U = -\mu_z \, H = -\gamma \, h \, H \, I_z \tag{2.37}$$

The allowed values of I_z are $m_I = I, I - 1, ...- I$, and $U = -m_I \gamma h H$.

In a magnetic field a nucleus with I = 1/2 has two energy levels corresponding to $m_I = \pm 1/2$ as shown in Figure 2.11.

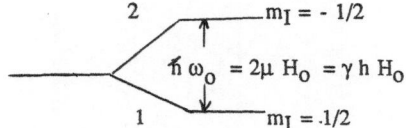

Figure 2.11   Energy level splitting of a nucleus of spin I = 1/2
in a static magnetic field H_0

If $h\omega_0$ denotes the energy difference between the two levels, then $h\omega_0 = \gamma h H_0$ or

$$\omega_0 = \gamma \, H_0 \tag{2.38}$$

This is the fundamental condition for magnetic resonance absorption.

2.4.1. Longitudinal and transverse relaxation times

The rate at which the magnetization component M_z approaches an equilibrium value M_0 is characterized by a spin-lattice relaxation time called the longitudinal relaxation time.

The equation of motion which governs the rate of change of angular momentum of a system is equal to the torque which acts on the system. The torque of a magnetic moment μ in a magnetic field H is $\mu \times H$

$$\hbar \, (dI \, / \, dt) = \mu \times H \tag{2.39}$$

$$(d\mu \, / \, dt) = \gamma \, \mu \times H \tag{2.40}$$

The nuclear magnetization M of the substance is the sum $\Sigma \, \mu_i$ over all the nucleus in an unit volume. If only a single isotope is important, we consider only a single value of γ, so that

$$(dM \, / \, dt) = \gamma \, M \times H \tag{2.41}$$

In thermal equilibrium, at temperature T, the magnetization, when the nucleus is placed in a static field $H = H_0 \, z$, will be along z :

$$M_x = 0 \quad M_y = 0 \qquad M_z = M_0 = \chi_0 \, H_0 = C \, H_0 \, / \, T \tag{2.42}$$

where $C = (N\mu^2 / 3 \, K_B)$ is the Curie constant. The magnetization of a system of spins with $I = 1/2$ is related to the population difference $N_1 - N_2$ of the lower and upper levels in Figure 2.11 :

$$M_z = (N_1 - N_2) \, \mu \tag{2.43}$$

where the N's refer to a unit volume. The population ratio in thermal equilibrium is just given by the Boltzmann factor for the energy difference $2\mu \, H_0$

$$(N_2 \, / \, N_1)_0 = \exp \, (- \, 2\mu \, H_0 \, / \, k_B T) \tag{2.44}$$

When the magnetization component M_z is not in thermal equilibrium, it approaches equilibrium at a rate proportional to the departure from the equilibrium value M_0 :

$$(dM_z \, / \, dt) = [(M_0 - M_z) \, / \, T_1] \tag{2.45}$$

T_1 is the longitudinal relaxation time or the spin lattice relaxation time.

The dominant spin-lattice interaction of paramagnetic ions in crystals is by the phonon modulation of the crystalline electric field.

Taking account of (2.45) the equation of motion (2.41) becomes

$$(dM_z \, / \, dt) = \gamma \, (M \times H_0)_z + [(M_0 - M_z) \, / \, T_1)] \tag{2.46}$$

where $(M_0 - M_z) / T_1$ is an extra term in the equation of motion, arising from interactions not included in the effect of the magnetic field H_0 ; besides precessing about the magnetic field, M will relax to the equilibrium value M_0.

In a static field $H_0 z$ if the transverse magnetization component M_x is not zero, then M_x will decay to zero, and similarly for M_y the decay occurs because at thermal equilibrium the transverse components are zero. The rate of transverse relaxation is :

$$(dM_x / dt) = \gamma (M \times H_o)_x - (M_x / T_2) \qquad\qquad (2.47a)$$

$$(dM_y / dt) = \gamma (M \times H_o)_y - (M_y / T_2) \qquad\qquad (2.47b)$$

T_2 is called the <u>transverse relaxation time</u>. The magnetic energy - $M \cdot H_0$ does not change as M_x and M_y change, provided that H_0 is along z. No energy need flow out of the spin system for relaxation of M_x and M_y to occur, so that the condition which determines T_2 may be less strict than for T_1.

The time T_2 is a measure of the time during which the individual moments that contribute to M_x and M_y remain in phase with each others. Different local magnetic fields at different spins will cause them to precess at different frequencies. If initially the spins have a common phase, the phase will become random in the course of time and the values of M_x, M_y will become zero. T_2 is the dephasing time.

Experimentally in addition to the static magnetic field applied in the z direction a rf magnetic field is usually applied along the x or y axes.

The frequency of free precession of the spin system in a static field $H_a = H_0 z$, deduced from eqs. (2.47a) and (2.47b), is $\omega_0 = \gamma H_0$ with a characteristic time $T' = T_2$

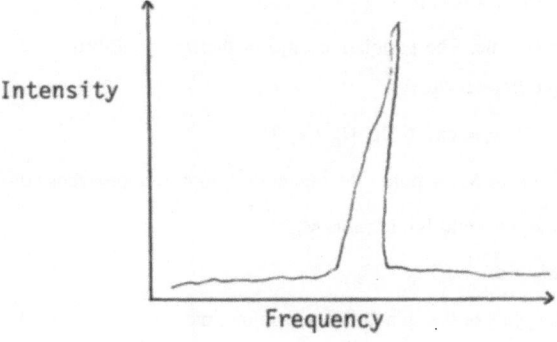

Figure 2.12

The motion is similar to that of a damped harmonic oscillator in two dimensions. The spin system will show resonance absorption from a driving field near the frequency $\omega_0 = \gamma H_0$, and the frequency width of the response of the system to the driving field will be $\Delta\omega \approx 1/T_2$. A typical spectrum is shown in Figure 2.12

As for the other spectroscopy methods NMR is used in attempts to determine structural properties but it is also adequate for studying the ionic motion. We shall examine here one example of each one of such investigations.

2.4.2. Structural models for borate glasses deduced from NMR

NMR has been used for many years to study structure and bonding in glasses with particular attention to borate glasses [32]. The ^{11}B isotope has proved useful for structure and chemical bonding studies since its electrical quadrupole moment interacts with any electric field gradient experienced by the nucleus. The electric field gradient arises principally from electrons in the chemical bonds formed by the boron atom, and the quadrupole interaction, which determines the shape and the structure of the NMR resonance, is extremely sensitive to changes in the chemical bonding and atomic environment.

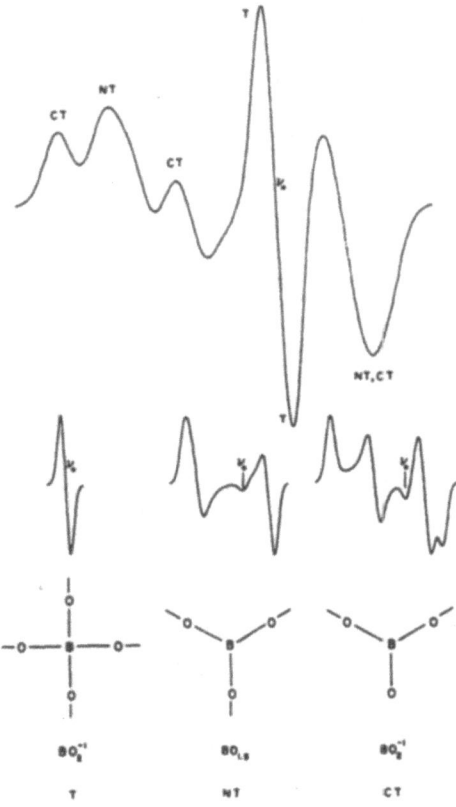

Fig. 2.13 The upper portion of the figure shows a typical ^{11}B NMR derivitive spectrum for a borate glass having a high alkali content. The spectrum is a superposition of the middle portion of the figure. The lower portion depicts the elemental sub-unit of the borate groups corresponding to the NMR responses shown above. These are the tetrahedral sub-units (T) containing 4 bridging oxygens, the neutral trigonal sub-unit (NT) containing 3 bridging oxygens, and the charged trigonal sub-unit (CT) containing 2 bridging and 1 non-bridging oxygen.

In Figure 2.13 is shown the ^{11}B spectrum, (the first derivative of the NMR absorption spectrum) for boron in various bonding configurations.

Matching the computer-simulated and experimental spectra yield three quantitative measurements :N_4, the fraction of boron in tetrahedral units ; N_{3S} the fraction of boron in symmetric trigonal units ; N_{3A}, the fraction of boron in asymmetric trigonal units. The ^{11}B spectra by themselves yield only the amounts of tetrahedral and charged or neutral trigonal bonding configurations.

The quadrupole moment of ^{10}B is much larger than that of ^{11}B, so the coupling to an electric field gradient at the nucleus is much greater.

In Figure 2.14 are displayed ^{10}B NMR spectra for pure B_2O_3 and several glasses in the system Na_2O - B_2O_3.

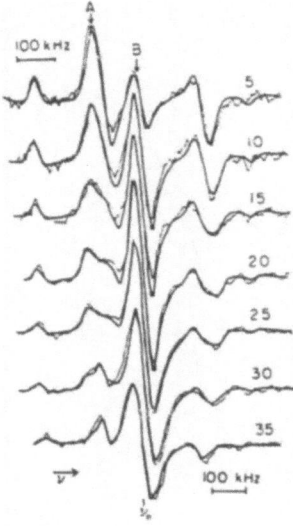

Fig. 2.14 ^{10}B NMR spectrum for pure B_2O_3 (top) and seven glasses in the system Na_2O - B_2O_3. The molar % of Na_2O is indicated to the right of each trace. The darker line in each trace is a computer simulated spectrum.

A detail from the spectrum shown in Figure 2.14 is displayed on expanded scale in Figure 2.15 along with a particular computer simulation that provided an acceptable fit to the experimental spectrum.

This simulation is obtained simply by adding together the spectra for the structural groups (boroxol, diborate, etc...) from the crystalline compounds in this system. Weightings for each component spectrum are not arbitrary, they are linked by constraints, so that there is actually only one variable parameter.

For glasses below the metaborate composition the parameter is chosen to be the member of diborate structural groups in the glass. When that value is set for a glass, the weightings for all other groups follow from the constraints.

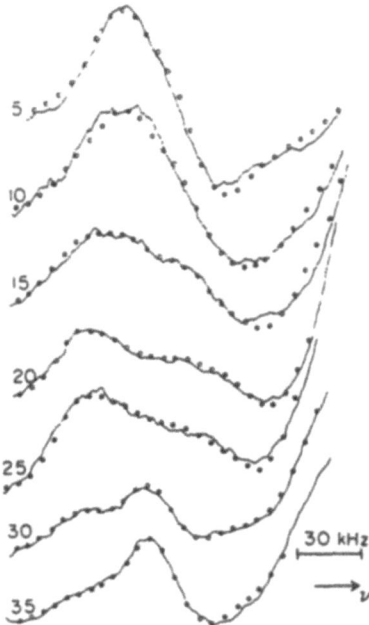

Fig. 2.15 Experimental ^{10}B NMR spectra displaying feature A of fig. 2.14 on an expanded scale. Circles show simulated spectra.

2.4.3. Use of NMR for studying motions of ions

There have been some strong claims [33] [34] that nuclear magnetic resonance can be a powerful tool for studying motion of ions and therefore has had extensive application in the field of superionic conductors. In superionic conductors a large number of ions are involved in the conduction process ; therefore correlated motion is a major aspect of the problem and it is of interest to see what can be learned about this from NMR. To understand the features of the dynamics which can be delineated and how NMR might be useful in distinguishing correlated from independent particle motion, we consider this technique in conjunction with bulk conductivity and infrared or Raman spectroscopy.

The most important property of superionic conductors, at least for technical applications, is the dc ionic conductivity σ which may be expressed as [34]

$$\sigma = (v/6)\ (N/V)\ e^2\ l^2\ /\ k_B T = (N/V)\ e^2\ D/k_B T \qquad\qquad (2.48)$$

where N is the number of mobile ions of charge e contained in volume V, T is the absolute temperature, and v^{-1} is the average time for the ion to hop an average distance l.

If the number of carriers is known, then the conductivity measures the product $v\ l^2$ and can thus determine the hopping frequency v only if the jump mechanisms are sufficiently well understood that the mean jump distance l is known. The hop may be to nearest-neighbor sites only, and such determination can be made, but this is not always the case.

The advantage of NMR is that it can often provide a direct estimate of v. The reason is that the NMR relaxation process is governed by the rate at which an interaction, such as coupling to an electric field gradient or to spin of a nearby ion, fluctuates. This generally depends on how long the nucleus remains at a given site and not on how far it travels from the site once it jumps. Conduction, on the contrary, clearly depends both on how frequent and how far the jumps are.

For the processes where the Arrhenius relation

$$v = v_0\ \exp\ (-\ E_A\ /\ k_B T) \qquad\qquad (2.49)$$

is obeyed, the activation energy E_A and the "attempt frequency" v_0 can both be extracted from NMR relaxation data. E_A is obtained from the overall temperature dependence of the conductivity, and once it is known, v_0 can be infered from the temperature at which v reaches a critical value associated with a maximum spin-lattice relaxation rate ($v \approx \omega_0$, the NMR frequency) or the onset of motional narrowing of the linewidth ($v \approx 1/T_{20}$, where T_{20} is the rigid lattice spin-spin relaxation time). A fundamental question is whether v_0 is related to a vibrational frequency v of the mobile ion residing in a potential well between hops. The question is what is the relation of v with the frequency of the metal ion vibration in the borate glass determined by infrared spectroscopy as show in Figure 2.6. Simple classical theory predicts that $v_0 = v$, which can be tested by combined NMR and infrared or Raman measurements. Hopping produces a broadening but, at least at sufficiently low temperatures, should not interfere with the spectroscopic determination of v. The interesting value of NMR here is that spin relaxation is insensitive to the rapid vibrations v, so that vibration between hops normally do not interfere with measurements of v as long as $v \sim \omega \ll v$.

It has been generally tempting to believe that in the simplest case of discrete, nearest neighbour, classical hopping, the activation energy and attempt frequency inferred from conductivity and NMR should agree, and the attempt frequency should correspond to a measured optical phonon mode. This is in fact not the case .In most of the cases there are vast discrepancies, by many orders of magnitude, between these three sets of measurements. There is a fundamental physical misunderstanding of the basic processes here. In the case of optical modes, the vibrating ion is bound to the lattice, and has the force constant of the compound, whereas the concept of an attempt frequency v_o concerns a free ion. The discrepancy originates from an improper application of simple hopping ideas to the theory of NMR. Another explanation may be that the difference is ascribed to correlation effects.

That correlated hopping may cause a discrepancy can be seen by noting that the v of eq. 2.48 from the measured conductivity is the true, correlated frequency which describes the collective motion of all the ions. But v deduced from NMR relaxation is related to the motion of a single ion, distinguishable by its spin label. If correlation effects are important these frequencies can be quite different. A gross effect occurs in the one dimensional diffusion of particles which cannot hop to occupied sites. Here, the bulk diffusion coefficient D of all the particles is well defined and the second equality in 2.48 holds. However, the motion of a single labeled particle is not diffusive : its mean square displacement x^2 is proportional to $t^{1/2}$ rather than to t for long times t. One would not, therefore, expect NMR inferred dynamics to agree with those obtained from bulk conductivity.

A good example of the difference between results obtained from bulk conductivity and NMR could be seen in the inequivalent site model shown in Figure 2.16.

A particle jumps rapidly from a b-type to an unoccupied a-type well, but an a   b type process is much

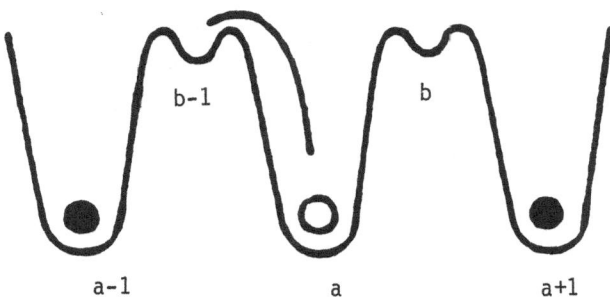

Fig. 2.16 Inequivalent site model with deep a-type and shallow b-type wells. Rapid hopping back and forth of nucleus among sites b - 1, a, b gives rise to a high frequency component in NMR relaxation ; but there is no net conduction unless particles are displaced from a - 1 or a + 1.

less probable. If the neighboring a wells are initially occupied, a nucleus will hop back and forth between the nearest b sites until one of the other a sites, is vacanted. This motion makes no contribution to the conductivity and yet can be effective in relaxing the nuclear spin if it experiences different interactions at a and b sites. In this case there is a true difference between distinguishable particle and total dynamics.

For all these reasons NMR deduced attempt frequencies and activation energies have to be handled with great care.

We shall not intend here to develop the general theory of magnetic relaxation in terms of correlation functions and interactions. Rather we shall try to present some experimental results and discuss their contribution to the understanding of the dynamics of fast ion conduction.

The general features of the longitudinal relaxation time T_1 and the transverse relaxation time T_2 can be described as functions of temperature. T_1^{-1} will have a maximum as a function of temperature T and very rapidly on each side T_2^{-2} will decrease nearly monotonically with increasing temperature since the narrowing becomes more complete with decreasing hopping time τ_c. These features of the T_1 minimum and the motionally narrowed T_2 are nearly universally observed in superionic conductors.

Examining some experimental results it appears that the most easily verified feature of NMR relaxation is the activation energy. For processes characterized by an activation energy E_A one should expect

$$T_1^{-1} \alpha \exp(-E_A / k_B T)$$

on the low temperature side of the minimum.

A good example is ^7Li in Li Al Si O_4 [36].

Line narrowing has been used frequently to obtain activation energies. Activation energies extracted from T_2 data generally agree with those from T_1 data and then correlate well with measurements of dc conductivity.

As for the attempt frequency the situation is quite different. I shall present here only one example which seems to be typical : Li Al SO_4. The ν_o obtained by NMR from line narrowing is 5.10^7 s^{-1} ; as determined by conductivity measurements it is 5.10^{14} s^{-1}, yet the optical mode frequency gives 5.10^{12} s^{-1}. It is difficult to admit that we are considering here the same physical quantity. In Na β-alumina on the contrary, the three methods yield practically the same quantity : 2.10^{11} s^{-1}.

These examples show that much care has to be exercised in assuming that τ measured is the true single-particle hopping time. Even more serious discrepancies occur in 1D and 2D systems.

REFERENCES

1. A. GUINIER, "X-Ray Diffraction", (W.H. FREEMAN, San Francisco 1963).

2. D.E. SAYERS and E.A. STERN, Phys. Rev. Lett. 27 : 120 (1971).

3. R.F. WALLIS and M. BALKANSKI, Many body aspects of Solid State Spectroscopy (ch 9) (North-Holland, 1986).

4. S.V. BERGER, Acad. Chem. Scand. 7 : 611 (1953).

5. A.H. SILVER and P.J. BRAY, J. Phys. Chem. 29 : 984 (1958).
   P.J. BRAY, J. Non-Cryst. Solids 73 : 19 (1985).

6. F.L. GALEENER, R.A. BARIO, E. MARTINEZ and R. ELLIOTT, Phys. Rev. Lett. 53 : 2429 (1984).

7. F.L. GALEENER in the Structure of non crystalline materials 1982, ed. P.H. GOSKELL, J.M. PARKER and E.A. DAVIS, (Taylor and Francis, London, 1983) p. 337.

8. W.M. RISEN Jr., J. Non-Crys. Solids 76 : 87 (1985).

9. P.J. BRAY and G.O'KEFFE, Phys. Chem. Glasses 4 : 37 (1963).

10. T.W. BRIL, Philips Res. Rep. Suppl., 2 (1976).

11. M. KBALA, Thesis, Bordeaux (1984).

12. M. BALKANSKI, A. AYYADI, P. CADET, M. JOUANNE, C. JULIEN, M. MASSOT, M. SCAGLIOTTI and A. LEVASSEUR, Solid State Commun. 57 : 41 (1986).

13. A. AYYADI, Thesis, Paris (1986).

14. KONIJNENDIJK, Philips Res. Rep. Suppl., 7 (1975).

15. M. IRION, M. COUZI, A. LEVASSEUR, J.M. REAU and J.C. BRETHOUS, J. of Solid State Chemistry 31 : 285 (1980).

16. P.M. BRIDENBAUGH, G.P. ESPINOSA, J.E. GRIFFITHS, J.C. PHILLIPS and J.P. REMEIKA, Phys. Rev. B20 : 4140 (1979).

17. J.A. ANOROVITZ, J.R. BANAVAR, M.A. MARCUS and J.C. PHILLIPS, Phys. Rev. 28 : 4454 (1983).

18. K.T. MURASE, K. FUKUNAGA, K. YAKUSHIJI and I. YUNOKI, J. Non-Cryst. Solids 59 : 160 (1984).

19. M. BALKANSKI, E. HARO, Z.S. XU and J.F. MORHANGE, Sixth Int. Conf. on Collective Phenomena ; Reports from the Moscow Refusnik Seminar Reprinted, Annals of the New York Academy of Sciences, 452 : 275 (1985).

20. BRESSER, W.J.P. BOOLCHAND, P. SURANYI and J.P. DE NEUFVILLE, Phys. Rev. Lett. 46 : 1689 (1981).

21. LUCOVSKY, F.L. GALEENER, R.H. GEILS and R.C. KEEZER, in Structure of Non-Crystalline Materials, H. Gaskell, ed.(Taylor and Francis, London, England, 1977) p. 127.

22. R. AZOULAY, J. Non-Cryst. Solids 18 : 33 (1975).

23. A. FONTANA, G. MARIOTTO and F. ROCCA, Phys. Stat. Sol. (b) 129 : 489 (1985).

24. E. KORDES, Glostech Ber. 32 : 267 (1959) ; 38 : 242 (1965).

25. B.D. McSWAIN, N.F. BORRELLI and G.J. SU, Phys. Chem. Glasses 4 : 1 (1963).

26. A. PAUL and R.W. DOUGLAS, Phys. Chem. Glasses 8 : 151 (1967) and 9 : 27 (1968).

27. H.R. PHILLIPS, Solid State Commun. 4 : 73 (1966).

28. G.H. SIGEL, Jr. J. Phys. Chem. Solids 32 : 2373 (1971).

29. E.I. KAMITSOS and W.M. RISEN, J. of Non-Cryst. Solids 65 : 333 (1984).

30. S. LEE and P.J. BRAY, Phys. Chem. Glasses 3 : 37 (1962) ; J. Chem. Phys. 39 : 2863 (1963).

31. D.L. GRISCOM, P.C. TAYLOR, D.A. WARE and P.J. BRAY, J. Chem. Phys. 48 : 5158 (1968).

32. P.J. BRAY, J. of Non-Cryst. Solids 75 : 29 (1985).

33. M.S. WHITTINGHAM and B.G. SILBERNAGEL, in "Solid Electrolytes : General properties Characterization, Materials Applications" ed. by P. HAGENMULLER and W. VAN GOOL (Academic Press, 1979).

34. P.M. RICHARDS, "Topic in Current Physics n° 15, Physics of Superionic Conductors" (Springer Verlag Berlin, 1979).

35. R. KUBO, J. Phys. Soc. Jpn. 12 : 570 (1957).

36. D.M. FOLLSTAEDT and P.M. RICHARDS, Phys. Rev. Lett. 37 : 1571 (1976).

X-RAY AND NEUTRON SCATTERING STUDIES OF SUPERIONICS

C.R.A. Catlow

Department of Chemistry
University of Keele
Keele, Staffs ST5 5BG

1.   Introduction

X-ray and neutron scattering techniques play a central role in current materials science research. The importance of crystallographic studies using Bragg scattering of X-rays and neutrons is obvious; but elastic diffuse scattering is also a valuable technique in the study of disordered materials.   Moreover, inelastic scattering of neutrons yields unique information on the dynamics of the system studied.   In the present chapter we review these techniques with emphasis on their application to superionics.   We will consider first a number of comparative issues in the field of diffraction: these issues will relate to the types of samples and types of sources that are used in contemporary studies.   We consider next a number of problems relating to analysis of diffraction data on high temperature materials, which we follow by a discussion of diffuse scattering and the information which we can derive from such data.   We then describe a range of applications of Bragg and diffuse scattering to the study of structural properties of superionics.

In addition to elastic scattering, inelastic or more particularly, quasi-elastic scattering of neutrons is of great value in studying superionics.   The value of this technique is nicely illustrated by the work of Hutchings et al[1] on the high temperature fluorite structured superionics, as will be illustrated in the final section of the chapter.

2.   Diffraction : Comparative Issues

The issues considered in this section relate first to the type of sample — single crystal or powder — and secondly to the type of radiation — X-ray or neutron.   For each type of radiation there are then choices to be made as to the source: in the case of X-ray studies there is the question of laboratory vs. synchrotron sources;   and for neutrons both reactor and pulsed sources are now available.

(2.1)   Powders vs. Single Crystals

Single crystal studies are the preferred diffraction technique when data of the highest quality are required.   Thus in studies of superionics, where we are endeavouring to determine accurately, low levels of defects, and where accurate parameters describing thermal motions are needed, single crystal experiments will where possible be performed.   Several elegant applications of the technique are given in the chapter of Wuensch, and in section (6) we summarise the results of single crystal studies of high temperature fluorites.   Three problems, however, arise with the use of single crystal methods: the first is the obvious question of sample availability;   single crystals of adequate size and quality can often not be grown.   Secondly, there may be difficulties with high temperature experiments.   Although several single crystal studies are reported for temperatures between room temperature and 1600 K, the type of furnace needed for very high temperature studies may be incompatible

167

with the 4-circle geometry of many single crystal instruments.   Thirdly, the <u>analysis of</u> <u>single crystal</u> data is commonly a lengthy procedure, and there is little prospect of these techniques becoming routine in studies of superionics.

Powder diffraction is inherently more limited in the quality of data and of the resulting crystal structure.   The scope of the method has, however, expanded enormously in the last 10–15 years.   One of the basic limitations of the technique is loss of information due to peak overlap, but this can be overcome to a large extent by using the technique of <u>profile refinement</u> developed originally by Rietveld[2].   The problem is illustrated schematically in fig (1) which illustrates the overlap of two peaks, which will commonly occur at high Bragg angles especially for structures with large low-symmetry unit-cells.   The intensities of these reflections will not be usable in structure refinement unless the components of the overlapping pattern can be separated.   Such separation is, however, possible given well defined line shapes.   The Rietveld technique takes advantage of the fact that peak shapes of Bragg reflections in neutron powder experiments, using reactor sources, are generally close to Gaussians;   and deviations from the Gaussian line shape can usually be adequately parameterized.   The variation of the line width i.e. the full width at half maximum, H, of the Gaussian peak is given by:

$$H = U \tan^2\theta + V \tan\theta + W \qquad\qquad (1)$$

Thus in Rietveld refinement the whole powder profile is fitted to the structual model and the parameters U, V and W are included as variable parameters.   Most work, reported to date, using this technique has employed neutron sources;   although we note that when pulsed sources are used (see below) more complex line shapes are observed.   The technique is beginning to be applied with X-rays, and, as discussed below, is proving to be particularly effective when synchrotron sources are used.   A good review of earlier work is given by Cheetham and Taylor[3], and recent applications are discussed by Cheetham and by Fitch in reference (4).   In section (4) we shall discuss a recent elegant study of Fitch and coworkers on proton conducting solids.

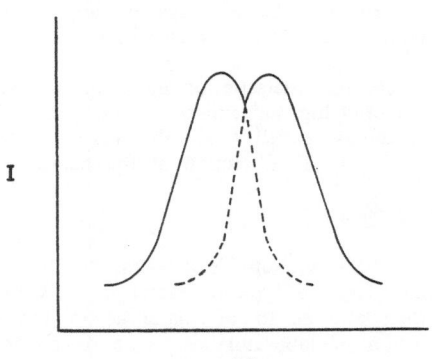

I

2θ

Figure 1.   Schematic illustration of pair of overlapping peaks in a powder diffraction pattern.   Dotted lines indicate component peaks.   Separation is possible given well defined peak shapes.

The problem of peak overlap can also be reduced by reducing the peak width. Narrow Bragg peaks can be achieved using neutron beams from pulsed sources and with synchrotron X-ray sources;   in the latter case, the high degree of collimation of the synchrotron beam allows very narrow line shapes and it is possible to achieve full widths at half maximum as low as 0.04˚ (in 2θ).   Data collected by McCusker and Glazer using the diffractometer on station (9.1) at the SRS Daresbury is illustrated in fig (2);   the material is silicalite – a porous, zeolite polymorph of SiO_2, with a complex crystal structure.   We note that most of the peaks are separated owing to the narrow line shape. In this case there is still sufficient overlap to require Rietveld refinement; but we can envisage for smaller unit cells, obtaining data in which the problem of overlap has been removed, allowing single crystal like refinement techniques to be used.

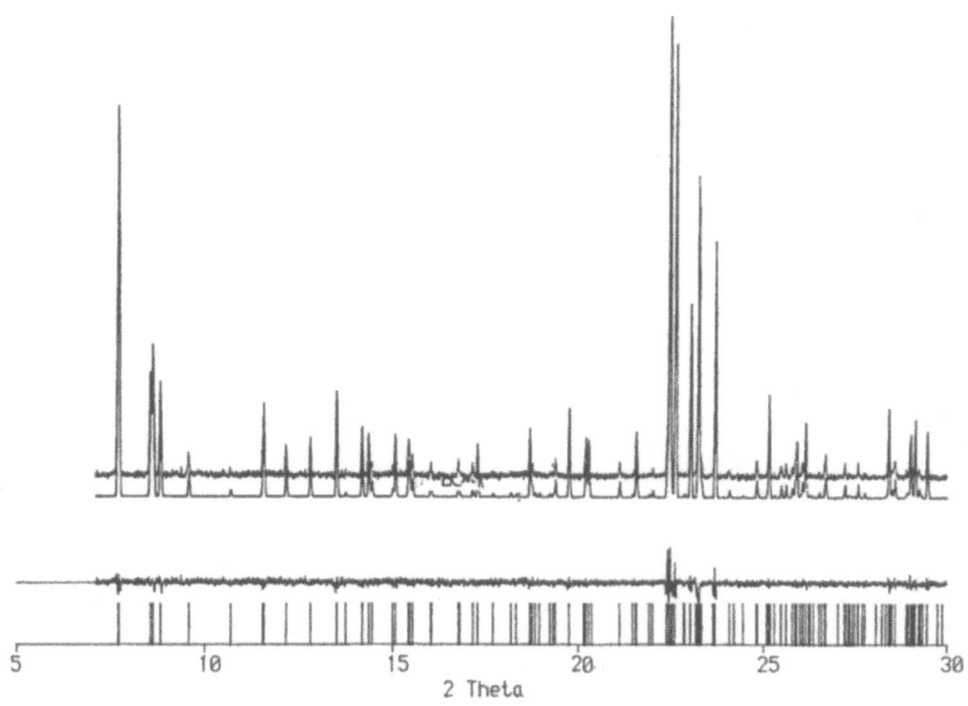

Figure 2. Powder diffraction pattern of silicalite. The pattern has been fitted using the Rietveld technique. Deviation between measured and fitted profiles are indicated in the lower part of the diagram.

Other problems with powder methods relate first to background subtraction; the procedure may be ambiguous when there is extensive peak overlap, although this difficulty is diminished by the availability of patterns with narrow line width discussed above. Secondly 'preferred orientation' i.e. the adoption of a non-random angular distribution of the crystallites in the powder sample will lead to errors. This problem is obviously more severe when samples are studied at high pressures, and when small samples are studied which is commonly the case in synchrotron powder diffraction. The effects can, however, partially be overcome if it is possible to spin the sample.

Finally we note that the Rietveld procedure used in analysis of powder data is a refinement rather than an ab-initio structure solving method. There has been some success by Cheetham and coworkers[5] in using 'direct methods' with powder data. But we would expect the use of such methods in powder studies to be limited in the near future. In practice this is not a major limitation in studies of solid electrolytes, for which we generally have reasonable approximate models and where our main need is usually to identify defect concentrations or the location of specific species e.g. protons or lithium ions. The work of Fitch and coworkers discussed in section (5) will illustrate the power of powder methods in these cases.

(2.2) X–Rays vs. Neutrons

X–rays are always the natural first choice in a diffraction experiment owing to the ready availability of laboratory X–ray sources and to the consi derably higher intensity of X–ray compared with neutron beams.   Neutrons have, however, a number of specific advantages in diffraction studies many of which are of particular relevance to the study of superionics.   These are as follows:

(i)   It is possible to detect light atoms in the presence of heavy ones.   Unlike the case for X–rays, the neutron scattering length of an atom does not vary systematically with atomic number.   Thus H atoms scatter neutron strongly (although due to the high incoherent scattering length of the proton, experiments with deuterated samples are preferred for diffraction studies), and neutron techniques are widely used in studies of the crystallography of hydrides and hydrates.   In addition lithium compounds are generally better studied by neutron methods and there are advantages in the studying of oxides and fluorides of heavy metals such as the lanthanides and actinides.

(ii)   Since neutrons are scattered by the nuclei, which have dimensions small compared with the neutron wavelengths used, there is no form factor for neutron scattering, i.e. no variation in the effective neutron scattering length with scattering angle.   This is not the case with X–rays, where since electrons effect the scattering, the scattering centres are of atomic rather than nuclear dimensions and consequently the X–ray form factor decreases markedly with the Bragg angle.   Neutron diffraction therefore yields more high angle data and can yield more highly resolved nuclear density maps.   Inversion of X–ray diffraction yields electron density maps, whose resolution into atomic components may be difficult. Neutron diffraction is therefore much better suited to the study of defective crystals where highly resolved nuclear density maps are needed.   A number of elegant illustrations of the use of neutron studies in identifying mobile ion distributions in superionics are given in the chapter of Wuensch in this book.

(iii)   Although high temperature X–ray studies are possible, it is generally easier to construct furnaces for use with neutron studies.   The need to use Be windows in X–ray equipment results in the construction of environmental cells being difficult.

(iv)   Traditionally, profile refinement of powder data has required neutron data owing to the well defined line shape of the latter.   However, X–ray Rietveld refinements are beginning to be reported, and with the increased availability of synchrotron powder instruments there will be a great growth in the use of X–ray Rietveld techniques.

Taken together, these factors have provided powerful incentives for the use of neutron diffraction in several studies of superionics.   Examples are given both in this chapter and in the chapter of Wuensch.

(2.3) X–Ray Sources : Laboratory vs. Synchrotron

Laboratory sources are again the natural first choice due to their greater accessibility and convenience.   Synchrotron radiation studies are, however, having an increasing impact on materials science studies.   Synchrotron radiation (SR) is emitted by a stored beam of electrons which have been accelerated to relativistic velocities and are constrained to a circular path by a set of powerful bending magnets, as shown schematically in fig (3).   The high energy electron beam throws off tangentially an intense white beam of electromagnetic radiation;   spectra for a variety of sources are shown in fig (4).   SR has the following unique features all of which can be exploited in materials studies;

(i)   Intensity, which is typically a factor of 10^4–10^5 greater than available with the best Laboratory sources (with rotating anode equipment).

(ii)   Tunability.   The SR spectrum ranges from the hard X–ray with wavelengths down to 0.1Å to the microwave region.   The spectrum can be changed by the use of 'insertion devices' e.g. 'Wiggler magnets' (see fig. 3) which alter the radius of curvature of the beam over a small region, and thereby increase the intensity of the beam at shorter wavelengths. And over the X–ray region of the spectrum, which is of greatest interest in the present context, high intensity is generally available at all required wavelengths.

(iii)   Collimation.   The SR beam is extremely well collimated in the plane of the synchrotron with typical divergence in the plane of the accelerator being ∼0.01˚.   In the present context, the most important application of this feature is that it leads to very narrow well defined line shapes in diffraction studies – a feature that has been exploited in

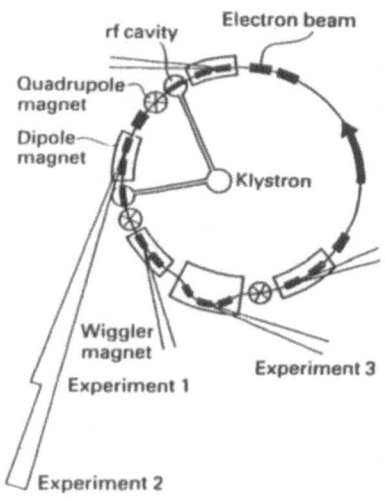

rf cavity

Electron beam

Quadrupole magnet

Dipole magnet

Klystron

Wiggler magnet

Experiment 1

Experiment 2

Experiment 3

Figure 3. A schematic representation of stored–beam synchrotron radiation source. Note that the energy lost by the emission of the radiation is replenished by radio frequency radiation produced by the central Klystron.

Rietveld refinement of SR powder diffraction data, and in the possibility of obtaining single crystal like data from powder experiments.

(iv)   <u>Polarisation</u>.   The SR is highly polarised in the plane of the synchrotron.   The feature can be exploited mainly in studies of magnetic structure.

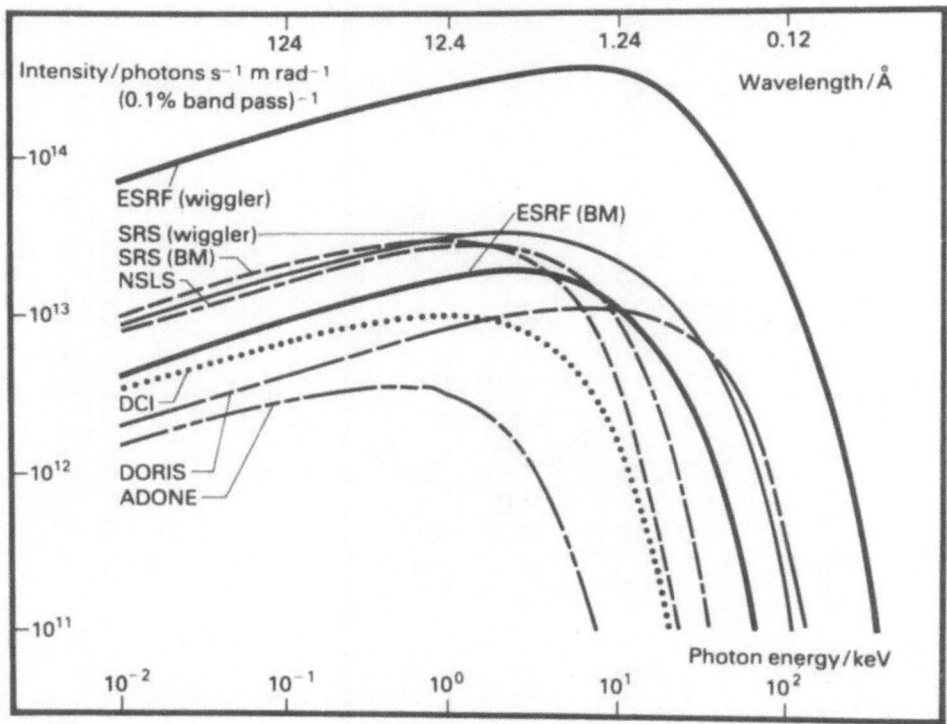

Figure 4. Electromagnetic spectrum from a variety of synchrotron radiation sources.

These unique features of SR give rise to several distinct advantages in diffraction studies.   First, the availability of a white beam opens up the opportunity of <u>energy dispersive diffraction</u>.   In the traditional angle dispersive diffraction, the Bragg condition: $n\lambda$ = 2d $\sin\theta$, is satisfied for different d spacings by varying θ, i.e. by scanning the detector over a range of angles;   in contrast, energy dispersive diffraction holds the detector arm fixed and the wavelengths of the scattered radiation are energy analysed, i.e. the Bragg equation is satisfied by varying λ: this can be achieved with a white X-ray beam and a solid state detector connected to a multi-channel analyser.   The resolution of the resulting data is lower than can be achieved with angle dispersive data collection, but the technique has major advantages in that it can be used relatively straightforwardly with special sample environments as in e.g. high pressure studies. Moreover, the need for only one detector arm setting is of great assistance in constructing environmental cells. Secondly rapid data collection is possible as the entire white beam of the synchrotron is being used simultaneously in the experiment, unlike the case with angle dispersive experiments where a monochromator must be used to select a narrow band of the synchrotron spectrum.

Rapid data collection is, however, possible in monochromatic work when a position sensitive detector (PSD) is used. Such a detector collects data over a wide angular range simultaneously. Again, there is a loss of resolution compared with the conventional scanning detector modes of data collection; but the resolution is superior to that obtained in energy dispersive experiments. Indeed the possibility of time resolved diffraction using both ED and monochromatic PSD techniques is one of the most exciting developments in current diffraction studies. The techniques open up the possibilities of making detailed structural studies of kinetic phenomena including intercalation, dehydration and other solid state reactions that are of direct importance in the field of solid state ionics.

A third important opportunity in the diffraction field that is opened up by synchrotron radiation is the effect of anomalous dispersion. The basis of this phenomenon is the change in the effective scattering power of an atom for X-rays that occurs close to X-ray absorption edges. This allows the determination of the contribution of individual atom types to the total structure factor. It is of particular importance in the study of amorphous materials where it enables the deconvolution of the total radial distribution function into partial terms arising from pairs of individual atom types. It is in this way very similar to the use of isotopic substitution techniques in neutron diffraction studies of liquid and amophous materials. There is clearly a considerable scope for the use of these methods in studying amorphous fast ion conductors.

The interested reader should consider references (6) and (7) for a general account of the nature and application of synchrotron radiation. We are confident that the techniques outlined above will be increasingly used in the materials science field in the next few years.

(2.4)   Neutron Sources : Reactor vs. Pulsed

Most neutron diffraction studies reported to date have used reactor sources, an excellent example being the High Flux Reactor at the Institut Laue-Langevin Grenoble, which was used for all of the neutron studies discussed subsequently in this chapter. Several research reactors are, however, available to the scientific community in sites distributed over the world. In recent years, however, a second type of neutron source has been developed. These are variously known as spallation or pulsed sources. The principle on which they work is again nuclear fission; but in this case the fission is produced by bombarding a heavy metal target with a high energy beam of protons. The proton beam is fired at the target in a series of pulses; and the resulting intense high energy beam of neutrons is also therefore pulsed. Spallation neutron sources are available currently at the ISIS facility of the SERC Rutherford and Appleton Laboratory, UK., and the Argonne and Los Alamos Laboratories in the USA. The main advantages of pulsed sources in the field of diffraction is that they have good intensities at low wavelengths giving increased data at short d-spacings and hence very highly resolved structures. Secondly, the white nature of the source together with the use of time-of-flight techniques in determinining the neutron energy, allows one to carry out energy dispersive diffraction experiments, but at high resolution. Indeed, the diffractometer HRPD on the ISIS facility is capable of yielding very high resolution, high quality powder diffraction data with data collection times of only a few hours. The chapter of David in reference (4) gives a good account of this instrument and of the field of high resolution powder diffractometry using pulsed neutron sources. A more general account of the use of these sources is given in reference (8).

3. Analysis of Diffraction Data : Special Problem for Superionics

The chapter of Wuensch in this book has addressed most of the central issues concerning data refinement in superionics. Refinement of crystallographic data invariably has two distinct stages: first, determination of the unit cell dimensions and symmetry, characterized by the space group; this information is obtained from the angles and systematic relationships between the Bragg peaks. For an unknown structure this process can be lengthy and laborious, although a number of automated indexing procedures are now available[9]. Secondly, the intensity data are analysed to yield the positions of the atoms in the unit cell. In this exercise we make use of the basic relationship between the intensity I_{hkl} and the structure factor f_{hkl}, of the reflection described by the Miller indices h, k and l:

$$I_{(hkl)} = |f_{(hkl)}|^2 \tag{2}$$

where the structure factor in turn is given by:

$$f_{(hkl)} = \int_0^V \rho(r)\, e^{-i\,\underline{G}_{hkl}\underline{r}}\, dr \tag{3}$$

in which $\rho(r)$ represents the distribution of scattering density over the unit cell; G_{hkl} is the reciprocal lattice vector corresponding to the $hjkl^{th}$ application, V is the volume of the unit cell over which the integration is taken. Equation (3) can be Fourier inverted to give the expression for the scattering density as a Fourier sum:

$$\rho(r) = \frac{1}{V} \sum_{\underline{G}_{hkl}} f_{(hkl)}\, e^{i\,\underline{G}_{hkl}\,\underline{r}} \tag{4}$$

One of the central problems in crystallography arises from the relationship between equations (2) and (4). The square root of the intensity of a Bragg reflection gives the modulus of the structure factor. The phase, which is required for use in Fourier syntheses is lost; use of equation (4) requires therefore some information on structures in order to fix the phases.

Refinements using diffraction data normally proceed by two procedures which may be used together:

(i) Least Squares Fitting in which a structural model containing a number of variable parameters is proposed: intensities are calculated using equations (2) and (3) and the structural parameters are varied via a least squares fitting procedure until the best agreement is obtained between calculated and experimental intensities. Assuming that cell dimensions have been determined from the positions of the Bragg peaks, the parameters to be fitted concern the atomic parameters, the coordinates of the atoms in the unit cell, the occupation numbers of each atomic site, which may deviate from unity in disordered crystals, and the thermal parameters which describe the distribution of scattering density around each atomic site due to atomic motion. The latter need careful attention in superionics as discussed in the chapter of Wuensch. Lattice vibrations in these systems have high amplitudes and are commonly anharmonic and anisotropic, in contrast to the harmonic isotropic motion that is commonly assumed in least squares refinements of crystallographic data. Wuensch elsewhere in this book has described the type of temperature factors that need to be used in refining data on superionics. Here we draw attention to the study of this problem reported by Kuhs[10], who describes the use of the very generalised expressions for the probability density functions $P(\underline{u})$ as a function of the displacement vector $\underline{u}$ around the lattice site.

$$P(\underline{u}) = P_H(u) \left[1 + \frac{1}{3!}\, c^{jkl}\, H(\underline{u})_{jkl} + \frac{1}{4!}\, c^{jklm} H(\underline{u})_{jklm} \right.$$

$$\left. + \frac{1}{5!}\, c^{jklmn}\, H(\underline{u})_{jklmn} + \frac{1}{6!}\, c^{jklmnp}\, H(\underline{u})_{jklmnp} \right] \tag{5}$$

where $P_H(\underline{u})$ is the simple Harmonic (i.e. Gaussian) function, where the $H(u)_{jkl}$ are the Hermite polynormals which are tabulated in the literature up to 6th order[11][12]. The $c^{jkl...}$ are known quasimoments[13]; and the expression may be recast in terms of alternative coefficients, $K^{jkl...}$ known as cumulants; hence the term cumulant expansion is often used for this type of expression. Equation (5) which goes to 6th order in the Hermite polynomial provides a highly flexible description of the distribution of scattering density about a site.

Kuhs[10] and coworkers examined a number of systems. Of particular interest is the study of $RbAg_4I_5$ where they demonstrated a very anisotropic distribution of Ag^+ ions. Fig. (5) shows the results of their refinement of the data using probability density functions for the Ag^+ ion of the type given in equation (5) but taken to 5th order. The pronounced anisotropy of the distribution is evident and the ability of the cumulant expansion approach to handle such complexities is demonstrated.

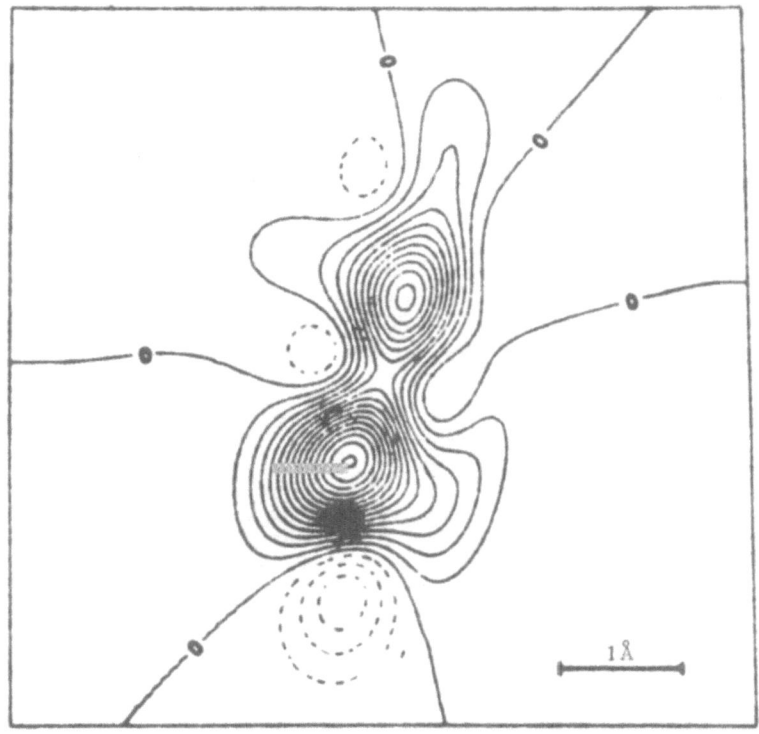

Figure 5. Distribution function for Ag^+ in $RbAg_4I_5$ (see reference 10).

Refinement of occupation numbers can also present serious difficulties as these are often highly correlated with the temperature factors. A good case study of this problem is produced by the high temperature fluorites which will be discussed in section (6) (and were discussed in our previous chapter). We recall that these systems show anion disorder at high temperature. Different workers have, however, produced, from diffraction studies, widely different estimates of the occupation numbers of lattice and interstitial sites in superionics. The differences are not due to differences in the data obtained, but to the treatment of the thermal parameters which in turn are reflected in the occupation numbers.

(ii)   Fourier Inversion

In view of the difficulties discussed above in determining thermal parameters and occupation numbers from least squares, fitting techniques, it is tempting to conclude that Fourier inversion of the data using equation (4) is the more appropriate approach for superionics. This will reveal directly the scattering density distribution. Moreover, the phase problem is commonly not a difficulty with superionics as sufficient of the structure (e.g. the immobile sublattice) is well known to fix all or most of the phases of the structure factors. Given knowledge of the phase angles there remain two main difficulties with Fourier techniques: the first is the need for highly accurate intensity data especially when small features of Fourier maps are to be interpreted in terms of defects. The second is the question of termination errors. A 'complete' Fourier sum will only be obtained if all Bragg intensities of appreciable intensity are included. Those omitted will normally be the low d spacing reflections beyond the Bragg cut–off corresponding to the wavelength of the radiation used. The effect of omitting these terms is at best to reduce the resolution of the structure revealed by the Fourier synthesis, and in some cases to give misleading and inaccurate features in the structure. In obtaining accurate Fourier maps it is clearly therefore desirable to use where possible short wavelength radiation to increase the number of terms in the synthesis. This factor favours the use of short wavelength neutron beams obtained from a hot source on a reactor or synchrotron radiation in the case of X–rays due to the high intensity available at short X–ray wavelengths.

In practice, many solutions of crystal structures from diffraction data make use of both least squares fitting and Fourier techniques. Examples will be given in section (5).

4. Diffuse Elastic Scattering

The Bragg condition for diffraction is derived by assuming that the contents of all unit cells in a crystal are identical. In a disordered crystal this condition does not hold, which is manifested by the appearance of scattering intensity, above the background level, between the Bragg peaks. The theory of diffuse scattering can be understood simply as follows. The differential scattering cross section ($d\sigma/d\Omega$) per unit solid angle per unit incident intensity of radiation is given by:

$$\frac{d\sigma}{d\Omega} = \sum_{m}^{N} \sum_{n=1}^{N} b_m b_n \exp[i\ \underline{Q}\ (\underline{R}_m - \underline{R}_n)] \tag{6}$$

where b_m and b_n are the scattering lengths of atoms m and n. The double summations are over all atoms in the unit cell. R_m and R_n are the vectors from an arbitrary constant origin to atoms m and n, and $\underline{Q}$ is the scattering vector: $\underline{Q} = \underline{K}_s - \underline{K}_i$ where $\underline{K}_s$ is the scattered and $\underline{K}_i$ is the incident wave vector. For a perfectly periodic crystal the Bragg condition $\underline{Q} = \underline{G}$ holds for scattering of appreciable intensity. But for a disordered crystal the condition is relaxed and scattering is observed for other values of Q.

Following the discussion of diffuse scattering given by Fender[14] we can recast equation (6) as follows:

$$\frac{d\sigma}{d\Omega} = \sum_{i}^{\nu} \sum_{j}^{\nu} \sum_{m}^{N} \sum_{n=1}^{N} \langle b_m^i \rangle \langle b_n^j \rangle \exp[i\underline{Q}\ (\underline{R}_m - \underline{R}_n)]\ +$$

$$\sum_{i}^{\nu} \sum_{j}^{\nu} \sum_{m}^{N} \sum_{n=1}^{N} (b_m^i b_n^j - \langle b_m^i \rangle \langle b_n^j \rangle) \exp[i\ \underline{Q}\ (\underline{R}_m - \underline{R}_n)] \tag{7}$$

in which we now sum over all atoms types ν, in each unit cell with N being the total number of unit cells. We have also introduced the terms $\langle b_m^i \rangle$ and $\langle b_n^j \rangle$ which represent the average scattering at a particular site in the unit cell. Of course in a perfectly ordered crystal the 2nd term in equation (7) will be zero as $b_m^i b_n^j = \langle b_m^i \rangle \langle b_n^j \rangle$. Only the 1st term remains, which of course gives rise to the Bragg condition. However, in disordered crystals, $b_m^i b_n^j \neq \langle b_m^i \rangle \langle b_n^j \rangle$ and the magnitude of the difference varies from cell to cell. The scattering arising from this term does not require that $\underline{Q} = \underline{G}$ and is manifested as the diffuse scattering between the Bragg peaks.

Extensive diffuse scattering is common in superionics as found in the study by Fender and Wright[15] of superionic AgI. The highly disordered nature of this crystal structure was manifested by a large diffuse background to the Bragg peaks. The modulated nature of the diffuse scattering was indicative of extensive short range order: a flat diffuse background would indicate a random distribution of defects.

Quantitative analysis of diffuse scattering can be achieved by fitting data to short range order models; alternatively Fourier transformation will yield defect-defect correlation functions. However, in solid state ionic systems, quantitative studies are difficult. The first difficulty arises from the fact that atomic vibrations give rise to a diffuse intensity known as thermal diffuse scattering: at any one instant thermal vibrations result in a deviation from perfect periodicity – the condition for observation of diffuse scattering. In superionics this effect may be large owing to the high amplitude of the thermal motions. The subtraction of this effect from that of a static defect distribution may be difficult. The second problem arising in neutron scattering is that certain nuclei (eg. the proton) give rise to extensive incoherent scattering which is again manifested as background and whose separation from the contribution due to diffuse scattering may be difficult. A third difficulty is that Bragg peaks will commonly be superimposed on the diffuse background. A partial solution to this problem is to use long wavelength radiation. In the following section we shall show how data collected on the cold neutron source at the Institut Laue-Langevin has proved to be of value in studies of superionics.

5. **Elastic Scattering : Examples**

On examining the proceedings of conferences held on fast ion conductors during the last ten years [16][17][18][19][20] the reader will find many examples of the use of X-ray and neutron diffraction in crystallographic studies and of the use of diffuse scattering in elucidating the nature of short range order. In addition, the chapter by Wuensch in this volume beautifully illustrates the role of high resolution single crystal studies and also of powder diffraction in determining the details of the mobile ion distribution in cation conducting superionics. In this section we will present results of two recent studies of topical superionics: the first will be our own studies of the high temperature, bismuth oxide oxygen-ion conductors; the second will concern the study of Fitch and coworkers on the isostructural proton conductors Hydrogen Uranyl Phosphate Tetrahydrate (HUP) and Hydrogen Uranyl Arsenate Tetrahydrate (HUAS).

(5.1) **The δ-Bi$_2$O$_3$ Oxygen Ion Conductors**

Bi_2O_3 has a complex crystal chemistry. There are 4 known phases of which the highest, the δ-phase, is stable above $740\,^{\circ}C$. On transformation to the δ phase the material becomes an exceptionally good oxygen ion conductor. Indeed, within the stability region of the δ-phase, the material shows probably the highest measured conductivity due to oxygen ion transport. The δ-phase can be stabilised by dopants e.g. Y_2O_3 and rare earth oxides. Unfortunately, however, although extending the stability field of the superionic phase to lower temperatures, the dopants lower the conductivity.

δ-Bi$_2$O$_3$ has an intriguing structure: it is based on the f.c.c. cation sublattice adopted by the fluorite structure; and the available crystallographic data indicates that the oxygen ion distribution is compatible with the Fm3m space group of the fluorite structure. But the oxygen/metal ratio of the compound is incompatible with a regular fluorite structured compound. In the simplest models the oxygen deficiency would be accommodated simply in terms of oxygen ions missing from regular lattice sites. In fact the structure is more complex with oxygen ion interstitials in addition to the expected vacancies.

Following earlier work of Harwig[21] there have been detailed powder neutron diffraction studies by Battle et al.[22][23]. This work examined the structure of pure and isovalent doped δ-Bi$_2$O$_3$. The diffraction pattern for the 40% Y_2O_3 doped material at room temperature is shown in fig. (6). The rapid decrease in intensity with 2θ should be noted as it is indicative of a high level of disorder. Analysis of the data (see table 1) reveals two types of oxygen interstitial displaced from the regular cubic interstitial site of the fluorite structure along both <110> and (111> directions; the concentrations of the latter species decrease with dopant concentration. In addition small <100> displacements of the cations in the doped crystals are observed. The results obtained for Er and Yb as dopants are very similar to these for the yttrium doped system.

There is as yet no clear interpretation of these results. Since the pure material has the highest conductivity and the highest concentration of <111> interstitials, it is likely that the latter play an import role in the conduction mechanism. The nature of this mechanism is, however, at present unknown.

Considerable insight into the structural nature of these complex disordered materials is provided by diffuse scattering studies undertaken by Battle et al[23] using the long wavelength neutron diffractometer, D7, at the Institut Laue–Langevin. Data are shown in figs. (7) and (8) for pure and 40% Y_2O_3 doped Bi_2O_3; similar results were obtained for

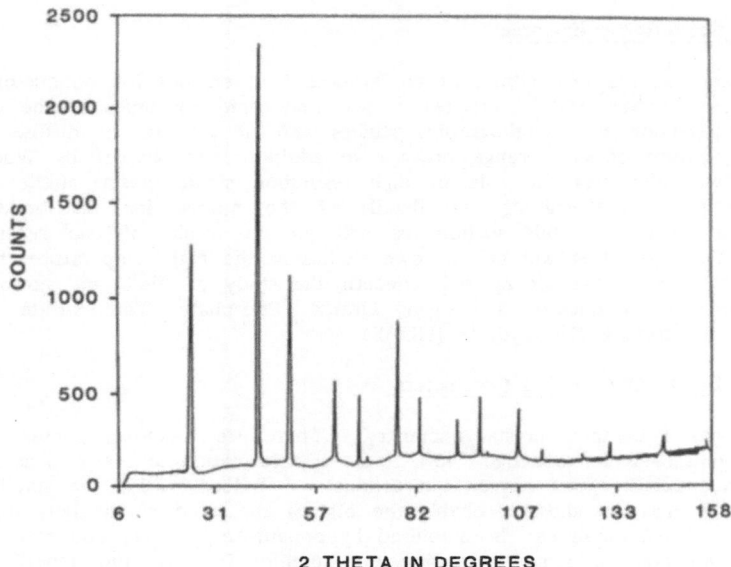

Figure 6. Neutron diffraction pattern of 40% Y_2O_3 doped Bi_2O_3 at room temperature.

the Er doped material ions. The data reveal little structure in the case of the pure material indicating little short range order. In contrast the doped material shows very pronounced structure which persists to high temperatures. Battle et al.[23] rationalise this behaviour in terms of the ordering by the dopants of the surrounding vacancy distribution. In particular they suggest the formation of <111> chains of oxygen vacancies around Y (see fig. 9). The pronounced structure of the diffuse scattering would suggest the existence of domains of such ordered regions extending over several unit cells. Further details of these models together with a discussion of the contrasting behaviour of Yb compared with Y and Er doped samples is given in references (22) and (23). The work demonstrates the value of Bragg and diffuse neutron scattering in revealing complex structural behaviour in apparently simple materials.

Table 1

Structural Parameters for $(Bi_2O_3)_{1-x}(Y_2O_3)_x$ as a function of concentration and temperature (after reference 22).

	Pure Bi_2O_3	27% Y_2O_3	27% Y_2O_3	34% Y_2O_3	40% Y_2O_3
Temperature (K)	1023	1023	298	298	298
Lattice parameter a (Å)	5.6485(6)	5.5466(2)	5.4784(2)	5.4650(1)	5.4469(1
Cation $(24e,x,0,0)$					
x	0.0	0.065(1)	0.048(4)	0.042(5)	0.046(3)
B ($Å^2$)	7.3(1)	0.3(2)	1.7(3)	2.1(4)	1.7(3)
Oxygen $(8c,\frac{1}{4},\frac{1}{2},\frac{1}{4})$					
B ($Å^2$)	11.8(7)	5.3(1)	5.3(2)	4.6(2)	4.6(2)
Occ. No.	1.72(9)	1.84(2)	2.01(6)	2.06(7)	2.33(6)
Oxygen $(32f,x,x,x)$					
x	0.354(3)	0.321(1)	0.319(3)	0.314(4)	0.320(4)
B($Å^2$)	9.9(9)	4.8(4)	5.3(2)	4.6(2)	4.6(2)
Occ. No.	1.28(9)	1.00(2)	0.83(6)	0.73(8)	0.55(6)
Oxygen $(48I,\frac{1}{2},v,v)$					
v	–	0.390(8)	0.316(10)	0.316(14)	0.288(23)
B ($Å^2$)	–	4.8(4)	5.3(2)	4.6(2)	4.6(2)
Occ. No.	–	0.16(2)	0.16(6)	0.20(11)	0.12(8)
R_I%	2.2	5.0	1.7	1.9	1.6

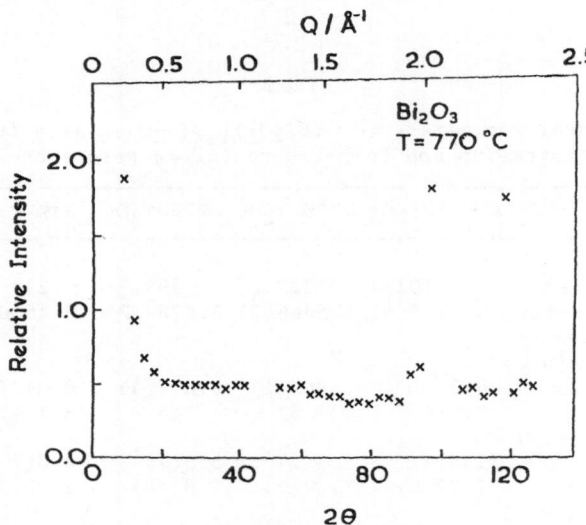

Figure 7. Diffuse neutron scattering for pure Bi_2O_3

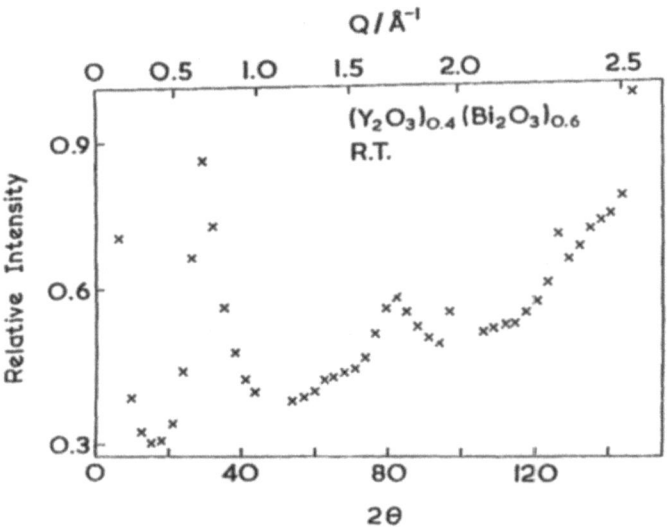

Figure 8. Diffuse scattering for 40% Y_2O_3 doped Bi_2O_3 at room temperature.

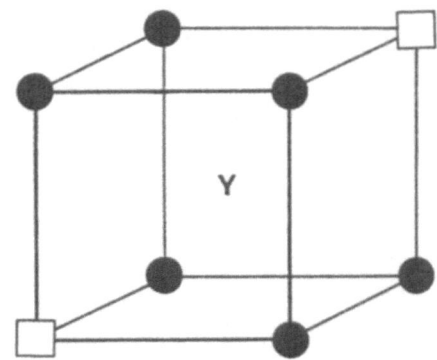

Figure 9. Postulated <111> distribution of vacancies in the Y coordination shell of Y doped Bi_2O_3.

(5.2) HUP and HUAs Proton Conductors

We describe here the detailed study of Fitch and coworkers on these intriguing superionic materials. The account we present is taken from the recent review of Fitch[24]. The isostructural layered hydrates $HUO_2AsO_4H_2O$(HUAS) and $HUO_2PO_4.4H_2O$ (HUP) have attracted attention owing to their promising proton conductivity at room temperature[25][26], and high hydrogen–ion mobility as demonstrated by solid–state NMR[27][28] and quasielastic neutron–scattering studies[29]. The structure of HUP was solved by single–crystal X–ray diffraction, which located the positions of all atoms except hydrogen[30][31]. HUP is tetragonal, P4/ncc, with a = 6.995(2), c = 17.491(4) Å.

The structure consists of layers composed from uranyl and phosphate ions, separated by layers composed of networks of hydrogen–bonded water molecules, as shown in figure 10.

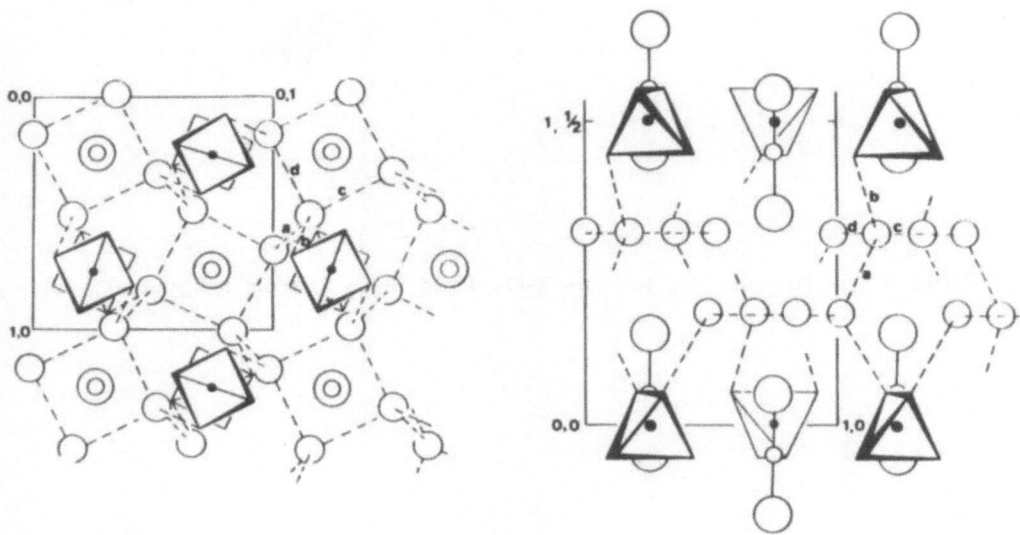

Figure 10. (left) Structure of HUP projected along (001); (right) along (100). Large open circles, uranyl oxygen; smaller open circles, water oxygen; small solid circles, P at the centre of the tetrahedral phosphate groups; small open circles, U. The water oxygens participate in three types of hydrogen bond; bridging bonds between squares, (a or H(3)); to the phosphate groups, (b or H(2)); and within the squares (c,d or H(1)) (From reference 31).

Each unit cell contains two layers of uranyl and phosphate ions, centered about $z = 0$ and $z = 1/2$, and two layers of water molecules, centered about $z = 1/4$ and $z = 3/4$. The uranyl ion is linear and lies parallel with the c–axis of the unit cell. The U atom is co–ordinated equatorially by four O atoms in a plane, each from a different phosphate group. Every oxygen of a phosphate group is hydrogen bonded by a hydrogen ion, H(2), to the oxygen of a water molecule. The water molecules are grouped into squares of side 2.81(2) Å. These squares, for the lower water layer in the unit cell, alternate about $t = \frac{1}{4}$ at $z = 0.1891(6)$ and $z = 0.3109(6)$. For the upper water layer in the unit cell, the water squares alternate about $z = 3/4$, at $z = 0.6891(6)$ and $z = 0.8109(6)$. All the water–molecule oxygens are related to each other by the symmetry operations of the tetragonal space group.

Not only are the oxygen atoms of the water molecules hydrogen–bonded to the oxygen atoms of the phosphate groups, they are also hydrogen–bonded to each other, via hydrogen ions both within the squares, H(1), and between squares at different values of z in the unit cell, H(3). These latter bridging hydrogen bonds represent the shortest oxygen–oxygen distance between water molecule oxygens, with an average distance of 2.56 Å. Within the unit cell there is a total of forty hydrogen–bond sites, but with four formula units per cell, only 36 hydrogen ions to fill them. It might be assumed that the observed high hydrogen–ion mobility should be related to the distribution of the 36 hydrogen ions over the 40 available sites.

A powder neutron diffraction pattern of deuterated HUAs was taken on the powder diffractometer DIA at the Institut Laue–Langevin, DIA at a wavelength of 1.909 A[32][33]. Early refinements confirmed the very close similarity between the layers of uranyl and phosphate/arsenate ions and the arrangement of the water molecules in HUP and HUAs. Fourier sections, (figure 11), were computed through the various hydrogen bonds, and the occupancies of the sites refined. These indicated:

1)   The hydrogen bonds to the arsenate oxygen atoms are fully occupied with 16 hydrogen ions per unit cell. The hydrogen ion is closely bonded to the oxygen atom of the water molecule, in accordance with IR data[34].

2)   The in–square hydrogens, H(1), are on average 3/4 occupied with 12 hydrogen ions distributed over the 16 sites per unit cell. There is only one apparent position for the H(1) ion between the water molecules, whereas earlier work had assumed that two nearly–equivalent sites might occur owing to the similar nature of the chemical environments[35].

3)   The bridging hydrogen sites, H(3), are fully occupied with 8 hydrogen ions per unit cell. The Fourier map, figure 11, reveals that these hydrogen ions are associated with a single maximum in scattering density on the two–fold axis midway between the two water oxygens. There is, however, a large anisotropy in the scattering density along the O–O direction.

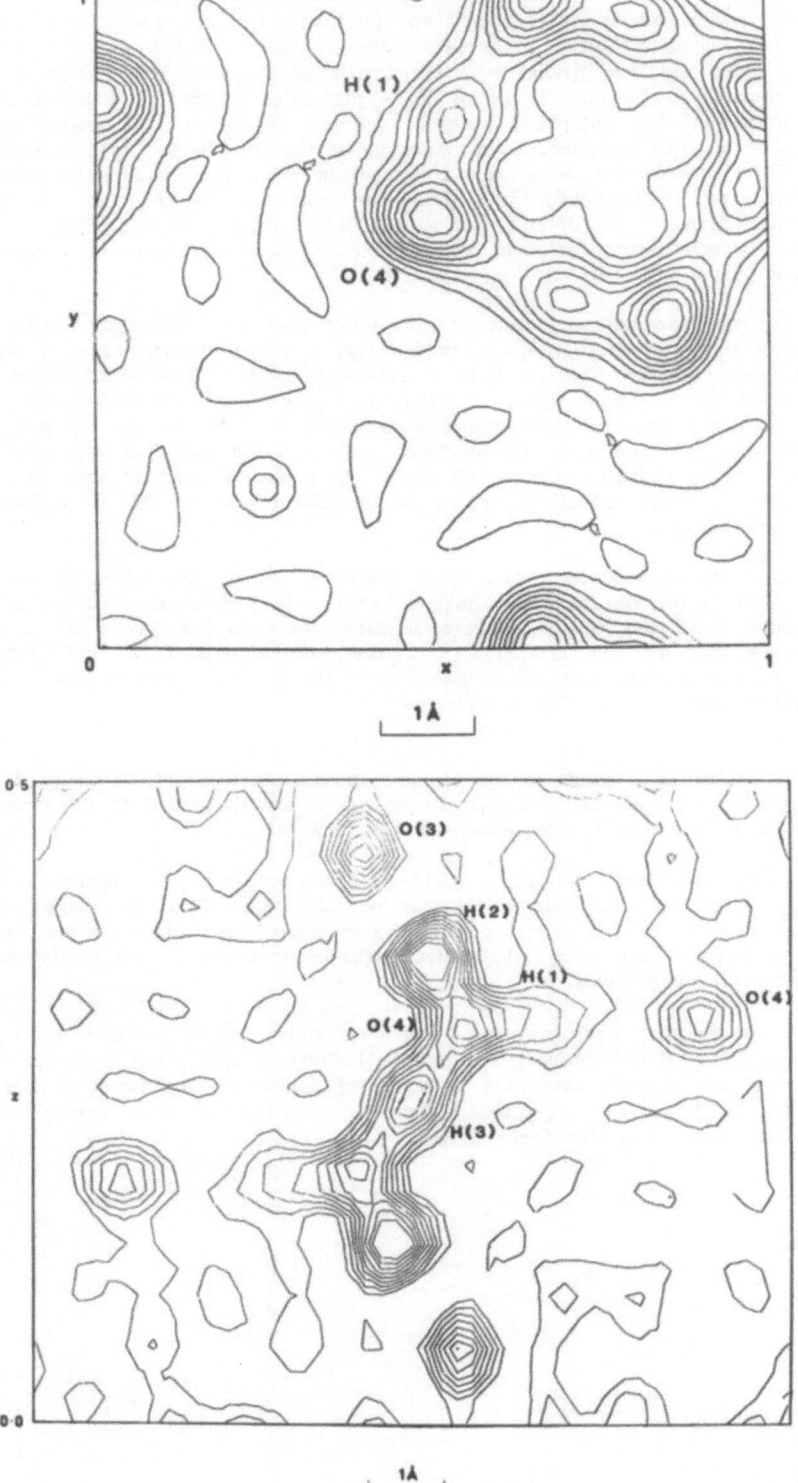

Figure 11. Fourier maps through the hydrogen bonds in HUAs; (a) through the water squares (top), and (b) through the bridging hydrogen bond (bottom).

Figure 11b shows the bridging hydrogen bond, which represents the shortest average O–O separation in the structure, (2.57(1) Å in deuterated HUAs), and suggests the presence of $H_5O_2^+$-type species. However, with the in-square hydrogen sites only 3/4 occupied, there is a probability of 1/2 that any one or other of the two in-square hydrogens associated with a pair of bridged water molecules will be vacant. In this case we have a water-molecule dimer H_4O_2. Thus the observed distribution of site occupancies suggests the presence of equal numbers of $H_5O_2^+$-type ions and water-molecule dimers H_4O_2. The observed scattering density of figure 2 represents the superposition of these species, and the O–O separation of 2.57(1) Å is the <u>average</u> O–O separation for these units. For the water-molecule dimers, the bridging hydrogen must be displaced away from the central position and is directly bonded to one or other of the water oxygens, resulting in the anisotropy in the H(3) density along the O–O direction. For the $H_5O_2^+$ species, it is possible that the bridging hydrogen is displaced away from the central position so that it is better described as a species such as $(H_3O^+.H_2O)$, or the H(3) ion may be in the central position, creating a true $H_5O_2^+$ ion. For the former we would expect the actual O–O separation to be about 2.46 Å, and for the latter about 2.40 Å. For the H_4O_2 units the O–O separation would be much longer, owing to the reduced polarity of H_2O compared to H_3O^+. Further evidence supporting the notion that the observed bridging distance is an average distance is found in the water-oxygen temperature factors, which are large, even in the X-ray study on HUP. This suggests the smearing out of the oxygen scattering-density, which would be expected if the observed position were the average of more than one individual position.

From this simple picture of the presence of equal numbers of water-molecule dimers and $H_5O_2^+$-type species, a basic mechanism for hydrogen-ion mobility can be proposed[32], involving proton jumps between filled and unfilled hydrogen-ion sites H(1) within the squares. This causes the interconversion of the H_4O_2 and $H_5O_2^+$-type species, and is followed by the coupled rotation of the two water molecules of a water-molecule dimer about their hydrogen bonds to the arsenate ions, thus transferring a proton from one square to another, (see figure 3 of reference (32)). To explain the proton conductivity, more-elaborate schemes have been suggested[37], along with alternative mechanisms involving migration of H_3O^+ ions[38], or conduction by surface water[39][40].

Low Temperature Phase

Both HUP and HUAs undergo a transition, the former around 274 K and the latter around 301 K, to a poorly-conducting low-temperature phase[34][41]. For HUAs[42] the transformation corresponds to a change from paraelectric to antiferroelectric behaviour. Whether HUP similarly becomes antiferroelectric has yet to be confirmed. NMR studies indicate a reduction in hydrogen-ion mobility, anad DSC gives an enthalpy of transition of less than 0.5 kJ per mole of water. Single crystals break up into domains with domain walls parallel to {110} and subdomain walls parallel to {010} of the original tetragonal unit cell[42][43]. Powder X-ray diffraction reveals the splitting of many of the peaks and a possible structure in the orthorhombic space group $P2_12_12$ was suggested[44]. This, however, was not compatible with the neutron powder diffraction pattern of deuterated HUAs taken at 4 K, owing to the presence of peaks corresponding to 100 or 010 neither of which is allowed in this spacegroup[45].

A further transition to a ferroelectric phase was reported at 253 K for HUAs by de Benyacar and de Dussel when an electric field was applied parallel with the c-axis of the room-temperature phase[42][43]. From detailed analysis of the domain and subdomain structure of crystals of HUAs below the transition, they suggested that for the antiferroelectric phase the point group is 1, 1, 2 or 222, with point group 1 for the additional ferroelectric phase. In the absence of the electric field, however, the transition to the ferroelectric phase could not be detected by powder X-ray diffraction[34] DSC[34], DTA[41], or powder neutron diffraction[45].

The physical characteristics of the transition clearly suggest that ordering of the protonic species is occurring, and a variety of models were tested in a number of spacegroups which are subgroups of the original disordered structure in P4/ncc. An additional problem in the refinement was the presence of 41 peaks from deuterated ice overlapping with the peaks of the main pattern. This ice resulted from a sample which had been kept slightly wet to ensure that the room–temperature structural determination was of the fully–hydrated product. Use was therefore made of the modified Rietveld procedure of Thomas & Bendall[(46)(47)] which allows the refinement of a diffraction profile containing overlapping contributions from more than one phase. By including a component to the profile from deuterated ice, whose structure is well known, these additional peaks were very successfully accounted for.

Only limited success was at first achieved using a variety of space groups, although the basic nature of the ordering was revealed. Refinements with even a large number of parameters still produced poor fits, and problems with the convergence and stability of the refinements were encountered. Non–physical temperature factors were also a problem, (e.g. in $P2_1/c$, with 72 atomic parameters, R_{wp}=16.1%; in $P2_1$ or Pc, with 135 atomic parameters, R_{wp}=12.8%, unstable refinements). A reduction in symmetry to the triclinic space group $P\bar{1}$ caused a dramatic improvement in the fit. With 135 atomic parameters (and a total of 161 variables in the refinement) the 2091 reflections could all be refined simultaneously and converged, with sensible temperature factor, to R_{wp}=4.2%, R_I=2.6%, R_E=3.0%. The fit is shown in figure 12.

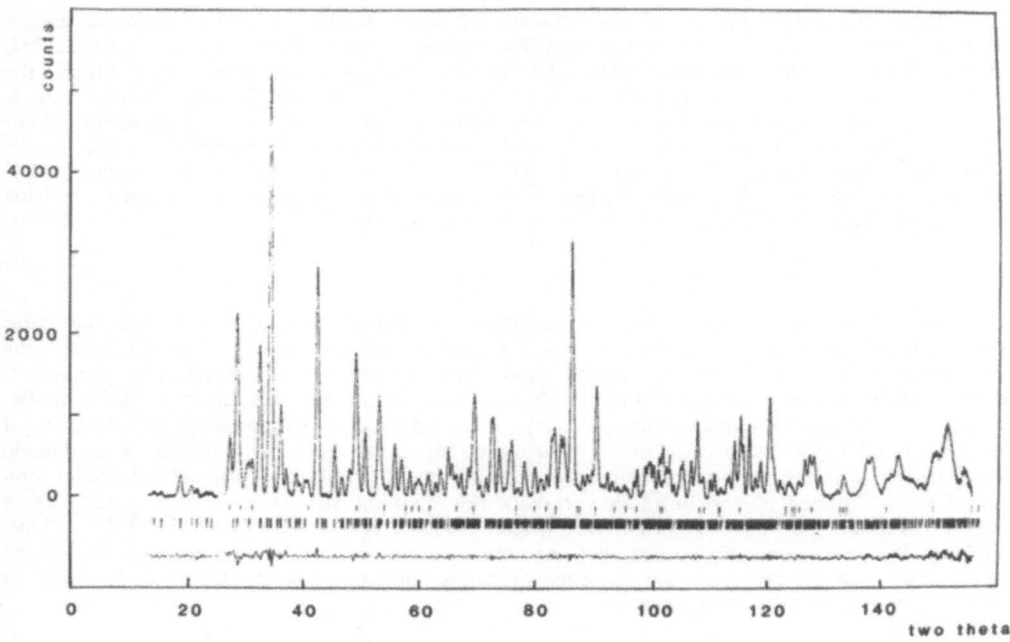

Figure 12. Observed (points), calculated (full curve), and difference profiles for deuterated hydrogen uranyl arsenate tetrahydrate at 4 K, in space group $P\bar{1}$. The positions of the contributing reflections are shown for deuterated HUAs (lower) and the contaminating ice (upper).

In the low–temperature structure, the layers of uranyl and arsenate ions are only slightly modified as compared with room temperature. In the water layers, the water–molecule species have ordered and formed themselves into hydrogen–bonded chains. Figure 13 shows the ordering of the hydrogen bonds within the water squared, and figure 14 shows the arrangement of the chains, in the lower half of the centrosymmetric unit cell.

186

One chain, composed of two types of water–molecule dimer, runs in the direction of the crystallographic a–axis, and is cross–linked by two sets of chains running in the b–direction, composed from $H_5O_2^+$–type species. For the first of these latter species, the O–O

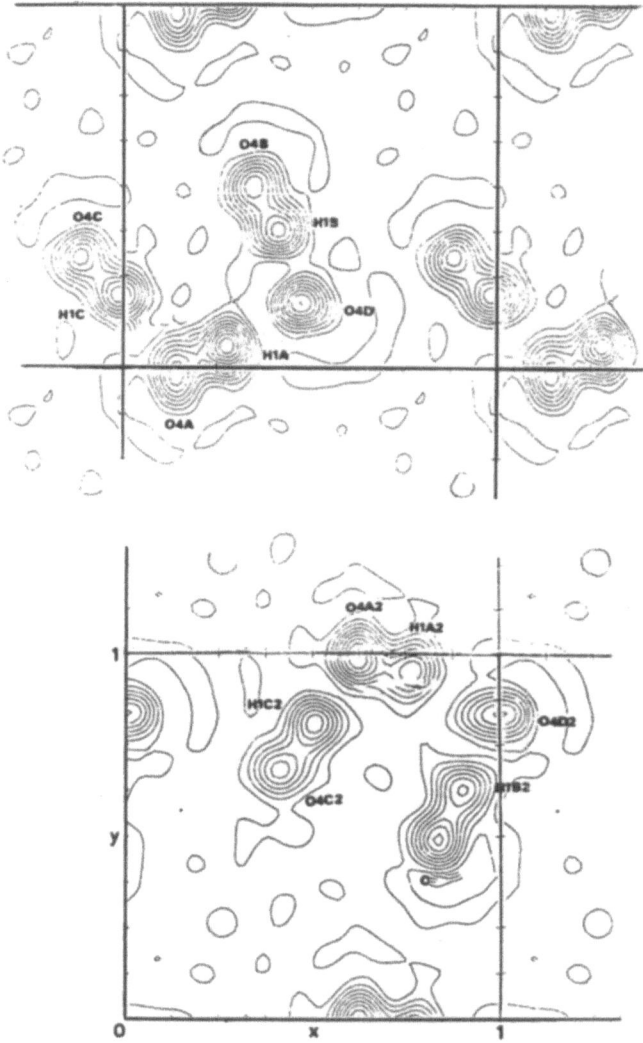

Figure 13. Fourier maps through the plane of the water squares at $z = 0.19$ (top) and $z = 0.31$ (bottom). The ordering of the three hydrogens over the four possible sites is clearly illustrated.

separation is 2.47(2) A, and the linking hydrogen ion is closely associated with one of the two water–molecule oxygen atoms. This is a species of the type $(H_3O^+H_2O)$. For the second cross–linking species, the O–O distance is 2.38(2) A and the bridging hydrogen ion is almost centrally situated between the two oxygen atoms, i.e. a true $H_5O_2^+$ unit.

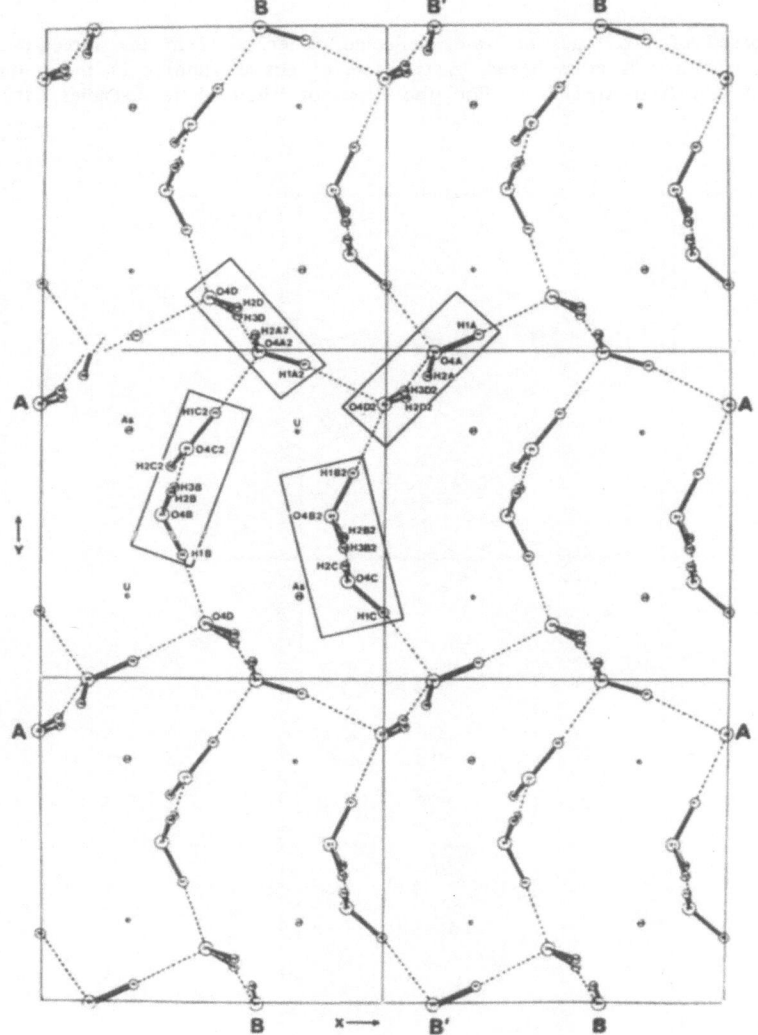

Figure 14. The hydrogen-bonded water-molecule chains that form in the low-temperature phase of deuterated HUAs. The chain running in the a direction is composed of water-molecule dimers. These are cross-linked by two chains in the b direction composed of $H_5O_2^+$-type species, (one an ($H_3O^+.H_2O$) and the other a true $H_5O_2^+$ ion).

The ordering of the hydrogen ions in this structure accounts for the decrease in hydrogen mobility at the transition, and the distortions to the unit cell from the original disordered tetragonal structure are consistent with the domain structure formed by single crystals through the transition. In addition, the structure is compatible with the observed antiferroelectric behaviour of this material. Thus the net polarization produced in a water layer by the ordering of the $H_5O_2^+$-type species below the transition is matched by an equal and opposite polarization in the next water layer owing to the centrosymmetric nature of the unit cell. An electrically-induced, small distortion could produce a non-centrosymmetric, ferroelectric structure, as reported by de Benyacar and de Dussel[42][43].

For the low-temperature antiferroelectric phase of HUAs, the ordering of the protonic species reveals the presence of two types of water-molecule dimer and two types of $H_5O_2^+$-type ion. For the H_4O_2 dimers, the O-O distances are 2.66(2) Å and 2.68(2) Å, respectively, whereas for the $H_5O_2^+$ species, one is of the type ($H_3O^+.H_2O$) with a non-central hydrogen ion and an O-O separation of 2.47(2) Å, and the other is a true

$H_5O_2^+$ with a nearly central linking hydrogen and an O–O separation of 2.38(2) A. The average O–O separation for these four species is 2.55 Å which is very close to the average O–O linking distance of 2.57(1) Å seen in the disordered room–temperature phase. This suggests that these species may also be present at room temperature, and supports the description of the average scattering density seen at room temperature as the superposition of equal numbers of disordered water–molecule dimers and $H_5O_2^+$–type ions. For the latter we were unable to conclude whether they are units of the type $(H_3O^+ \cdot H_2O)$ or true $H_5O_2^+$ ions. The results for the low–temperature phase suggest that, indeed, both types are present at room temperature. Further evidence in support of this description comes from a recent study on HUP using a combination of IR and incoherent inelastic neutron–scattering spectroscopies, which was found to be consistent with the presence of both H_3O^+ and $H_5O_2^+$ ions[48].

6.   **Quasielastic Neutron Scattering Studies**

Quasielastic (QES) neutron scattering may be observed in materials in which appreciable diffusion is occurring; this gives inelastic scattering but with only small energy loss or gain around the elastic peak. It yields both structural and diffusional information on the material studied and a series of elegant experiments have been undertaken on superionic fluorites by Hutchings and coworkers (see the excellent review of Hutchings et al.[1]). These workers have mainly been concerned with two aspects of the coherent QES: first the integrated diffuse QES, which they define as

$$S^D(Q) = \hbar \int {}^{QES}S^D(Q,\omega)\,d\omega - I(Q,\infty), \qquad (8)$$

where $S^D(Q,\omega)$ is the QES excluding all contributions from Bragg peaks. The integration over ω refers to the inelastic component of the scattering (for $\omega \neq 0$) but is confined to the quasielastic region of small ω, which omits any contribution from phonon scattering. $I(Q,\infty)$ is obtained from the purely elastic coherent cross section and yields a time averaged picture of the arrangement of the scattering nuclei. In contrast $S^D(Q)$ yields information on the instantaneous nuclear positions relative to this time averaged structure.

The second major piece of information concerns the width (in ω) of the QES peaks. This can be related to the life time of species, which assuming exponential decay of the latter with time, and assuming a Lorentzian line shape with a full width at half maximum of $\Gamma(Q)$ is given by

$$\tau_{coh} = \frac{4.14}{\Pi[2\Gamma(Q)]} \qquad (9)$$

The value of these studies will be apparent from the discussion in the next section which reviews the fascinating problems posed by the high temperature superionic fluorite structured crystals.

(6.1)   **High Temperature Superionic Fluorites**

At low temperatures all stoichiometric fluorite structure compounds e.g. CaF_2 are 'normal' ion conductors. Superionic behaviour appears however to be general to this class of compounds at temperatures close to the melting point. The unusual behaviour of high temperature fluorites was first documented in any detail by Dworkin and Bredig[49] who noted the general tendency for fluorites to show a diffuse phase transition, manifested by a specific heat excess within a few hundred degrees of the melting point. Moreover, it had long been known (see e.g. Faraday[50]) that at high temperatures certain fluorides e.g. PbF_2 had high ionic conductivities. The correlation between the diffuse phase transition and the high conductivity has been convincingly demonstrated first by O'Keefe and coworkers[51] and in greater detail by Chadwick's school[52-(57)]. Greatest detail is available for PbF_2 and $SrCl_2$ which have the lowest temperature phase transitions; and for the former compound conductivity and specific heat data are shown in fig. (15). Table (2) gives the values of the phase transition temperatures (identified by the maxima in the excess specific heat plots). Enthalpies of the phase transitions can be obtained from excess enthalpy studies, or, where good data are available, from integration of the specific heat excess; they are generally in the range 0.1–0.2 eV per formula unit.

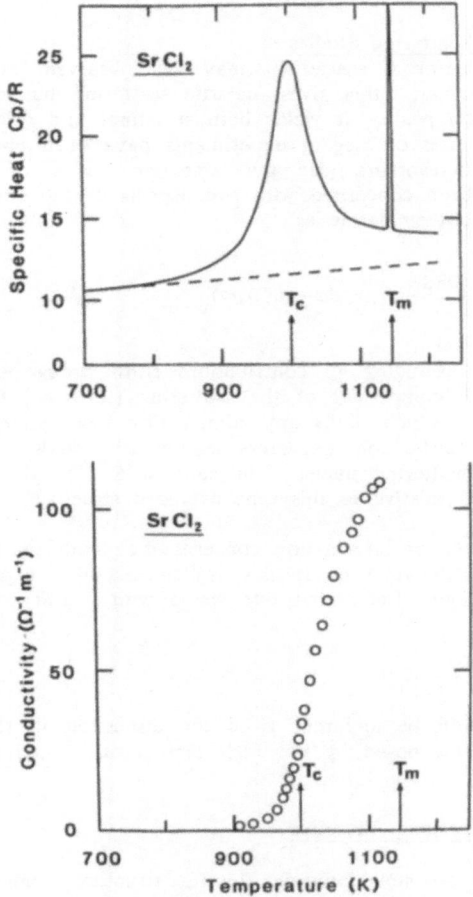

Figure 15. Specific heat and conductivity vs. temperature for $SrCl_2$ (taken from reference (1)).

Table (2)   **Superionic Phase Transition in Stoichiometric Fluorites**

Compound	Transition Temperature (K)†
CaF_2	1430 (1633)
SrF_2	1400 (1723)
BaF_2	1275 (1550)
PbF_2	712 (1158)
$SrCl_2$	1001 (1146)
UO_2	2770 (3100)

† Melting Point given in brackets

Three empirical features of the 'superionic transition' in fluorites deserve immediate emphasis: first, and most obviously the superionicity is due to anion tranport as has been shown by radiotracer measurements which have shown that cation mobility is low[58]; secondly magnitudes of the phase transition energies are small compared with the anion Frenkel energy.   Thirdly, after the rapid increase in the conductivity through the phase transition, the subsequent variation with temperature is small.

These observations cannot be explained easily.   The early theories of the transition[59][60] were based on the concept 'sub-lattice melting', implying a radical disordering of the anion sub-lattice on melting.   But it was pointed out by Catlow et al[61] and Catlow[62] that such models were not compatible with the low ratios of the energy of the phase transition to the Frenkel energy: we would expect the disordering to involve the creation of anion interstitials given that these are the dominant modes of disorder, and thus the relatively low values of the phase transition energy cannot be compatible with massive anion disorder.   This conclusion, based on simple arguments, was subsequently supported by a detailed computational, molecular dynamics study which was discussed in our earlier chapter.

What kind of models can we therefore advance to explain the nature of the superionic transition?   The arguments in the previous paragraph support a model based on the generation of limited disorder;   indeed the simplest predictions based on the ratios reported of the Frenkel to the phase transition energy would suggest that <5 mole % of anion Frenkel defects were present in the superionic phase.   These concentrations would, however be sufficient to lead to the observed superionic behaviour.   We need therefore to characterise the nature of the disorder and the features which cause its limited extent.   We need moreover to understand the dynamics of the superionic phase, and the qualitative differences between anion transport in the low temperature and superionic phases.

(6.1.1) Structural Data on the Superionic Phase

In the last ten years there have been several high quality single crystal neutron diffraction studies of high temperature superionic fluorites; table (3) summarises the most important features of the results of these studies.   Neutron rather than X-ray techniques are preferred because of the greater ease of high temperature experiments with neutron techniques referred to earlier and because the lighter O and F atoms will give rise to a much lower fraction of the total scattering when X-rays rather than neutrons are used. Moreover neutron diffraction is better for separating occupancy variation from the effects of thermal factors owing to the lack of form factor as discussed in section (2.2).   Single crystals rather than powder experiments, are needed because of the very high quality required in data which is to be used to detect low concentrations of interstitials.

Table (3)   Neutron Diffraction Studies of Stoichiometric Superionic Fluorites

Compound	Reference	Temperature	Fraction of Anion sites vacant
CaF_2	(74)	1430	23%
BaF_2	(74)	1275	34%
$SrCl_2$	(73)	1073	16%
	(74)		21%
PbF_2	(73)	711	49%
	(68)		<5%

Interpretation of the data is difficult and in many cases still controversial. Some of this difficulty and controversy arises from matters of definitions, in particular the question of when a relaxed lattice ion or an atom executing large ampitude thermal vibration about a lattice site is to be classified as an 'interstitial'. As noted diffraction measurements yield simply the distribution of scattering density in the average unit cell, and the interpretation of this information in terms of defect concentrations is in this case ambiguous. With these reservations in mind, we can nevertheless draw the following conclusions:

(i)   Interstitial concentrations depend on the interpretation of the data. But in some cases low concentrations are reported in accordance with our discussion earlier in this section.

(ii) There is some evidence, particularly in the studies of Hutchings, Hayes and coworkers (see for example the review of Hutchings et al.[1]) for the formation of clusters of the type illustrated in fig. (16). The evidence is based on the refinements of the Bragg scattering data yielding the two types of interstitial present in the cluster. Moreover, similar refinements of neutron data[63] on the doped anion excess alkaline earth fluorides have been successfully interpreted in terms of interstitial cluster models. Indeed, the cluster shown in fig. (17) is very close to the celebrated '2:2:2' cluster originally proposed by Willis[64] in a study of UO_{2+x}, and subsequently advanced by Cheetham et al.[63] to explain their neutron data on Y/CaF_2. Further support for the formation of this type of cluster was provided by calculations of Catlow and Hayes[65], who found a large binding energy for the interstitial–vacancy aggregate shown in fig. (16), as was discussed in our previous article.

The question of cluster formation in the high temperature superionics remains, however, controversial. Gillan[66][67] has argued persuasively that all the existing data can be rationalised without invoking the formation of these species. Moreover, analyses of this molecular dynamics simulations do not reveal any evidence for clustering. Again, the controversy may be partly semantic in origin. Clusters will certainly only persist for periods of a few atomic vibrations; and if their lifetime becomes of the order of 1–2 vibrations, the justification of describing them as a well defined cluster may become doubtful.

One point over which there can, however, be no doubt is that vibrational amplitudes are large in the superionic high temperature fluorites and that there is extensive anharmonicity. Indeed studies of PbF_2 by Bachmann and Schulz[68] have suggested that the structural changes occurring on transition to the superionic state can largely be accounted for by high amplitude anharmonic thermal motions. We consider that it is most likely that the transition to the superionic state involves the generation of limited disorder which is accompanied by a general 'loosening' of the F^- sublattice leading to the enhanced vibrational amplitudes.

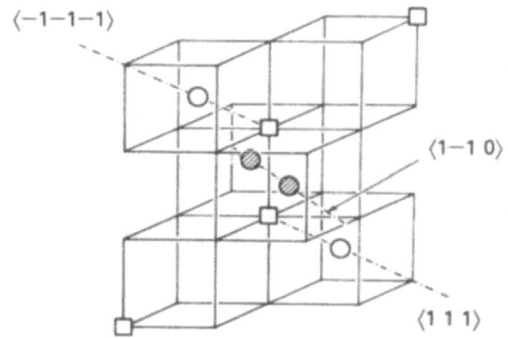

Proposed interstitial–vacancy cluster in high-temperature fluorites.

Figure 16

QES studies of Hutchings and coworkers have proved to be illuminating. We recall that the scattering cross–section $S^D(Q)$ yields information on the instantaneous nuclear positions relative to this time averaged structure, and can therefore yield information on e.g. short–lived clusters. Hutchings et al. can interpret their data in terms of the formation of vacancy–interstitial clusters of the type illustrated in fig. (16), in which an interstitial is stabilised by the relaxation of two neighbouring F^- ions; a compensating vacancy is located in the vicinity of the cluster. More complex variations on this theme are discussed by Hutchings et al.[1].

In addition we recall that from the width in energy of the QES, it is possible to obtain estimates of the life times of scattering centres. Hutchings et al interpret their results in terms of cluster decay for which they report life times of 2–4 psec for $SrCl_2$ and 1–7 psec for PbF_2 depending on the temperature. The results confirm the short life–time of the proposed clusters.

Finally we note that in $SrCl_2$, incoherent as well as coherent scattering has been studied for $SrCl_2$ (Dickens et al.[69]). The advantage of this latter techniques is that it allows single particle dynamics to be studied. Analysis of the results indicates a jump diffusion model and that vacancy diffusion predominates. These results are consistent with those obtained from the theoretical techniques discussed in the next section.

As noted molecular dynamics techniques have also been reported for high temperature fluorites. Work of Dixon and Gillan[70] and Gillan and Dixon[71] on superionic $SrCl_2$ has elegantly demonstrated the value of the MD technique when applied to superionics. Their simulations correctly demonstrate the mobility of the anion sublattice; and they calculate diffusion coefficients in reasonable agreement with experiment. No evidence is found for high levels of interstitial occupancy although large amplitude thermal vibrations are observed in the simulation. Anion transport takes place predominantly by a hopping mechanism as indicated by the incoherent neutron scattering study of $SrCl_2$

Analyses of the simulation has led to a wealth of detail on anion dynamics, and we refer the reader to the original papers and to the reviews of Gillan for details. We note, however, that, as discussed earlier, no evidence was found in the simulations for the clustering which had been proposed on the basis of the neutron scattering studies.

Nevertheless despite the extensive battery of theoretical and experimental techniques that have been brought to bear on the high temperature fluorites over the past ten years, the materials are not fully understood. It seems that limited interstitial generation occurs during the transition. But what is the nature of the high temperature defect structure, and how are we to understand the phase transition? In the opinion of the present author, the cluster models of the type developed by Hutchings et al. provide the most plausible answer to those questions. Moreover, we note that although they have not received support from MD studies, their stability was supported by static simulation studies of Catlow and Hayes[65]. In addition, analyses of conductivity data by Allnatt, Chadwick and Jacobs[72] are also in accordance with these models. A particularly attractive feature of the models is that they provide a qualitative explanation of the generation limited disorder: the clusters are large and at relatively low concentrations (5–10 mole %) interactions between them might be expected to suppress further cluster generation. Such models must, however, remain tentative until there is further experimental and theoretical evidence.

7. Summary and Conclusions

It is hoped that this article has shown the great range and scope of X-ray and neutron scattering methods in the study of superionics. Details of structure and dynamics can be probed and the level of information will improve with increased intensity and energy range of available sources. The methods are particularly powerful when used in conjunction with computer modelling techniques of the type discussed elsewhere in this book.

Acknowledgements

We are grateful to M.T. Hutchings, B.E.F. Fender and W. Hayes for many useful discussions. I would also like to thank Dr Andy Fitch for his helpful comments on this manuscript. The discussion of the HUP and HUAS proton conductors follows closely his account in reference (24).

References

(1)   Hutchings M.T., Clausen K., Dickens M.H., Hayes W., Kjems K.J., Schnabel P.G. and Smith C., J. Phys. C., 17, 3903 (1984).
(2)   Rietveld H.M., J. Appl. Crystallog. 2, 65 (1969).
(3)   Cheetham A.K. and Taylor J.C., J. Solid State Chem. 21, 253 (1977).
(4)   High Resolution Powder Diffraction (ed Catlow C.R.A.), Materials Science Forum vol 9 (1986).
(5)   Cheetham A.K. in 'High Resolution Powder Diffraction' (ed Catlow C.R.A.), Materials Science Forum vol 9 p 103 (1986).

(6)   Catlow C.R.A. and Greaves G.N. Chemistry in Britain, $\underline{22}$, 8061 (1986).

(7)   Skelton E.F., Physics Today $\underline{37}$, 44 (1984).

(8)   Windsor C.G. 'Pulsed Neutron Scattering' (Taylor anad Francis; London), 1981.

(9)   Werner P.E., Erickson L. and Westdahl M., J. Appl. Cryst. $\underline{18}$, 367 (1985).

(10) Kuhs W.F. Acta Cryst. $\underline{A39}$, 148 (1983).

(11) Johnson C.K. and Levy H.A. International Tables for X-Ray Crystallography vol IV, p 311, Birmingham; Kynoch Press (1974).

(12) Zucker Y.H. and Schulz H., Acta Cryst. $\underline{A38}$, 563 (1982).

(13) Kuznetsov P.I., Stratonovich R.L. and Tikhonov V.I., Theory Probab. Its Appl. (USSR) $\underline{5}$, 80 (1960).

(14) Fender B.E.F. in 'Thermal Neutron Scattering' (ed Willis B.T.M.)(AERE Harwell) p 250 (1972).

(15) Fender B.E.F. and Wright A., J. Phys. C. $\underline{10}$, 2261 (1977).

(16) Mahan G.D. and Roth W.L. (eds) Superionic Conductors' (Plenum Press; New York) (1977).

(17) Vashishta P., Mundy J.N. and Shenoy G.K., (eds), Fast Ion Transpsort in Solids (North Holland; Amsterdam) (1979).

(18) Bates J.B. and Farrington G.C. (eds) 'Fast Ion Transport in Solids' (North Holland; Amsterdam) (1981).

(19) Kleitz M., Sapoval B. and Chabre Y. (eds) 'Solid State Ionics-83' (North Holland; Amsterda,) (1983).

(20) Solid State Ionics. vol. 18/19 (1986).

(21) Harwig H.A., Z. Anorg. Allg. Chem. $\underline{444}$, 151 (1978).

(22) Battle P.D., Catlow C.R.A., Drennan J. and Murray A.D. J. Phys. C. $\underline{16}$, L561 (1983).

(23) Battle P.D., Catlow C.R.A., Heap J.W. and Moroney L.M., J. Solid State Chem. $\underline{63}$, 8 (1986).

(24) Fitch A.N. in High Resolution Powder Diffraction (ed Catlow C.R.A.) Materials Science Forum vol $\underline{9}$, p 113 (1986).

(25) Shilton M.G. and Howe A.T., Mat. Res. Bull. $\underline{12}$, 701 (1977).

(26) Howe A.T. and Shilton M.G., J. Solid State Chem., $\underline{34}$, 149 (1980).

(27) Gordon R.E., Strange J.H. and Halstead T.K., Solid State Comm. $\underline{31}$, 995 (1979).

(28) Halstead T.K., Boden N., Clark L.D. and Clarke C.G., J. Solid State Chem. $\underline{47}$, 225 (1983).

(29) Poinsignon C., Fitch A.N. and Fender B.E.F., Solid State Ionics $\underline{9}$ &$\underline{10}$, 1049 (1983).

(30) Morosin, B., Acta Cryst., $\underline{B34}$, 3732 (1978).

(31) Morosin B., Phys. Lett., $\underline{65A}$, 53 (1978).

(32) Bernard L., Fitch A.N., Howe A.T., Wright, A.F. and Fender B.E.F., J. Chem. Soc. Chem. Comm. 784 (1981).

(33) Fitch A.N., Bernard L., Howe A.T., Wright A.F. and Fender B.E.F., Acta Cryst. $\underline{C39}$, 159 (1983).

(34) Shilton M.G. and Howe A.T., J. Solid State Chem. $\underline{34}$, 137 (1980).

(35) Howe A.T. and Shilton M.G., J. Solid State Chem. $\underline{28}$, 345 (1979).

(36) Thomas J.O. and Liminga R., Acta Cryst. $\underline{B34}$, 3686 (1978).

(37) Bernard L., Fitch A.N., Wright A.F., Fender B.E.F. and Howe A.T., Solid State Ionics, $\underline{5}$, 459 (1981).

(38) Kreuer K.D., Rabenau A. and Weppner W., Angew. Chem. Int. Ed. Engl. $\underline{21}$, 208 (1982).

(39) Kreuer K.D., Rabenau A. and Messer R., Appl. Physics, $\underline{A32}$, 155 (1983).

(40) Krogh Andersen E., Krogh Andersen I.G., Simonsen K.E., Shau E., Lundsgaard J.S. and Malling J. in Proceedings of the 2nd Conference on Solid State Protonic Conduction, eds Goodenough J.B., Jensen J. and Kletz M., Odense University Press, Odense, 253 (1983).

(41) de Benyacar M.A.R. and de Abelado M.J., Am. Mineral. $\underline{102}$, 763 (1983).

(42) de Benyacar M.A.R. and de Dussel H.L., Ferroelectrics $\underline{9}$, 241 (1975).

(43) de Benyacar M.A.R. and de Dussel H.L., Ferroelectrics $\underline{17}$, 469 (1978).

(44) Shilton M.G. and Howe A.T., J. Chem. Soc. Chem. Comm., 194 (1979).

(45) Fitch A.N., Wright A.F. and Fender B.E.F., Acta Cryst. $\underline{B38}$, 2546 (1982).

(46) Thomas M.W. and Bendall P.J., Acta Cryst. $\underline{A34}$, S351 (1978).

(47)   Bendall P.J., Fitch A.N. and Fender B.E.F., J. Appl. Cryst. 16, 164 (1983).

(48)   Kearley G.J., Fitch A.N. and Fender B.E.F., J. Mol. Struc. 125, 229 (1984).

(49)   Dworkin A.S. and Bredig M.A., J. Phys. Chem. 72, 1277 (1968).

(50)   Faraday M.   Experimental Researches in Electricity (R. & J.E. Taylor, London), vol. I, p 426 (1839).

(51)   Derrington C.E., Navrotsky A. and O'Keele M., Solid State Comm. 18, 47 (1976).

(52)   Chadwick A.V., Radiation Effects 74, 17 (1983).

(53)   Carr V.M., Chadwick A.V. and Figueroa D.R., J. Phys. (Paris) 37, C7–337 (1976).

(54)   Figueroa D.R., Chadwick A.V. and Strange J.H., J. Phys. C. 11, 55 (1978).

(55)   Carr V.M., Chadwick A.V. and Saghafian R., J. Phys. C. 11, L637 (1978).

(56)   Chadwick A.V., Kirkwood F.G. and Saghafian R., J. Phys. (Paris) 41, C6–216 (1980).

(57)   Azimi A., Carr V.M., Chadwick A.V., Kirkwood F.G. and Saghafian R., J. Phys. Chem. Solids. 45, 23 (1984).

(58)   Matzke, Hj. J. Mater. Sci., 5, 831 (1970).

(59)   Rice M.J., Strassler S. and Toombs G.A., Phys. Rev. Lett., 32, 596 (1974).

(60)   Huberman, B.A., Phys. Rev. Lett. 32, 1000 (1974).

(61)   Catlow C.R.A., Comins J.D., Germano F.A., Harley R.T. and Hayes W., J. Phys. C. 11, 3197 (1978).

(62)   Catlow C.R.A., Comments on Solid State Physics, 9, 157 (1980).

(63)   Cheetham A.K., Fender B.E.D. and Cooper M.J., J. Phys. C. 4, 3107 (1971).

(64)   Willis B.T.M., Proc. Brit. Ceram. Soc. 1, 9 (1963).

(65)   Willis B.T.M., J. Phys. (Paris) 25, 431 (1964).

(66)   Gillan M.J., J. Phys. C. 19, 3391 (1986).

(67)   Gillan M.J., to be published.

(68)   Bachman R. and Schulz H., Solid State Ionics 9/10, 521 (1983).

(69)   Dickens M.H., Hayes W., Schnabel P., Hutchings M.T., Lechner R.E. and Renkner B., J. Phys. C. 16, L1 (1983).

(70)   Gillan M.J. and Dixon M., J. Phys. C. 13, 1901 (1980).

(71)   Dixon M. and Gillan M.J., J. Phys. C. 13, 1919 (1980).

(72)   Allnatt A.N., Chadwick A.V. and Jacobs P.W.H. Proc. Roy. Soc. A410, 385 (1987).

(73)   Dickens M.H., Hayes W., Hutchings M.T. and Smith C., J. Phys. C. 15, 4043 (1982).

(74)   Hutchings M.T., A.I.P. Conf. Proc. No. 89, 209 (1982).

ELECTROCHEMICAL MEASUREMENT TECHNIQUES

Werner Weppner

Max-Planck-Institut
für Festkörperforschung
Heisenbergstr.1
D-7000 Stuttgart 80
Fed. Rep. Germany

INTRODUCTION

Phase relations, thermodynamic and kinetic data are among the most fundamental materials parameters. They provide information about the existing stable phases, stability ranges, non-stoichiometries, atomic disorder, energies of formation, and electronic and ionic transport properties. Many other materials properties are closely related to these quantities. In spite of the importance of these fundamental parameters, little information is presently known as may be easily recognized by looking at the small number of investigated phase diagrams of multinary systems. In addition, the precision of the parameters is getting more and more important since even very small variations of the composition may have a large influence on the materials properties. The activities of the components change rapidly for most solids as a function of the composition. Furthermore, the present tendency of miniaturization requires more carefully the consideration of stability requirements [1].

Electrochemical techniques have shown many advantages for studying phase equilibria, and fundamental thermodynamic and kinetic properties. The ionic conductor acts as a probe for chemical potentials, partial pressures, energies of formation, atomic disorder, stoichiometric deviations, phase equilibria, partial conductivities, diffusion coefficients, mobilities and reaction rate constants. Often, solid electrolytes are presently used in order to cover a wide temperature range, but liquid electrolytes may be employed in all cases as well.

The electrochemical techniques show a variety of important general advantages compared to other methods, e.g.:

- Electrical quantities are readily measurable with high precision.
- The fundamental thermodynamic and kinetic quantities are directly transduced into cell voltages and electrical currents without any intermediate step of conversion.
- The composition may be precisely changed in-situ. The data may be determined in this way as a function of composition. This allows to reduce drastically the number of samples to be employed.

- The experimental arrangement is generally very simple. Certain requirements of stability of the various parts of the galvanic cell are required, however.

The electrochemical techniques are described in the following and are mainly illustrated by the application to multinary systems. These materials are becoming more and more important because of the additional degrees of freedom and the possible variation of the materials properties by changes of the composition. Both predominantly ionically and electronically conducting solids will be considered. In order to discuss the electrochemical methods coherently, some fundamental principles underlying all techniques will be briefly considered first.

GENERAL PRINCIPLES OF ELECTROCHEMICAL CELL TECHNIQUES

Any voltage measurement, even the application of an electrometer with a very high input impedance, requires a small current. A chemical reaction will occur in the galvanic cell as a result of the transfer of ions. The chemical energy of this process at constant total pressure and temperature, the Gibbs energy ΔG of reaction, corresponds to the electrical work that accompanies the current flux:

$$\Delta G = - nqE \tag{1}$$

where n, q and E are the number of charges that are transfered for the considered reaction, the elementary charge and open cell voltage (emf), respectively. The variation of the Gibbs energy with the change of the number of any type of atoms in the electrodes corresponds to the chemical potential μ of this component. Accordingly, the transfer of ions A^{z+} corresponds to the cell voltage

$$E = (zq)^{-1}(\mu_A^l - \mu_A^r) \tag{2}$$

(l and r stand for the left and right hand electrode, respectively)
The chemical potential μ may be expressed by the activity a or the equilibrium partial pressure p

$$\mu = \mu^o + kT \ln a = \mu^o + kT \ln p \tag{3}$$

μ^o, k, T are the chemical potential in the standard state, Boltzmann's constant and absolute temperature, respectively. Substitution of Eq. (3) into (2) results in the following relation

$$E = (kT/zq)\ln(a_A^l/a_A^r) = (kT/zq) \ln (p_A^l/p_A^r) \tag{4}$$

It should be pointed out that the voltage is determined by the reaction that really occurs by the application of the voltmeter. This may not always agree with the thermodynamically most favourable reaction.

In the case of an electrical current I across the galvanic cell, the emf is superimposed by various polarization voltages, e.g., the IR drop, concentration, transfer or crystallization polarization. The observed polarization voltages may be used to analyse the related phenomena. If the diffusion of ions or electrons in a compositional gradient is rate determining, the flux density j of the predominantly mobile ions is given by Fick's first law

$$j = - \tilde{D} \frac{\partial c}{\partial x} \tag{5}$$

c is the concentration of the mobile species, $\tilde{D}$ is the effective chemical diffusion coefficient. The flux is converted into an electrical current with the density i through the galvanic cell:

$$i = zqj \tag{6}$$

Diffusion processes may be studied by comparing the current to a given (known) concentration gradient (Eqs. (5) and (6)) or by studying transient processes and comparing the experimentally observed concentration as a function of time or location with the solution of Fick's second law

$$\partial c/\partial t = \tilde{D} \; \partial^2 c/\partial x^2 \tag{7}$$

with the appropriate initial and boundary conditions.

Another general aspect is the relation of the concentrations or chemical potentials of the k components of a compound according to Gibbs-Duhem's equation for any change of the concentrations

$$\sum_{i=1}^{k} \mu_i \; dc_i = 0 \tag{8}$$

It should also be recalled that the number of independently variable chemical potentials is decreased in case of an equilibrium between several phases and given according to Gibbs' phase rule. The number of thermodynamic degrees of freedom f is for the presence of p phases:

$$f = k - p + 2 \tag{9}$$

ELECTROCHEMICAL DETERMINATION OF PHASE EQUILIBRIA

The electrochemical technique for the determination of phase equilibria is based on the coulometric titration which was originally developed for the variation of the stoichiometry of binary compounds [2]. A charge flux $\int I dt$ through the galvanic cell

counter-		electrolyte			
	electrode		sample	(I)
reference-		for A^{z+} ions		

changes the mass m_A of the component A in the sample according to Faradays's law

$$\Delta m_A = (M_A/zF) \int I \; dt \tag{10}$$

(M_A and F are the atomic weight of A and Faraday's constant, respectively) This corresponds to a change in the stoichiometric number x of the component A in the sample by

$$\Delta x_A = \sum_{i=1}^{k} x_i M_i /mzF \tag{11}$$

The summation runs over all atomic weights multiplied by the stoichiometric numbers x_i. m is the total mass of the sample at the starting composition.

The current is interrupted from time to time, and the equilibrium cell voltage (emf) is determined and plotted as a function of the variation of the composition. Since the emf is an indication of the chemical

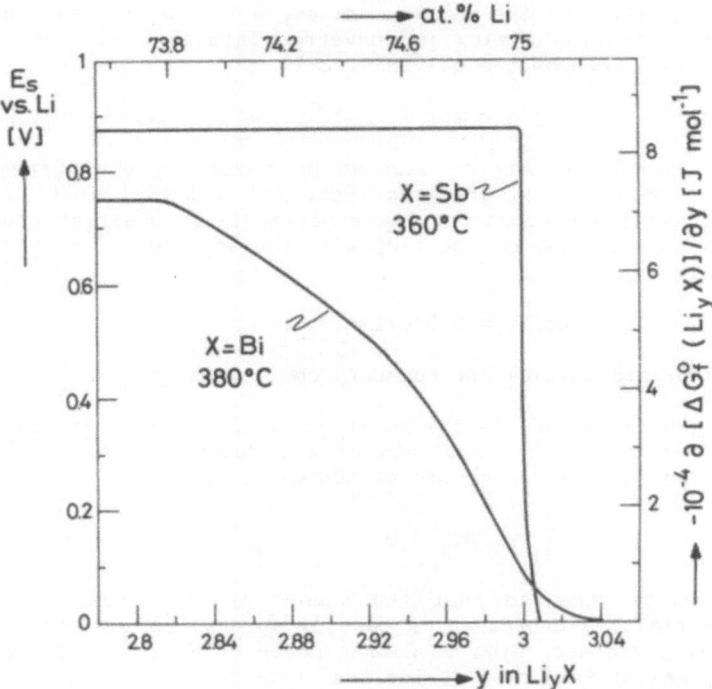

Fig. 1 Coulometric titration curve for the systems Li-Sb and Li-Bi using a molten salt electrolyte for lithium ions. The equilibrium cell voltage is plotted with reference to elemental lithium as a function of the lithium concentration.

potential of the electroactive component according to Eq. (2), a constant composition independent value is observed if the number of co-existing phases corresponds to the number of components. As soon as the overall composition of the sample is in the range of the presence of a smaller number of phases, a variation of the emf with the composition must occur because of the increased number of degrees of freedom according to Gibbs' phase rule (9). The cell voltage has to change monotonously since the minimum of the Gibbs energy for any given composition may not be fulfilled otherwise.

Fig. 1 shows as an example the voltage drop at the occurrence of the single phases Li_3Sb and Li_3Bi of the binary systems Li-Sb and Li-Bi. The employed experimental arrangement is illustrated in Fig. 2. Elemental antimony and bismuth are used as starting materials which is inserted as the working electrode into a molten salt lithium electrolyte. A eutectic mixture of LiCl and KCl is used in this case. Two-phase mixtures of LiAl and Al are employed both for a large counter and a small reference electrode. This electrode material has the advantage of being solid at the experimental temperature and having a much smaller electronic leakage current in contrast (in both regards) to the application of elemental lithium [3]. Small voltage steps are observed at the compositions Li_2Sb and LiBi. The voltage is otherwise independent of the composition within the 2-phase regions. A large decrease of the voltage over wide stoichiometric ranges are found around the ideal compositions Li_3Sb and Li_3Bi.

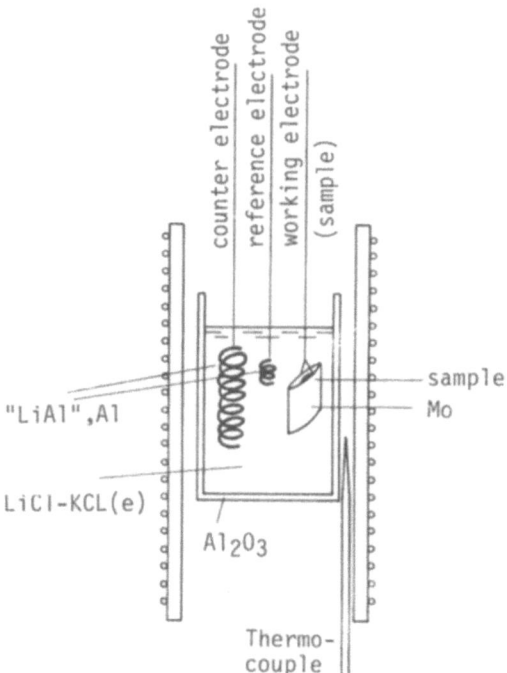

counter electrode
reference electrode
working electrode
(sample)

"LiAl",Al

LiCl-KCL(e)

sample
Mo

Al₂O₃

Thermo-
couple

Fig. 2   Schematical representation of the experimental arrangement to
         investigate phase equilibria of lithium systems. 2-phase equili-
         bria LiAl,Al are used for the reference and counter electrodes
         which are wound to spirals to accommodate easily for the large
         expansion in the course of the charge and discharge processes.

        The electrochemical technique requires only a very small number of
sample preparations. A single sample has been sufficient to study the
phase diagram of the system Li–Bi. A ternary system requires only that
many samples as required conventionally for studying a binary system.
Since the cell voltage also provides thermodynamic information (as will
be discussed later), these data may be taken into consideration and the
number of sample preparations may be even further reduced.

        An example of the determination of a ternary phase diagram is shown
in Figs. 3 and 4 for the system Cu–Ge–O. The composition of the sample
follows a straight line in the Gibbs triangle which connects the starting
composition with the corner of the electroactive component which is oxygen in the present case by using zirconia solid electrolytes at elevated
temperatures [4,5]. Several voltage plateaus are observed which correspond to 3-phase equilibria of the ternary system. These plateaus are
separated by small 2-phase regions. Data are shown for the 4 different
Cu:Ge ratios 4:1, 3:2, 1:1 and 1:3. Plateaus of the same cell voltage
belong to the same 3-phase region for all samples with different Cu:Ge
ratios. The results of the coulometric titrations are translated into
Gibbs' phase diagram (Fig. 4). Only one ternary phase, $CuGeO_3$, occurs at
the experimental temperature of 900°C. 3-phase regions which have one
side in common with one of the binary legs of the electroactive component

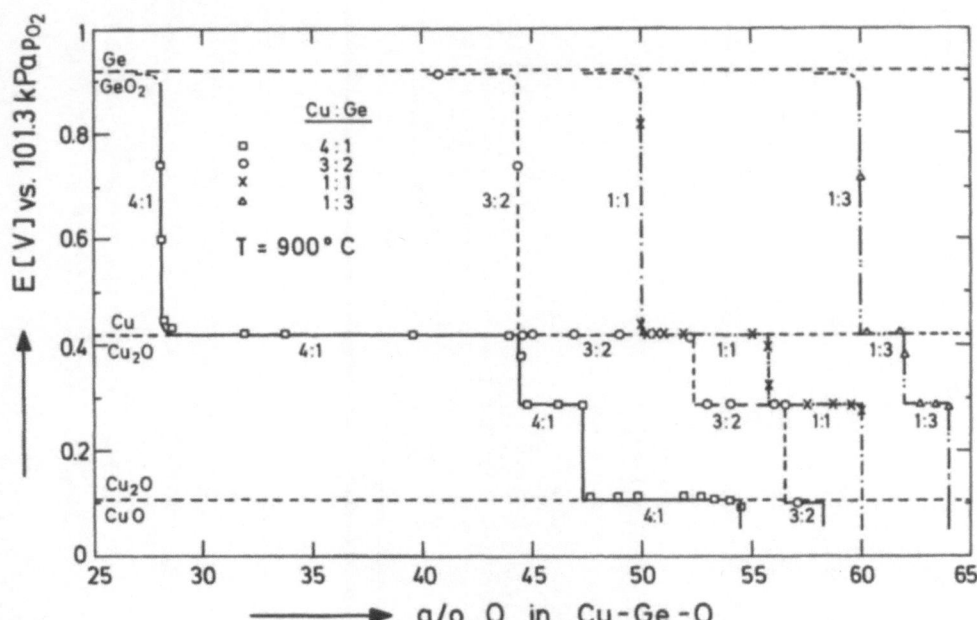

Fig. 3 Coulometric titration curves for the ternary system Cu-Ge-O for samples with different Cu-Ge ratios using a solid oxygen ion conductor. The voltages related to the equilibria of the binary oxide systems are shown as horizontal broken lines.

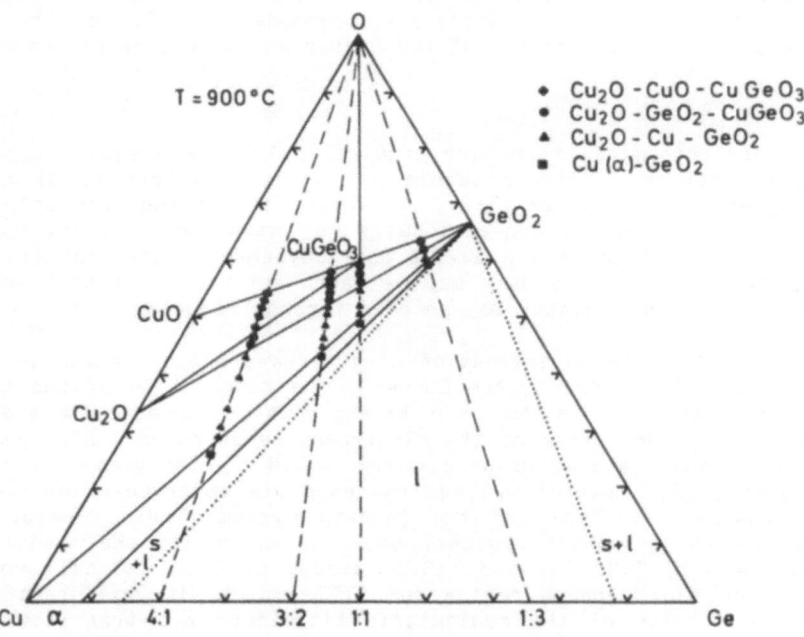

Fig. 4 Phase diagram (Gibbs triangle) of the ternary system Cu-Ge-O. The ranges of plateau voltages in Fig. 3 are translated into this diagram and regions of identical voltages for different Cu:Ge ratios are combined to 3-phase regimes indicated by the same sybols.

(Cu-O or Ge-O) show the same emf as for the presence of the binary system only. The third phase has no (or only little) influence since it does not become involved in the cell reaction (with the exception of some dissolution of the third component in the binary phases). Only the relative ratios of the binary compounds are changed. It should also be verified for the construction of the phase diagram that the voltage along any (thought or experimentally investigated) straight line through the corner of the electrochemically transfered component in the Gibbs triangle changes monotonously. This may be helpful to exclude the presence of metastable states. Very helpful is also the comparison of the titration results in both directions, at longer annealing times and as a result of temperature variations.

The resolution for the various regions of the phase diagram is extremely high as a result of the very sensitive measurement of currents and time intervals. An experimentally easily controllable current of 1 µA for the period of 10 s corresponds to a mass change of the sample which may be not achievable for any balance, e.g., 8.3×10^{-10} g for oxygen or 7.2×10^{-10} g for lithium. Even much smaller charge fluxes are easily accessible. Compounds which are commonly called line phases may be readily resolved with regard to the stoichiometric width. As an example, the stoichiometric width for lithium in $LiAlCl_4$ was found to be 2×10^{-6} [6,7].

The electrochemical technique does not require quenching samples and may be therefore applied also for the investigation of such system which may not be conventionally analysed because of phase transitions during quenching. The investigation of the material is in-situ and does not require the destruction of the sample. The equilibration process of the sample is always visible by looking at the time dependence of the emf. There is no need to prepare identical samples which are successively analysed after different periods of time for annealing. Also, annealing for unnecessary long periods of time is avoided.

The observed emf during the titration provides fundamental information on the thermodynamics of the system in addition to the phase equili-

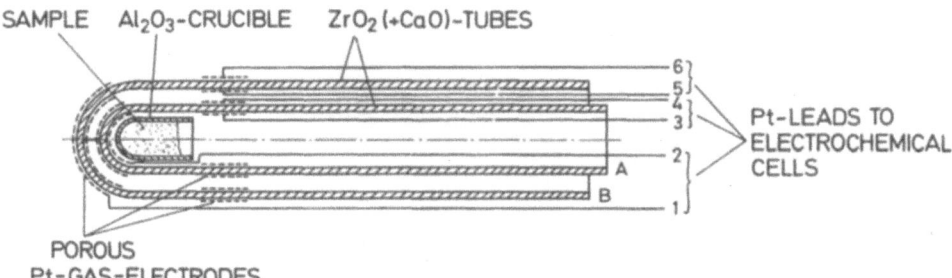

Fig. 5   Experimental arrangement for the coulometric titration and emf measurements of the ternary systems Y-Fe-O and Dy-Fe-O. A double electrolyte cell construction is employed to avoid oxygen permeation to the inner electrolyte tube. The oxygen partial pressure in the intermediate gas space is automatically controlled by a potentiostat at the same pressure as over the sample. A current is passed to that purpose through the outer electrolyte.

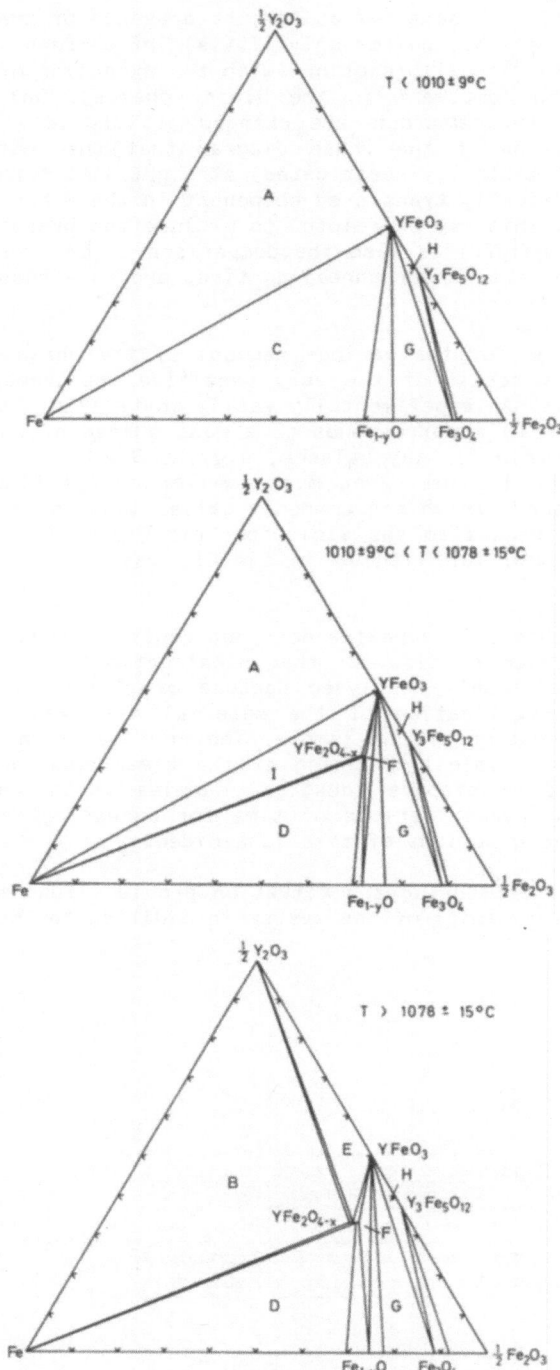

Figs. 6-8 Section of the Gibbs triangle of the system Y-Fe-O for different temperature regimes. The phase YFe_2O_{4-x} occurs only above 110°C. The coexistence of the phases $Fe-YFeO_3$ changes at 1078°C in favour of the equilibrium between Y_2O_3 and YFe_2O_{4-x}. The letters in the various 3-phase regions refer to the thermodynamic data reported in ref. [8].

bria. The knowledge of the emf as a function of stoichiometry allows the determination of thermodynamic data with high resolution as a function of the composition of the sample.

A variety of other systems have been investigated similarly. Experimentally, one may observe many differences, however. In the case of the systems Y-Fe-O and Dy-Fe-O the long equilibration times have required drastic reduction in the oxygen leakage rate of the electrolyte. This has been performed by a double electrolyte cell with a gas space inbetween as shown in Fig. 5. The oxygen partial pressure in the gap has been kept at the same value as in the inner compartment which contains the sample. This was done experimentally with the help of a potentiostat and a current flux through the outer electrolyte to remove the oxygen that permeates through this electrolyte from the counter/reference electrode. The activity gradient is kept zero across the inner electrolyte. There has been therefore no driving force for oxygen permeation through the inner electrolyte and leakage was suppressed. Depending on the temperature quite different phase equilibria were found (Figs. 6 - 8). It is seen that the phase YFe_2O_{4-x} occurs only above 1010°C. Above 1078°C a change of the co-existence of the phases $Fe-YFeO_3$ occurs toward the equilibrium between Y_2O_3 and YFe_2O_{4-x} [8]. Instead of carrying out coulometric titrations in small temperature steps it is also possible to find the transition temperatures from the intersection and branching of the straight lines of the emfs for the various 3-phase regions if the open cell voltage is measured as a function of temperature.

An example of the great potential of electrochemical techniques to provide detailed knowledge even in the case of very complex systems is illustrated in Figs. 9 and 10 for the ternary system Li-In-Sb. Two new ternary phases, Li_3Sb_2In and Li_6Sb_3In are observed along the quasibinary section between Li_3Sb and InSb. The 3-phase regions are separated in this system by very wide 2-phase regions [9,10].

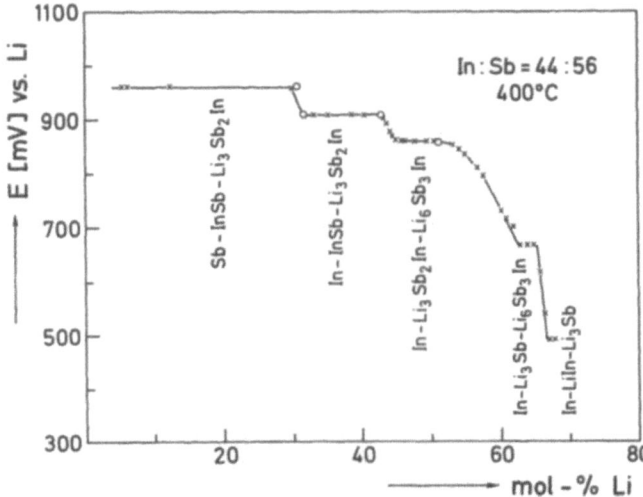

Fig. 9   Coulometric titration curve of the reaction of lithium with a mixture of InSb and Sb with an atomic In/Sb ratio of 44:56. The plateaus correspond to the indicated 3-phase equilibria.

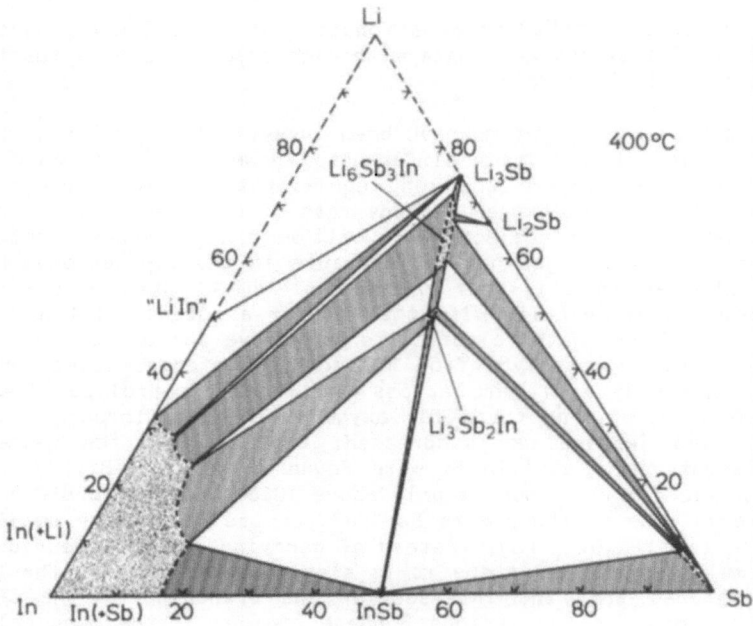

Fig. 10   Phase diagram (Gibbs triangle) of the ternary system Li-In-Sb.
Two new ternary phases are found, Li_3Sb_2In and Li_6Sb_3In, which
show large stoichiometric ranges.

So far, mixed conducting systems have been considered. Phase equilibria of ionically conducting compounds may be easily studied electrochemically by employing these materials themselves as ionic conductors in
galvanic cells [6,7]. The phase equilibria of the ternary system Li-Al-Cl
were verified by equilibration of the lithium ion conductor $LiAlCl_4$ at
one side with a mixture of LiCl,Al and at the other side with a mixture
of $AlCl_3$,Al. The measured emf allows one to determine the Gibbs energy of
formation of $LiAlCl_4$ which then allows calculation of the thermodynamically most favourable equilibria (see [1]). The calculated activities of
all components should show the highest value for any composition for any
assumed 3-phase equilibrium.

The ternary systems Li-N-Cl, Li-N-Br and Li-N-I show several ternary
fast ion conductors along the quasibinary sections Li_3N - lithium halide
(Fig. 11). The Gibbs energies of formation were determined by electrochemical decomposition of the electrolyte. The decomposition voltages then
allow calculation of the thermodynamically most favourable phase relations [1,11,12].

The discussed electrochemical techniques may be employed for the
determination of phase diagrams with any number of components. A general
treatment is available from the literature [13]. The determination of
thermodynamic data from the coulometric titration curves will be illustrated in the next chapter.

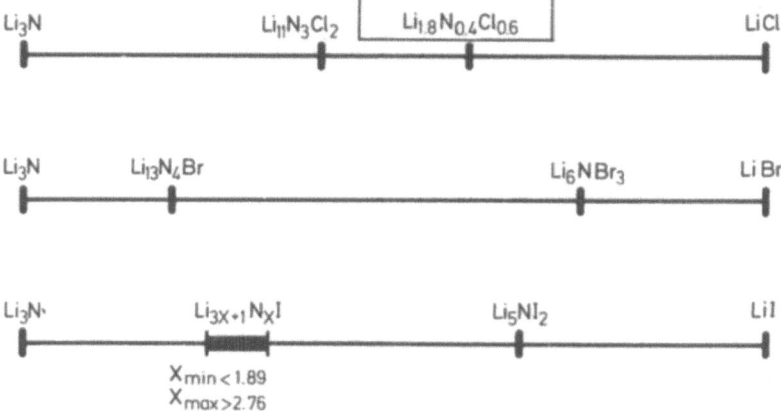

Fig. 11   Thermodynamically stable phases along the quasibinary sections
           Li_3N-lithium halide.

ELECTROCHEMICAL DETERMINATION OF THERMODYNAMIC PROPERTIES

The emf multiplied by the charge $z\delta q$ of a small current of δ ions through the galvanic cell is equivalent to the Gibbs energy $\Delta_\delta G$ of the corresponding chemical reaction. Accordingly, the Gibbs energy of formation for any composition may be obtained by integration of the emf along the path of coulometric titration [13]:

$$\Delta G = zq \int_{x_o}^{x} E(x) \, dx + \Delta G_o \qquad (12)$$

The integration runs from the stoichiometry of the starting composition (x_o) to the stoichiometry of the sample under consideration (x). ΔG_o is an integration constant corresponding to the Gibbs energy of formation of the starting material. This relation allows one to determine the Gibbs energy of formation as a function of stoichiometry.

In case of an equilibrium of k phases of a system which consists of k components, the cell voltage may be derived from the Gibbs energies of formation of the k phases which exist in thermodynamic equilibrium. The 'determinant formula' turned out to be very useful in this regard to calculate the emf E [13]:

$$E = - \frac{1}{zqd} \sum_{i=1}^{k} (-1)^i \, d_{i1} \, \Delta G_f^o \quad (A_{x_i} B_{y_i} \cdots) \qquad (13)$$

assuming the application of an ionic conductor for A^{z+} ions. The voltage is given with reference to the pure (elemental) electroactive component under standard conditions. d is the determinant formed by the k stoichiometric numbers x_i, y_i, $\cdots$ of the k compounds which are in equilibrium with each other, e.g., in the case of a ternary system

$$d = \begin{vmatrix} x_1 & y_1 & z_1 \\ x_2 & y_2 & z_2 \\ x_3 & y_3 & z_3 \end{vmatrix} \qquad (14)$$

d_{i1} is the minor that is formed by eliminating the i-th row and first column which consists of the stoichiometric numbers of the conducting component, e.g., in the case of the ternary system

$$d_{11} = \begin{vmatrix} y_2 & z_2 \\ y_3 & z_3 \end{vmatrix}, \quad d_{21} = \begin{vmatrix} y_1 & z_1 \\ y_3 & z_3 \end{vmatrix}, \quad d_{31} = \begin{vmatrix} y_1 & z_1 \\ y_2 & z_2 \end{vmatrix} \quad (14a)$$

Eq. (13) turned out to be very useful especially in the case of multicomponent systems and complicated stoichiometric ratios. Measurements of the emfs of the regions of equilibrium of k phases allow in reverse to determine the Gibbs energies of formation of all phases in the case of small stoichiometric widths of all phases, i.e., if the stoichiometries of the compounds and accordingly the Gibbs energies are approximately the same for all k-phase equilibria in which the compound is involved.

The temperature dependence of the Gibbs energy of formation allows one to determine the entropy of formation according to the Gibbs-Helmholtz relation $\Delta S_f^0 = -(\partial \Delta G_f^0 / \partial T)$, i.e., the temperature dependence of the cell voltage relates to the entropy of the cell reaction, $\Delta S = nq(\partial E / \partial T)$. The enthalpies of formation and cell reaction follow according to $\Delta H = \Delta G + T\Delta S$. Knowledge of these fundamental thermodynamic quantities allows one to derive comprehensively the thermodynamics of the system.

These theoretical considerations of the electrochemical determination of thermodynamic data will be illustrated by a few examples. The standard Gibbs energy of formation for any composition along the binary system Li-Sb is shown in Fig. 12 as determined from the integration of

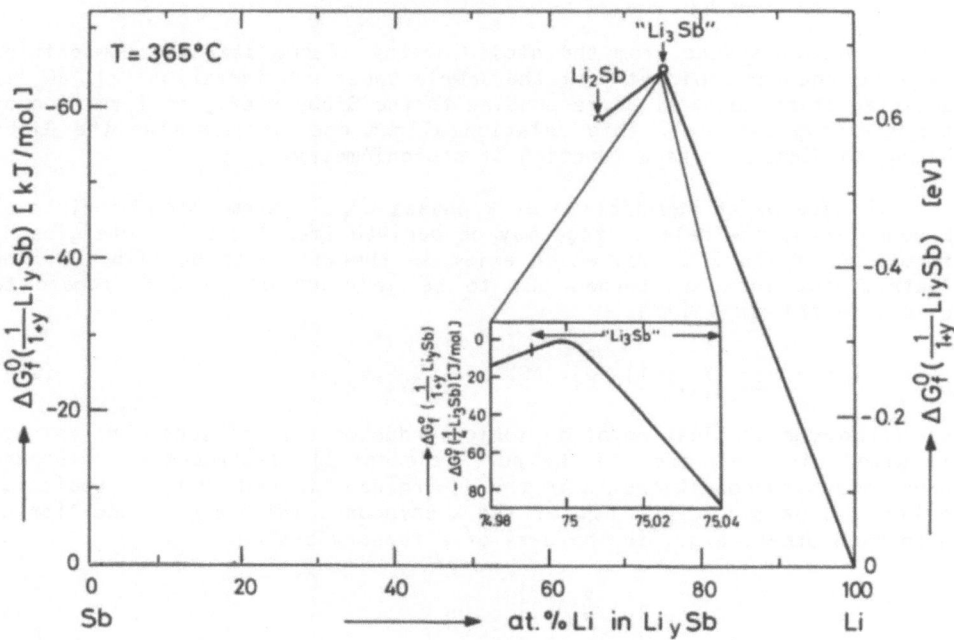

Fig. 12   Standard Gibbs energy of formation as a function of composition for the binary system Li-Sb.

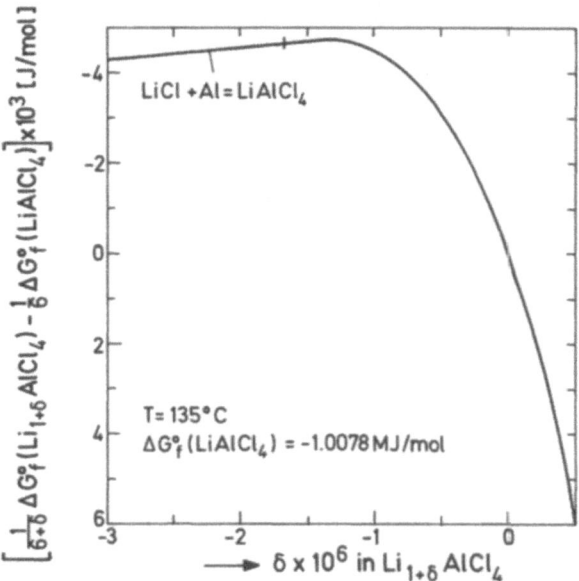

Fig. 13. Gibbs energy of formation of $Li_{1-\delta}AlCl_4$ from the elements under standard conditions as a function of the deviation from the ideal stoichiometry. The change of the Gibbs energy of formation is plotted compared to the value for the stoichiometric compound.

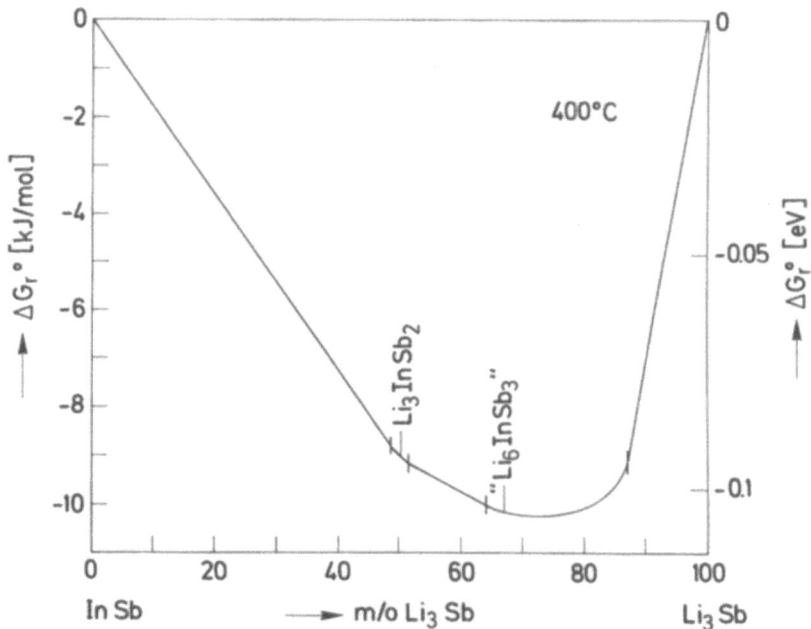

Fig. 14   Determination of the standard Gibbs energy of formation along the quasibinary section $InSb-Li_3Sb$.

209

the Coulometric titration curve. The enlargement of the region around the composition Li_3Sb shows the high resolution of the technique.

The variation of the Gibbs energy of formation of a phase with a very narrow stoichiometric width, $LiAlCl_4$, was determined by integration of the coulometric titration curve and is shown in Fig. 13. The data are shown relative to the ideal stoichiometry. The resolution is well below 1 J/mol and is hard to be matched by other techniques. The technique allows one to obtain the Gibbs energy of formation for any composition. This may be used to calculate the Gibbs energy of formation also for any other directions of the Gibbs triangle. Results for the quasibinary section $InSb-Li_3Sb$ of the ternary system $Li-In-Sb$ are shown in Fig. 14.

Another example is the determination of the emfs of the 3-phase equilibria of the ternary system Y-Fe-O (see Figs. 6 - 8) which are plot-

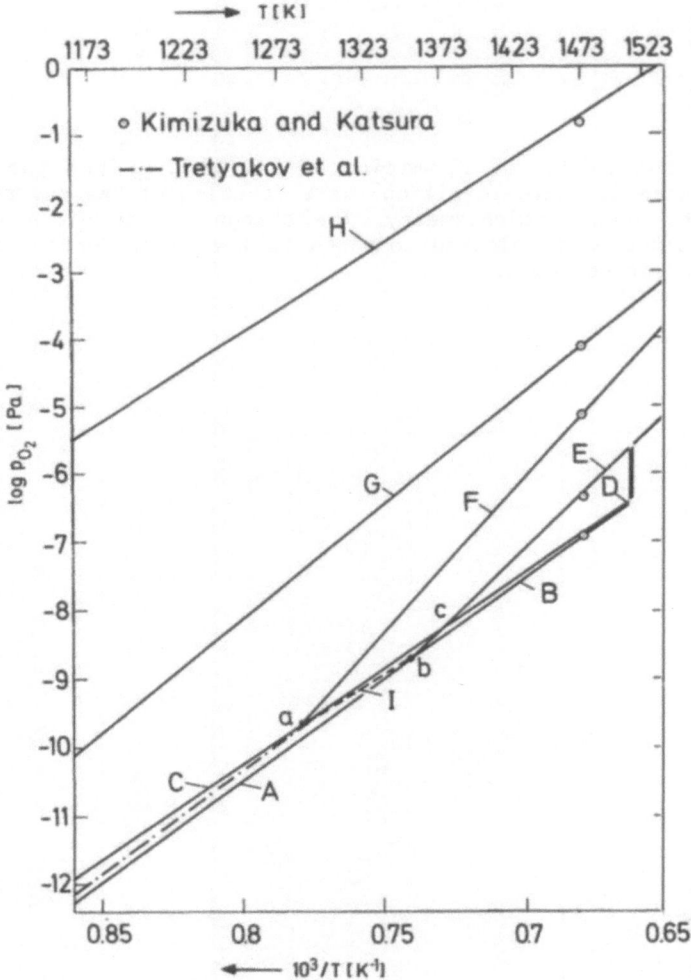

Fig. 15   Temperature dependence of the oxygen partial pressure of the various 3-phase equilibria of the system Y-Fe-O (cf. Figs. 6-8). The branching shows the formation of new phase equilibria at the corresponding temperature.

ted as a function of temperature in Fig. 15. The emf has been converted into the oxygen partial pressure according to Eq. (4). These data made it possible to calculate the Gibbs energies of formation of all ternary phases in that system [8]. Several other examples of the electrochemical detection of stability ranges of ionic conductors are given in another contribution to this volume [1].

It may be possible that metastable phases occur in the course of the coulometric titrations because it is necessary to employ finite current densities. Since the emf corresponds to the Gibbs energy of the reaction which really occurs, this information may be used in this case for the determination of the Gibbs energy of the metastable phase. The formation of the thermodynamically stable phases has to provide of course the lower Gibbs energy of reaction and therefore the higher emf. Fig. 16 shows as an example the electrochemical formation of 'Li_2S_2' at 400°C by the reaction of FeS_2 with Li. The formation of the metastable phase shows a voltage which is 79 mV smaller than the steady state value which appears after a sufficiently long equilibration time. The standard Gibbs energy of formation of 'Li_2S_2' is determined to be −310.1 kJ/mol at 400°C. The thermodynamic reaction of the formation of Li_2S and S is, however, more stable by about 119 kJ/mol.

Of special interest with regard to practical applications are those systems which include a gaseous component. The emf depends in this case also on the partial pressure of the gas. The galvanic cell may be employed as a gas sensor. The gaseous component is not necessarily in agreement with the electroactive component of the electrolyte. As an example, the voltage across the lithium ion conductor $LiAlCl_4$, to which LiCl or $AlCl_3$ is added at the face which is exposed to the gas shows a well defined emf characteristic as a function of the chlorine partial pressure [14].

So far, nearly exclusively the properties of solids in thermodynamic equilibrium or under metastable conditions have been analysed. In the

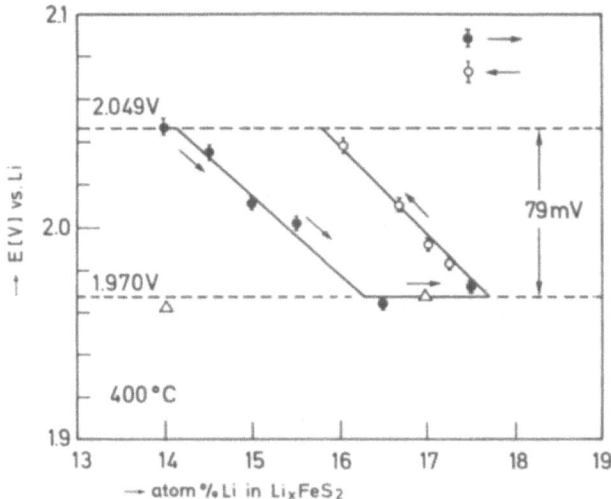

Fig. 16   Thermodynamic equilibrium voltages for the titration of lithium into FeS_2 (upper curve) and metastable apparent equilibrium voltage as a result of the formation of the metastable compound 'Li_2S_2' from Li and FeS_2 (lower curve).

following we will focus on transient phenomena which are similar to many technical processes and natural phenomena. Most important for these processes are the transport properties of the two predominantly mobile species, in most cases one type of ions and electrons or holes. The motions of these species are coupled by local electroneutrality conditions. Rate determining is generally, in this case, the type of species which provides the second largest contribution to the transport process. It will be shown that electrochemical techniques allow one to obtain very elegantly a large amount of fundamental information on kinetic properties.

ELECTROCHEMICAL TECHNIQUES FOR KINETIC STUDIES

Conductivities of predominantly mobile species are commonly investigated by impedance spectroscopy which is discussed elsewhere [15]. The investigation of the kinetics of the second fastest species requires other electrochemical techniques. The conductivity of the minority charge carriers may be derived from polarization measurements with the sample sandwiched between an inert ionically blocking electrode and a reversible electrode [16,17]. The current is carried by the electronic minority charge carriers in the steady state. The driving force for the electrons and holes is the concentration gradient which is generated by the application of the voltage in the case of high disorder

$$E = (zq)^{-1} (\mu_A^l - \mu_A^r) = q^{-1} (\mu_{e-}^l - \mu_{e-}^r) = q^{-1} (\mu_{h+}^r - \mu_{h+}^l) \qquad (15)$$

The effect of electrical fields is neglegible since these are suppressed by the presence of the highly conducting ions. Any possibly generated electrical field will be destroyed immediately by the motion of ionic charge carriers. This holds, however, only in the case if the partial lattice of the mobile ions shows a large number of mobile defects in order to avoid the build-up of a major concentration gradient of the ions.

The steady state current as a function of the applied voltage may be derived by substitution of the chemical potential or concentration difference of the electronic charge carriers as given by Eq. (15) into Fick's first law

$$i = i_{h+} + i_{e-} = q(j_{h+} - j_{e-}) = q \ (D_{h+} \frac{c_{h+}^r - c_{h+}^l}{L} - D_{e-} \frac{c_{e-}^r - c_{e-}^l}{L}) \qquad (16)$$

assuming concentration independent diffusion coefficients and ideal dilute solutions for the holes h^+ and electrons e^-. L is the thickness of the sample. The result reads:

$$i = \frac{kT}{qL} \{\sigma_{e-}^{rev} [\exp(\frac{Eq}{kT}) - 1] + \sigma_{h+}^{rev} [1 - \exp(- \frac{Eq}{kT})]\} \qquad (17)$$

Eq. (17) contains one term that approaches a plateau of the current and another term that shows an exponential increase of the current with increasing voltage. The relative contributions depend on the magnitudes of the prefactors which are the partial conductivities at the activity imposed by the reversible electrode. The shape of the log (current) versus voltage curves allows one to determine the type of minority charge carriers and to derive their conductivities from the plateau value, and the intersection with the log (current) axis of the extrapolated straight line.

Fig. 17 shows results obtained for ZrO_2 doped with 10 mol% Y_2O_3. σ changes with p_O with the fourth power. This dependence corresponds to divalent oxygen vacancies as the predominant type of disorder. Under very low oxygen partial pressures the material becomes mainly electronically conducting before it is decomposed. The temperature dependence of the

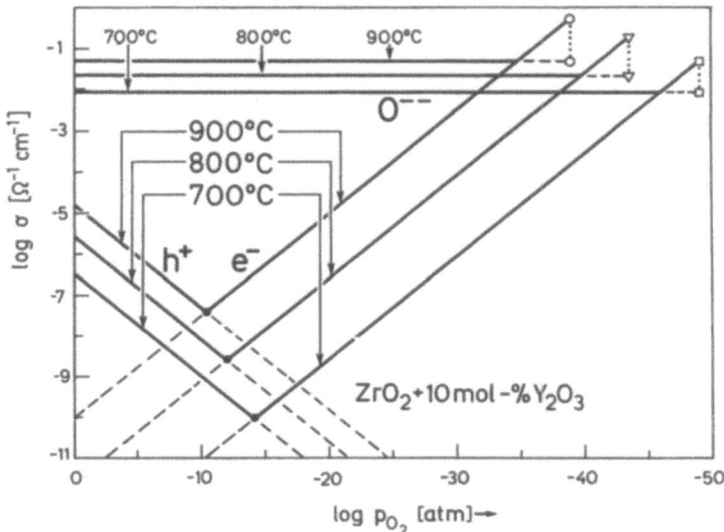

Fig. 17   Oxygen ion, excess electron and hole conductivity of ZrO_2 (+ 10 mol% Y_2O_3) as a function of the oxygen partial pressure at three elevated temperatures.

electronic concentration at the intrinsic composition allows one to determine the band gap of the material which shows close agreement with optical data (4.1 eV) [14].

The application of the cell voltage produces a linear concentration gradient of the electronic minority charge carriers under steady state conditions. The resulting current is given by the product of the electrical mobility and the concentration. It is of interest to obtain information on both parts separately. Transient measurements may be used in order to distinguish between these two contributions to the conductivity. The time dependence of the relaxation process for the electronic concentration gradient after switching off the applied voltage is indicative of the mobility alone. The transport occurs again exclusively by diffusion under the influence of a concentration gradient. The mobile ionic species will shield any possibly formed electrical field. The homogenization process is rate determined by the diffusion of the electronic species which follows Fick's second law with the appropriate initial and boundary conditions. The time dependence of the concentration of electrons and holes at the phase boundary with the inert electrode is measured by the cell voltage. Comparison with the solution of Fick's second law provides information about the diffusivity or mobility [18,19]. This procedure is illustrated in Fig. 18. The results are shown in Fig. 19 as a function of temperature. It is seen that the diffusivity of the electronic charge carriers is very small compared to that in typical semiconductors and metals. Similar low values have been observed for all ionic conductors investigated so far which suggests that small polarons exist in this type of material and the conductivity occurs via a hopping process. It should be noted that Hall measurements may not be performed because of the low mobility of electrons and holes.

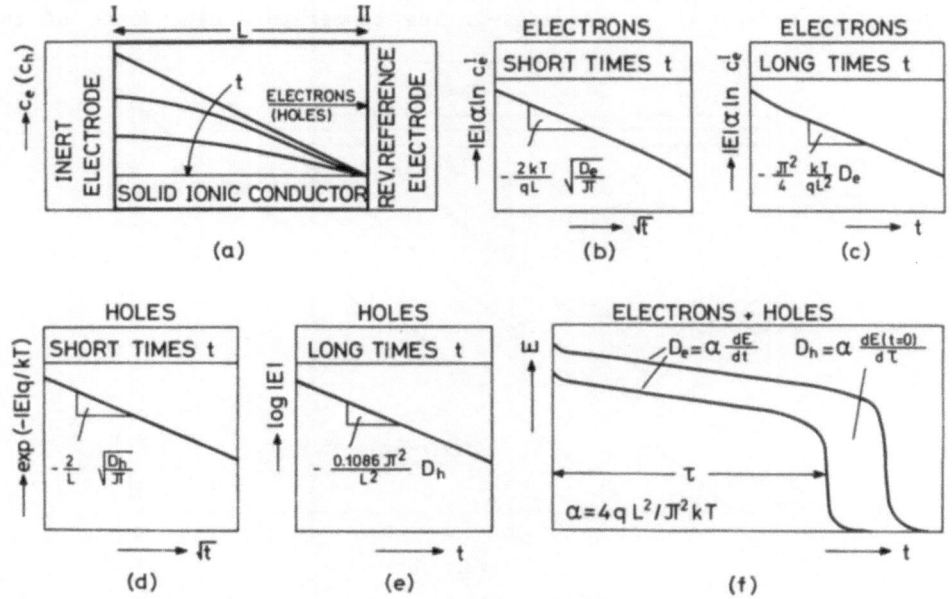

Fig. 18   Principle of the voltage relaxation technique for the determination of the mobilities or diffusion coefficients of solid ion conductors.

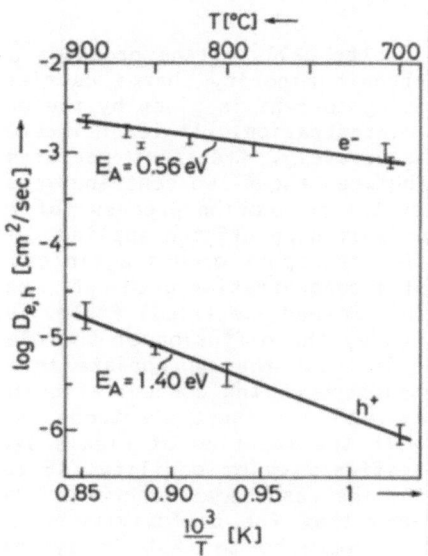

Fig. 19   Diffusion coefficients of the electrons and holes of the solid oxygen ion conductor ZrO_2 (+ 10 mol% Y_2O_3) as a function of temperature.

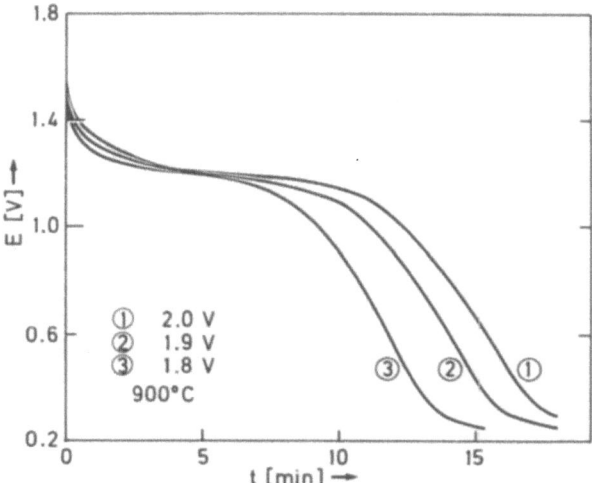

Fig. 20   Voltage relaxation of the solid electrolyte ZrO_2 (+ 10 mol%
           Y_2O_3) which has been polarized between a reversible and ioni-
           cally blocking electrode. An electronic p-n junction has been
           formed during the polarization process which is responsible
           for the formation of the plateau during the relaxation.

        Fig. 20 illustrates the relaxation behavior for three different
voltages which have generated predominant excess electron conductivity at
the inert electrode whereas predominant hole conductivity exists at the
side of the reversible air electrode. Application of the voltage has
therefore induced an electronic p-n junction in a previously homogeneous
material with p-type conduction. During the relaxation process it is
necessary that the electrons pass through the p-n junction in the blocking direction. The voltage remains therefore at a high level until the
junction has moved out of the sample. The initial decrease of the voltage
provides an information about the diffusivity of excess electrons whereas
the length of the plateaus is an indication of the hole diffusivity.

        For the measurement of diffusion coefficients and other kinetic
quantities of mixed ionically and (predominantly) electronically conducting materials, a variety of electrochemical techniques has been developed in recent years which have meanwhile found widespread application
by a large number of investigators. The mixed conducting sample is employed as an electrode which is in contact with an ionic conductor for
one of its components. Depending on the experimental conditions it may be
necessary to use a separate reference electrode in addition to the counter electrode. An enforced variation of the cell voltage or the application of a current forces ions to be delivered to or taken off from the
sample. A given voltage controls the activity of the electroactive component at the phase boundary electrolyte/sample and the current is measured
which is necessary to maintain this activity. Application of a current on
the other hand controls the diffusion of ions in the sample immediately
at the phase boundary with the electrolyte and therefore the concentration gradient at this location according to Fick's first law. The activity which is necessary to maintain this condition at the phase boundary is
measured by the cell voltage. A variety of other signals may be also
applied if a suitable solution of Fick's second law is available. In any

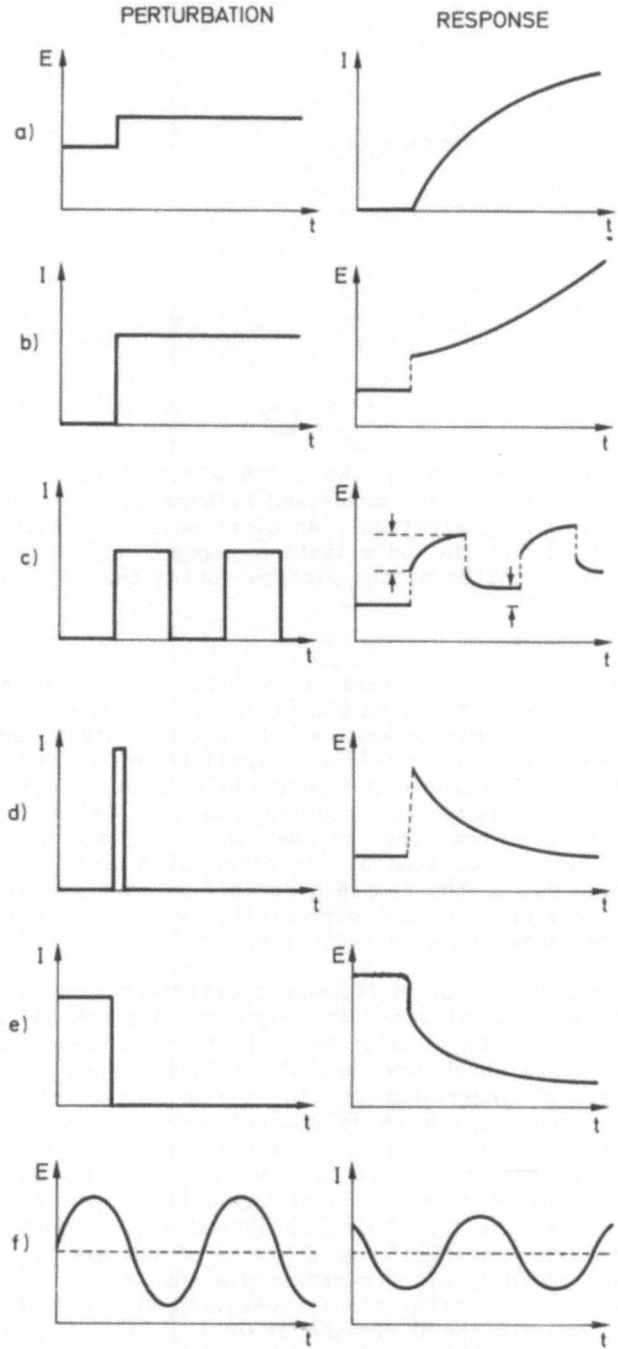

PERTURBATION   RESPONSE

Fig. 21   Most common perturbations and responses of electrochemical
           relaxation measurements for the determination of chemical
           diffusion coefficients and other kinetic quantities.

case, a perturbation is initiated with the help of an electrolyte which serves at the same time to monitor the time dependence of the response. Several examples are shown in Fig. 21.

A general complication arises from the fact that the emf determines the activity of the electroactive component in case of non-ideal behavior whereas the diffusion laws require the knowledge of the concentration. This problem was solved by making use of the coulometric titration curves which provide a relation between the activity (emf) and the concentration (stoichiometry).

The most frequently applied geometries and relaxation techniques are compiled in Fig. 22. The potentiostatic technique starts with an originally homogeneous sample, and a new slightly different voltage is imposed. The current is measured as a function of time. If the diffusion in the bulk of the sample toward the interface is rate determining, a proportionality between the current and the reciprocal square root of time is observed for the initial period of time ($t \ll l^2/\tilde{D}$).

$$ I = \frac{zqN_A}{V_M} \frac{\Delta E}{\pi \; dE/d\delta} \; \tilde{D} \tag{18} $$

N_A and V_M are Avogadro's number and the molar volume, respectively.

The chemical diffusion coefficient may be determined from the slope. For long periods of time when the finite length l of the sample shows an influence, the current changes logarithmically with time ($t > l^2/4\tilde{D}$):

$$ \log I = \log \left(\frac{2zqN_A}{l \; V_M} \frac{\Delta E}{dE/d\delta} \; \tilde{D}\right) - \frac{1.071 \; \tilde{D}}{l^2} \; t \tag{19} $$

The slope and the intersection with the logI axis allow again a determination of the chemical diffusion coefficient. After the sample has reached a new homogeneous concentration according to the new applied voltage, another voltage step may be applied. The chemical diffusion coefficient may be obtained in this way successively with high resolution as a function of stoichiometry.

A disadvantage of the potentiostatic technique is a theoretically infinite current which is required at the beginning of the process. Even by the application of a separate reference electrode it is practically impossible to avoid IR polarization completely. This IR drop across the electrolyte changes rapidly because of the fast change of the current with time. The boundary conditions for the solution of Fick's second law can be therefore only fulfilled to some extent. From this point of view it is advantageous to provide a constant current and to measure the cell voltage (galvanostatic technique). The IR drop is constant in this case and does not change the shape of the time dependence of the cell voltage. The solution of Fick's second law for the galvanostatic technique is for short periods of time ($t \ll l^2/D$):

$$ E = E(t=0) = \frac{2V_M}{zqL} \frac{i_o \; t}{\pi \; \tilde{D}} \frac{dE}{dx} \tag{20} $$

A plot of the voltage as a function of $\sqrt{t}$ provides the chemical diffusion coefficient if the slope of the coulometric titration curve dE/dx has been measured separately. To avoid nonlinearities of the titration curve it may be necessary to consider only small changes of the voltage or restrict the evaluation to the initial period of the relaxation process. A review of the kinetic measurements is available from the literature [20].

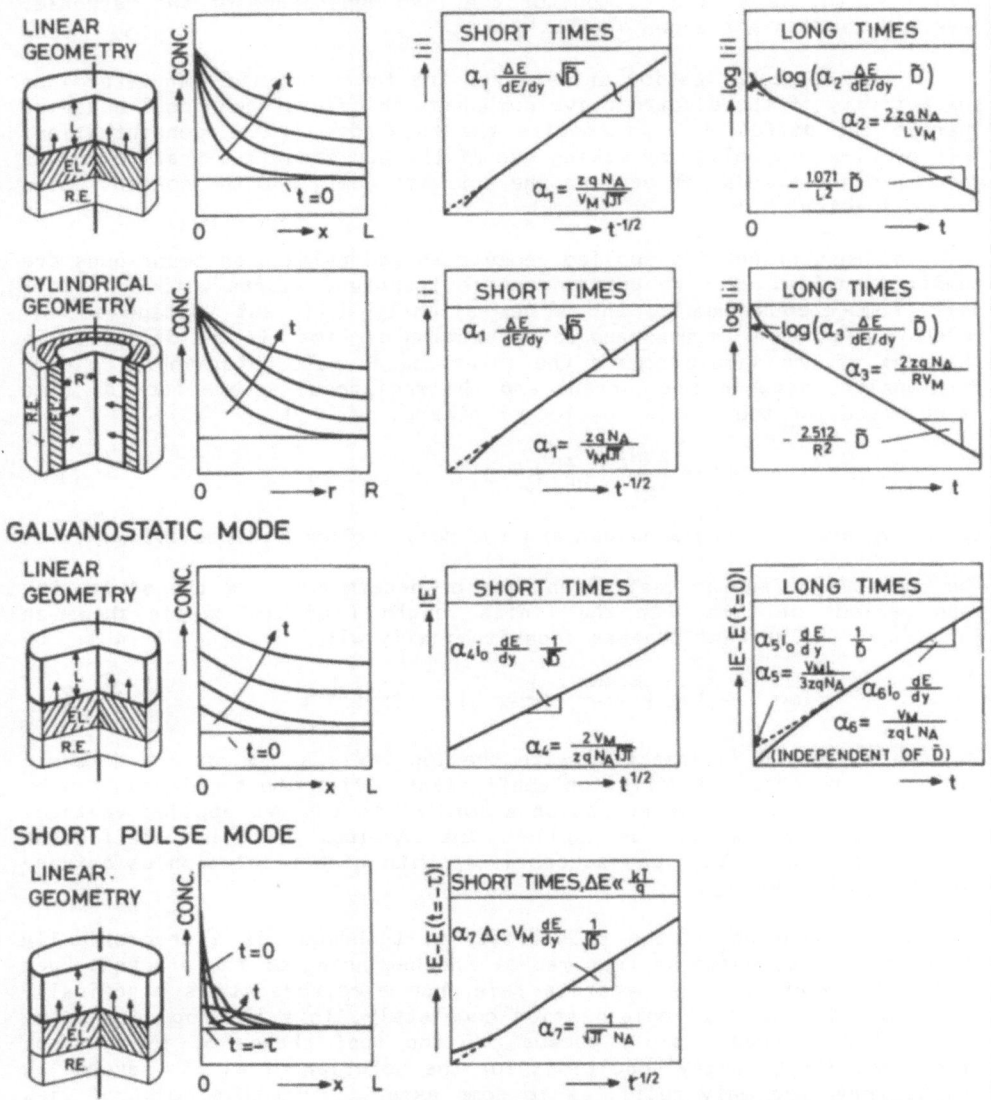

Fig. 22 Compilation of frequently used electrochemical relaxation measurements for the determination of chemical diffusion coefficients.

An elegant combination of the galvanostatic procedure with simultaneous determination of the coulometric titration curve and using a simple mathematical evaluation of the chemical diffusion coefficient as a function of stoichiometry has been developed and is known as the galvanostatic intermittent titration technique (GITT) [21,22]. This technique is illustrated in Fig. 23. A constant current is applied for the period of time τ after the sample has equilibrated. An IR drop is observed instantaneously to which a voltage change according to the $\sqrt{t}$ law is superim-

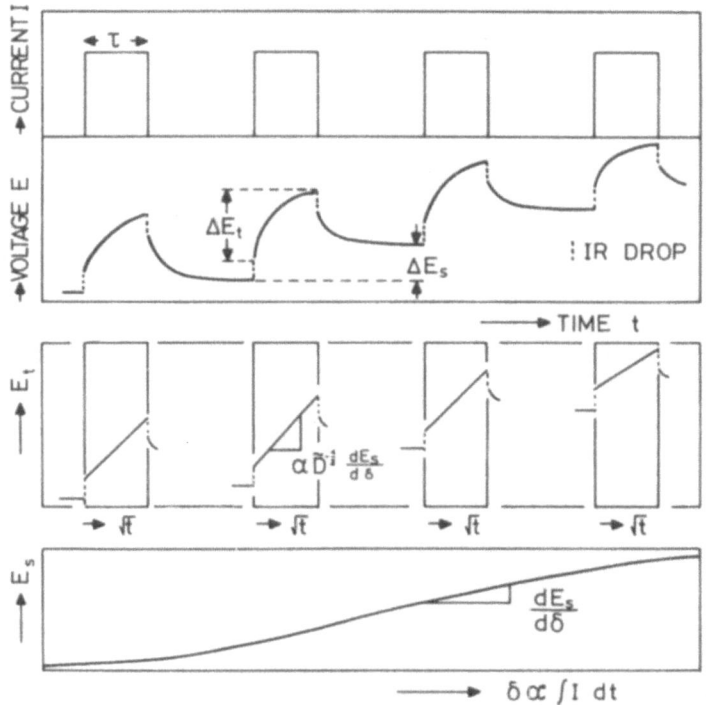

Fig. 23   Schematic representation of the time dependence of the current
           and voltage for the application of the galvanostatic intermit-
           tent titration technique (GITT), the analysis of the chemical
           diffusion coefficient and the determination of the coulometric
           titration curve.

posed. It should be verified that this dependence holds during the entire
period τ. Otherwise, shorter intervals or smaller currents have to be
chosen. The total change of the transient voltage without the IR drop
(which is easily detectable by a fast recorder) is ΔE_t. When the current
is switched off, the sample equilibrates according to the new concentration of the electroactive component which has been changed by the current
flux. The change of the steady state voltage compared to the previous
value is ΔE_s. The chemical diffusion coefficient $\tilde{D}$ for the given stoichiometry is easily determined from the ratio ΔE_t and ΔE_s

$$\tilde{D} = \beta \frac{\Delta E_s}{\Delta E_t} \; ; \; \beta = \frac{4}{\pi \tau} \left(\frac{nV_M}{zS}\right) \tag{21}$$

where S is the interface area with the electrolyte.

A large variety of other kinetic quantities may be easily derived
from ΔE_t and ΔE_s such as the diffusivity (or component diffusion coefficient) D, the electrical mobility u, the general mobility b, Wagner's
enhancement factor W and the partial ionic conductivity σ of the mobile
type of ions in the mixed conductor:

$$D = \frac{\beta_1}{c} \frac{\Delta E_s}{(\Delta E_t)^2} ; \; \beta_1 = -\frac{4kTmV_M I_o}{\pi z^2 q^2 S^2} \tag{22}$$

$$u = |z|qb = \frac{\beta_1|z|q}{kTc} \frac{\Delta E_s}{(\Delta E_t)^2} \tag{23}$$

$$W = -\frac{zqcV_M}{kTL} \frac{dE}{dx} \tag{24}$$

$$\sigma_A = \beta_2 \frac{\Delta E_s}{(\Delta E_t)^2}; \quad \beta_2 = -\frac{4nV_MI_0}{\pi S^2} \tag{25}$$

The process is repeated after the voltage has reached a steady state value until the range of existence of the phase is exhausted. As an example, Fig. 24 shows the variation of kinetic parameters as a function of stoichiometry for $Li_{3+\delta}Sb$ and $Li_{3+\delta}Bi$. A jump in the diffusion coefficients may be easily recognized which gives an indication of different diffusion mechanisms.

The knowledge of the kinetic parameters as a function of stoichiometry allows one to calculate the growth (or decomposition) of materials under the influence of activity differences (e.g., in the case of oxidation or reduction processes). This can be done for any combination of boundary conditions. The parabolic growth of the reaction product at different given activities of the components is described by Tammann's parabolic tarnishing rate constant k_t and may be calculated from the integration of the diffusion constants [23] over the sample

$$k_t = \frac{1}{c} \left| \int \tilde{D} \, dc \right| = \left| \int Dt_e \, d \ln a \right| \tag{26}$$

$\bar{c}$ and t_e are the average concentration of the mobile component of the sample and the tranference number of the electrons, respectively.

Fig. 25 shows the parabolic tarnishing rate constant for the growth (or decomposition) of Li_3Sb as a function of the lithium activity if the material is at the other side in contact with (a) Li_2Sb, (b) elemental lithium and (c) a phase with an intermediate stoichiometry which has a lithium activity of $10^{-3.5}$. Conventionally it would be necessary to perform a separate experiment for the determination of k_t for each pair of boundary conditions. Electrochemically, the GITT technique allows one to obtain the information from a single sample. Favourable reaction parameters may be chosen with such information available. If, e.g., fast growth is required, a high lithium activity is necessary. It is not important in this case whether Li_2Sb or Li_3Sb with an intermediate lithium activity are used.

The composition may deviate substantially during transient processes from any composition that might be expected from the overall composition of the sample in the case of multinary systems. The surface of the sample follows a "kinetic path" in the multinary phase diagram which may result in the formation of unexpected and metastable phases. With the knowledge of the cell voltage for the various regions in the multinary phase diagram, the observed galvanic cell voltage in the course of a transient process may also provide information on the path of composition in this case.

ELECTROCHEMICAL DETERMINATION OF KINETIC PATHS

For a binary system it is unimportant which of the activities of component A or B is changed in a chemical diffusion coefficient since both are related by Duhem Margules' equation

$$c_A d \ln a_A + c_B d \ln a_B = 0 \quad (p,T = const.) \tag{27}$$

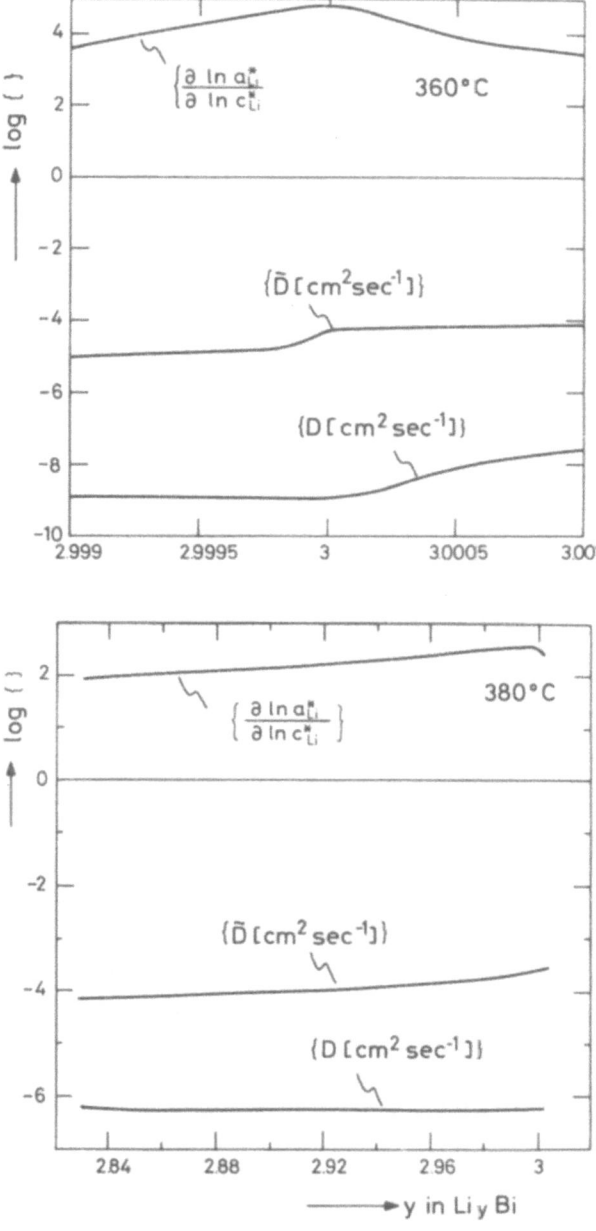

Fig. 24   Wagner factor (thermodynamic factor) W, chemical diffusion coefficient D̃ and diffusivity (component diffusion coefficient)
D as a function of the stoichiometric deviation for the compound Li₃Sb.

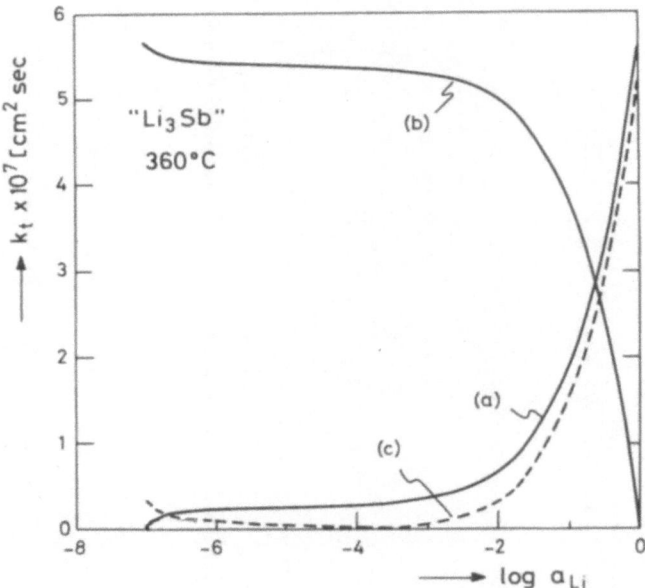

Fig. 25   Parabolic (Tamann's) rate constant of Li_3Sb as a function of
the lithium activity for three different lithium activities of
the substrate: (a) Li_2Sb, (b) Li and (c) an intermediate lithium activity of $10^{-3.5}$.

For the description of the composition as a function of location and time
it is unimportant which of the components is predominantly mobile since
the motion may be characterized by the ratio of the two concentrations.
Assuming local thermodynamic equilibrium, the composition of the sample
at any time and location agrees with the compositions that are observed
for infinitely slow processes.

A ternary system has an additional degree of freedom and the local
composition may deviate substantially from the sequence of equilibrium
compositions between the starting and final compostion in the course of a
relaxation process. The chemical potentials of all components (A, B and
C) are again related,

$$c_A \, d\,\mu_A + c_B d\mu_B + c_C d\mu_C = 0 \ (p,T = const.) \tag{28}$$

but not the two independent ratios of concentrations of the three components. Variation of one of the chemical potentials results in the variation of the chemical potentials of both other components (Fig. 26). A
driving force exists for any of the components and the predominantly
mobile species will change the composition whereas the ratio of the concentrations of the other components remains nearly unchanged during the
initial period of time until a steady state equilibrium is reached. If
the predominantly mobile species differ from the component which has been
forced to change, a deviation will occur from the sequence of composi-

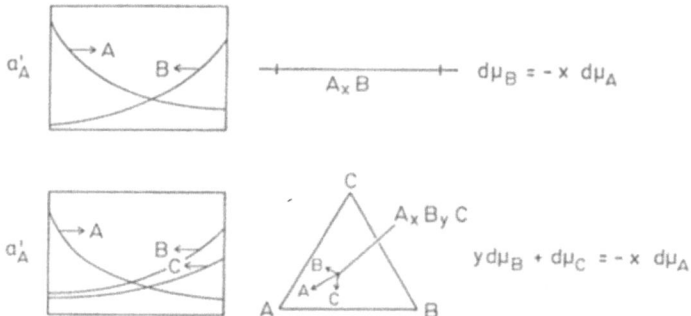

Fig. 26   Schematic representation of the variation of the chemical
           potentials and the driving forces for the motion of the diffe-
           rent components for a binary and ternary system. The composi-
           tions of a binary phase belong to thermodynamic equilibrium
           compositions of the system. Variation of the activity A in a
           ternary system may result in large deviations of the local
           composition (e.g., at the surface) from the thermodynamic path
           of the straight line that connects the original composition
           with the A corner. If B or C show higher mobilities, the
           composition will deviate in the directions of the corners of
           these components.

tions which would be observed under permanent quasi-thermodynamic equilibrium of the sample. Quite different stoichiometries and phases may be
observed which have other thermodynamic, kinetic, catalytic and other
physicochemical properties.

As for the determination of kinetic quantities, an electrolyte may
be used to probe the surface composition of multinary systems. It is
especially easy to recognize the temporary equilibrium between k phases
of the k component system. Fig. 27 shows the kinetic path of the reduction and reoxidation of $CuGeO_3$. Application of a current of 10^{-4} A/cm^2 at
900°C by employing a zirconia solid electrolyte shows immediately voltage
plateaus corresponding to the 3-phase equilibrium $Cu-Ge-GeO_2$ without
showing any shoulder corresponding to the other 3-phase equilibria along
the thermodynamic path (broken line). This may be explained by predominantly mobile copper ions in $CuGeO_3$. Their activity is increased at the
surface with the decrease of the oxygen activity. Copper diffuses into
the sample and leaves the surface enriched in germanium. The equilibrium
between Ge and GeO_2 determines the oxygen activity. Reoxidation shows an
opposite behavior. Increasing oxygen activity results in a decrease of
copper activity at the surface. Copper becomes enriched compared to germanium and the Cu-O equilibria determines the oxygen activity. Voltage
plateaus corresponding to the equilibria $Cu-Cu_2O(-GeO_2)$ and
$CuGeO_3-Cu_2O-GeO_2$ are observed. The kinetic path is shown in Fig. 27 as a
dotted line. It is obvious that the surface layers are important for the
properties of the materials. The germanium oxides control the reduction
processes whereas the copper oxides are determining the oxidation processes.

Reactions under real conditions may tend to form metastable phases
which may be easily recognized from voltage plateaus which may not be
attributed to equilibria of the thermodynamic values of the phase dia-

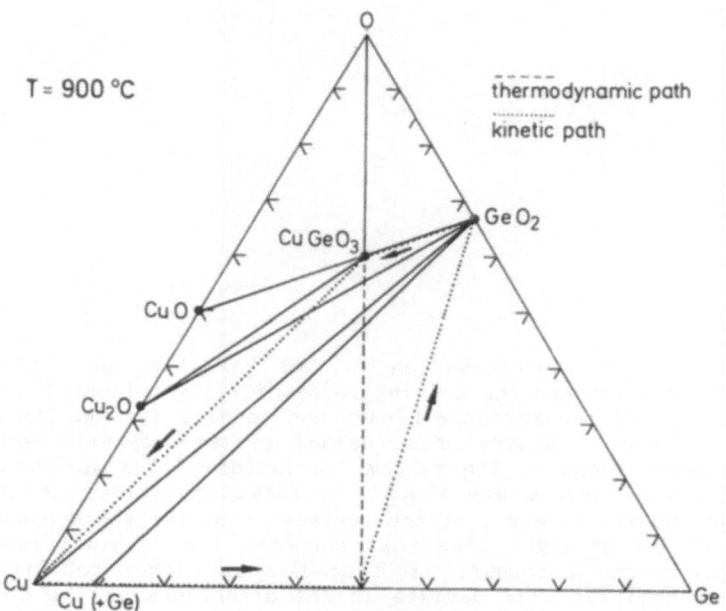

Fig. 27   Schematic representation of the path of surface composition of a CuGeO$_3$ sample during reduction and subesequent oxidation. The composition tends to move away or toward the copper corner because of the higher mobility of copper ions.

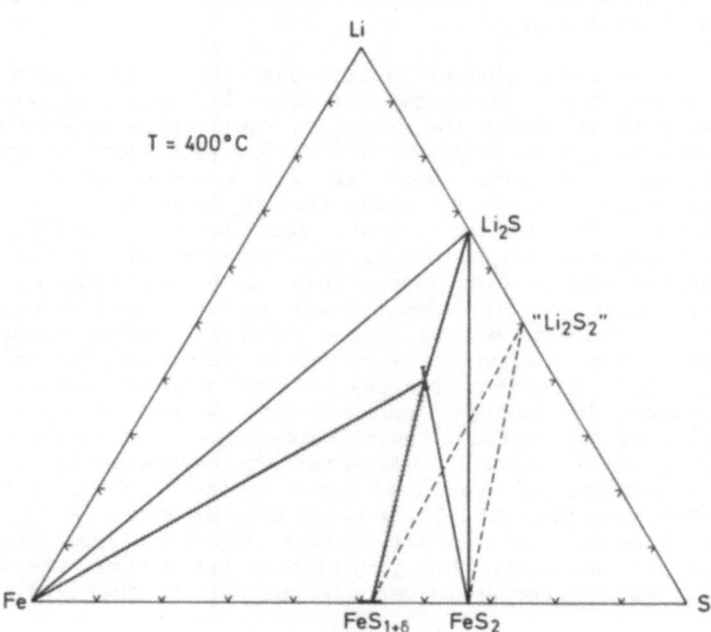

Fig. 28   Phase equilibria of the ternary system Li-Fe-S including the metastable phase 'Li$_2$S$_2$'.

gram. This behavior has been observed in the case of the ternary system Li-Fe-S which shows a voltage plateau 79 mV below the thermodynamic value of 2.049 V at 400°C upon titration of lithium into FeS_2 (Fig. 16). Titration of lithium in the reverse direction shows again the upper voltage. The upper voltage is also observed after waiting for thermodynamic equilibrium for several hours. This behavior is explained by the formation of "Li_2S_2" which is temporarily formed as a metastable compound and a quasi-equilibrium exists between FeS, FeS_2 and Li_2S_2 (Fig. 28).

REFERENCES

[1]   W. Weppner, Phase Stability of Crystalline Fast Ion Conductors, this volume
[2]   C. Wagner, J. Chem. Phys. 21, 1819 (1953)
[3]   W. Weppner and R.A. Huggins, J. Electrochem. Soc. 125, 7 (1978)
[4]   Chen Li-chuan and W. Weppner, Naturwiss. 65, 595 (1978)
[5]   W. Weppner, Chen Li-chuan, and A. Rabenau, J. Solid State Chem. 31, 257 (1980)
[6]   W. Weppner and R.A. Huggins, in: Fast Ion Transport in Solids; Electrodes and Electrolytes (P. Vashishta, J.N. Mundy, and G.K. Shenoy, Eds.), Elsevier North Holland Inc. New York, N.Y., 1979, p. 475
[7]   W. Weppner and R.A. Huggins, Solid State Ionics 1, 3 (1980)
[8]   W. Piekarczyk, W. Weppner, and A. Rabenau, Z. Naturforsch. 34a, 430 (1979)
[9]   W. Sitte and W. Weppner, Appl. Phys. A 38, 31 (1985)
[10] W. Sitte and W. Weppner, Z. Naturforsch. 42a, 1 (1987)
[11] P. Hartwig, W. Weppner, and W. Wichelhaus, in: Fast Ion Transport in Solids; Electrodes and Electrolytes (P. Vashishta, J.N. Mundy, and G.K. Shenoy, Eds.) Elsevier North Holland Inc., New York, N.Y., 1979, p. 487
[12] P. Hartwig, W. Weppner, W. Wichelhaus, and A. Rabenau, Angew. Chem. 92, 72 (1980; Angew. Chem. Int. Ed. Engl. 19, 74 (1980)
[13] W. Weppner, Chen Li-chuan, and W. Piekarczyk, Z. Naturforsch. 35a, 381 (1980)
[14] W. Weppner, Novel Solid State Gas Sensors, this volume
[15] E.J.L. Schouler, in: Solid State Protonic Conductors III; for Fuel Cells and Sensors (J.B. Goodenough, J. Jensen, and A. Potier, Eds.) Odense Univ. Press Odense, Denmark, 1985, p. 16
[16] M. Hebb, J. Chem. Phys. 20, 185 (1952)
[17] C. Wagner, Proc. 7th Meeting CITCE, Lindau, 1955, Butterworth, London, 1957, p. 361
[18] W. Weppner, Z. Naturforsch. 31a, 1336 (1976)
[19] W. Weppner, Electrochim. Acta 22, 721 (1977)
[20] W. Weppner and R.A. Huggins, Annual Review of Materials Science 8, 269 (1978)
[21] W. Weppner and R.A. Huggins, J. Electrochem. Soc. 124, 1569 (1977)
[22] W. Weppner and R.A. Huggins, J. Solid State Chem. 22, 297 (1977)
[23] W. Weppner and R.A. Huggins, Z. Physikal. Chem. N.F. (Frankfurt) 108, 105 (1977)

PHASE STABILITY OF CRYSTALLINE FAST ION CONDUCTORS

Werner Weppner

Max-Planck-Institut für
Festkörperforschung
Heisenbergstr. 1
D-7000 Stuttgart 80
Fed.Rep.Germany

INTRODUCTION

Fast ionic transport in solids has received much attention in recent years. This is not only due to the curiosity in this phenomenon but also caused by several potential important technological applications [1]. A variety of concepts relates these materials to new approaches in the fields of energy storage devices (high energy and power density secondary batteries), direct energy conversion (fuel cells), alternative fuels (hydrogen production; water electrolysis) and environmental protection (timers). Some other interesting possible applications are in connection with other recent technological advances, especially in the fields of microelectronics and ceramics. Examples are chemotronic devices, timers, memory elements, integrated batteries, or electrochromic windows and displays. In addition, on a more scientific level, fast solid ionic conductors became very important for precise measurements of fundamental thermodynamic and kinetic data and for the determination of equilibrium phase diagrams [2].

The fast ionic conductors are employed in all these applications in galvanic cells, i.e., in contact with electrode materials. These generate electrical voltages in case of the presence of different chemical potentials of the mobile component in both electrodes. Alternatively, voltages may be applied to the galvanic cell and the resulting currents may produce changes of the chemical potential of the mobile component in the electrodes and chemical reactions. The contact between the phases should be generally as unpolarizable as possible. That means, the transfer of the ions should not be impeded. The fast exchange and motion of the mobile component favours the equilibration between the phases. If these are not thermodynamically in equilibrium, this results often in several undesirable effects:

- The formation of new electronically conducting intermediate phases which have different activities of the mobile component and lower the cell voltage. The electrolyte and electrode may be eventually completely consumed by the chemical reaction.

- The formation of a kinetic barrier for the transfer of the ions by the reaction product, especially if the new compound is a poor ionic conductor. The current density will be drastically decreased.

Even thin reaction layers may cause important changes in the cell voltage or the resistance of the cell.

The galvanic cells are commonly multicomponent systems which may result in a large number of possibly formed intermediate phases. A 'kinetic' stability because of extremely slow reaction (diffusion) rates is generally not sufficient in the case of electrolyte/electrode contacts. The operation of a galvanic cell depends commonly on the fast motion of the ions in all phases. Only electrochemically inactive components which may be present to form a specific structure or conductivity may be possibly present under 'kinetic' stability conditions, but even the formation of very thin reaction layers of these components may result in a blocking of the transport of the mobile ions.

The stability requirements are quite different depending on the type of application of the galvanic cell. This depends mainly on the required depth of discharge and the cyclability of the cell. A battery, e.g., has to make use of electrodes with a sufficiently large difference in the activity of the electroactive component in order to generate a large voltage. In addition, the composition of the electrodes has to vary largely in the course of the required deep discharge process. The electrolyte has to be stable against reaction with the electrodes under all these different conditions. In contrast, other applications require only minor changes in the composition of the electrodes, e.g., capacitors, timers, or electrochromic windows or displays. The composition of the electrodes is not changed at all in the case of fuel and elctrolysis cells. Other practical galvanic cells require only a small stability window of the electrolyte since the activity of the counter electrode may be chosen arbitrarily and may be therefore close to the activity of the working electrode. Primary cells may be in some cases sufficiently tolerant to a slow formation of undesirable reaction products (as in the case of many primary lithium batteries based on liquid organic electrolytes) whereas secondary (rechargeable) cells will approach equilibrium more and more with each cycle and deviate from the desirable phase relations.

Thermodynamic data are also closely related to other materials properties than the stability. A change in the activity of the components corresponds to a change of the stoichiometry of a compound which causes a variation of the number or even the type of ionic defects and often very sensitively the electronic properties. The phase width is strongly dependent on these parameters and has to be very narrow in the case of ionic conductors. As will be shown later, the thermodynamic properties are also closely related to the ionic transport quantities and to the kinetic enhancement factor.

STABILITY OF BINARY ELECTROLYTES

The thermodynamic stability requirement is comparatively transparent in the case of binary and quasibinary systems. The limits of stability of the electrolyte A_yB are given by the equilibria with the two neighbouring phases A_xB and A_zB $(x<y<z)$. The equilibrium between two binary phases corresponds to an activity a_A of the electroactive component A which is determined by the difference of the Gibbs energies of formation ΔG_f° (related to 1 molecule) of the compounds which are in thermodynamic equilibrium

$$A_xB + (y-x)A = A_yB; \quad \Delta G_f^O(A_xB) + (y-x)kT \ln a_A = \Delta F_f^O(A_yB) \tag{1}$$

The activities of the electroactive component have to be in both electrodes within the range that is given by the two neighbouring phases

$$\exp\{\frac{\Delta G_f^O(A_yB)-\Delta G_f^O(A_xB)}{(y-x)kT}\} \leq a_A \leq \exp\{\frac{\Delta G_f^O(A_zB)-\Delta G_f^O(A_yB)}{(z-y)kT}\} \tag{2}$$

In the case of gaseous electrodes, e.g., in fuel cells or sensors, the activity difference corresponds to the applicable partial pressure range. The electrolyte opens a "stability window" [3] for the activity of the electroactive component in the electrodes. But the electrodes do not only consist of the electroactive component in most cases, and the phase stability of a system with three or more components has to be considered. Side reactions may occur which will be illustrated by a few examples.

The fast silver ion conductor AgI may be stable in an activity range of silver which includes the stability range of the compound Ag_2S but a reaction occurs nevertheless due to the presence of sulfur. The ternary compound Ag_3SI is formed at the interface. This phase is a fast silver ion conductor and does therefore not influence the voltage and kinetics of the galvanic cell. The electrolyte is a series connection of two phases in that special case. The growth of the intermediate layer is very slow because of the low electronic conductivity. The simultaneous motion of the silver ions is required for the net transport of neutral components across the reaction product layer.

Zirconia is one of the most stable oxides and a good oxygen ion conductor at elevated temperatures when it is stabilized in the cubic or tetragonal modification by a small amount of dopant such as CaO, MgO or Y_2O_3. The decomposition voltage is 2.22 V with reference to 1 atm oxygen partial pressure at 1000°C. This large stability range allows one to apply the material both under high oxidizing and reducing conditions, e.g., in fuel cells, in high temperature water electrolysis cells or in oxygen sensors down to an oxygen partial pressure of 10^{-36} atm [4]. Porous platinum is commonly used as a "chemically inert" electrode material which provides the electronic lead to the electrolyte. No reaction takes place between ZrO_2 and Pt when heated in air. If the oxygen activity is small, however, a reaction between zirconia and platinum will take place because of the increased zirconium activity in accordance with the Gibbs–Duhem relation. The electroactive component is only indirectly involved in the reaction.

The Pt-Zr compounds were formed and investigated in-situ using galvanic cells which employ zirconia as solid electrolyte:

Inert gas, Pt(Pt-Zr) | ZrO_2(+10 m/o Y_2O_3) | air, Pt

Originally, only a thin platinum layer was present at the left hand side. Applying a positive voltage E to the galvanic cell reduces the chemical potential $\mu_{O_2}^!$ or the oxygen partial pressure $p_{O_2}^!$ at the left hand phase boundary where the electrolyte is in contact with the Pt electrode according to Nernst's equation

$$E = \frac{1}{4q}(\mu_{O_2}^" - \mu_{O_2}^!) = \frac{kT}{4q} \ln (p_{O_2}^"/p_{O_2}^!) \tag{3}$$

$\mu_{O}^"$ and $p_{O}^"$ refer to the reversible air electrode. k, T and q are Boltzmann's constant, the absolute temperature and elementary charge, respectively.

The decrease in oxygen chemical potential is coupled to an increase in the zirconium chemical potential μ_{Zr} in the zirconia:

$$d\mu_{O_2} + d\mu_{Zr} = 0 \tag{4}$$

At a sufficiently low oxygen partial pressure the zirconium chemical potential will be high enough to form Pt–Zr compounds. The oxygen which is formed during this reaction is transported as ions through the electrolyte. The electrons come from the Pt lead. Once the compound has been formed, the emf of the cell is determined by the zirconium activity at the left hand side. According to Gibbs' phase rule, the chemical potential of Zr becomes independent of the composition when two phases are present and in equilibrium. The emf of the galvanic cell is therefore independent of the composition in this range at the left hand electrode. The voltage at which the formation of Pt–Zr compounds ($ZrPt_y$ and $ZrPt_x$) occurs is related to the Gibbs' energy of the cell reaction

$$\frac{x}{y-x} ZrPt_y + ZrO_2^{cub.} = \frac{y}{y-x} ZrPt_x + O_2(g) \tag{5}$$

which is a combination of the standard Gibbs energies of formation of the two Zr–Pt compounds and of zirconia:

$$4q(E + E_o) = \frac{x}{y-x} \Delta G_f^o(ZrPt_y) + \Delta G_f^o(ZrO_2^{cub.}) - \frac{y}{y-x} \Delta G_f^o(ZrPt_x) \tag{6}$$

E_o relates to a standard reference electrode of $p_{O_2} = 1$ atm.
The formation of Pt–Zr compounds starts at voltages as low as 1.225 V with reference to 1 atm oxygen partial pressure (or at a corresponding oxygen partial pressure of 4.1×10^{-20} atm) at 1000°C.

Eq. (6) allows one to determine the Gibbs energy of formation or the decomposition voltage of the electrolyte ZrO_2 if data for the formation of the compounds $ZrPt_3$ and $ZrPt_5$ are taken from the literature [5]. The following equations for the standard Gibbs energy of formation ΔG_f^o and the decomposition voltage E^* are obtained for the temperature range between 900 and 1100°C [4]

$$\Delta G_f^o (ZrO_2^{cub}) \text{ [eV]} = -11.379 + 1.96 \times 10^{-3} T \text{ [K]} \tag{7}$$

$$E^* \text{ [mV]} = 2844.77 - 0.49 T \text{ [K]} \tag{8}$$

This is an example of an indirect measurement of the stability of an electrolyte since a determination by direct electrochemical decomposition is not possible in this specific case because the material becomes electronically conducting before the decomposition voltage is reached.

From a practical point of view, the formation of Pt–Zr compounds may be not important for the performance of the cell. Rather than pure platinum, a Pt–Zr alloy acts as the electronic lead under these circumstances. The specific phases which are formed will be adjusted by the oxygen partial pressure. The Pt–Zr alloy does not determine the voltage in this case in the presence of oxygen but may have an influence on the kinetics.

In view of the complications with the chemistry of additional components of the galvanic cell by the electrodes, it is often helpful to restrict stability considerations to one electrode side. This may be easily achieved by selecting electrolytes which are stable with the pure electroactive component such as in the case of Ag/AgI, Li/LiI or Li/Li$_3$N. The disadvantage may be, however, the difficulty in handling some very reactive materials such as elemental lithium. In the case of the lithium battery based on a LiI solid electrolyte a similar approach as for the anode has been applied to the cathode side. Elemental iodine (which is combined with an electronically conducting charge transfer material) is used.

$$Li \mid LiI \ (+Al_2O_3) \mid I_2, \ poly-2-vinylpyridine$$

Such a cell in which the electrodes consist of the two elemental components of the binary electrolyte has also the advantage to be selfhealing in case of some mechanical fracture of the electrolyte. Contact of both electrode materials will immediately result in an in-situ formation of the electrolyte. The disadvantage of LiI is the low ionic conductivity at ambient temperatures and batteries based on the system have only found practical use in low power density pace maker cells. Other anionic components of ionic conductors than iodine are even less suitable for pure elemental cathode materials.

A general problem in the application of binary solid electrolytes are related to such observations that all investigated materials have either high stability or high conductivity but not both properties simultaneously, if high temperatures for high conductivities should be avoided. The interest has therefore been directed to look into compounds with three or more components. The increased number of components increases the number of degrees of freedom, however, and the stability requirements may become quite complex.

PHASE STABILITY OF MULTINARY SYSTEMS

A major breakthrough in the search for fast ion conductors has been the discovery of Ag_4RbI_5 which is stable at room temperature in contrast to α-AgI which requires heating to above the phase transition temperature of 147°C. The additional cation does not influence the stability against elemental silver which may be therefore used as an anode material.

A similar approach of 'modification' of the lattice of the conducting cations shows quite different results in the case of lithium compounds. LiCl and $AlCl_3$ react in an equimolar ratio to form the fast ion conductor $LiAlCl_4$. This compound may be easily prepared by heating the two binary salts to about 160°C in a common test tube in the glove box. The ternary compound melts at 146°C. Both the solid and the melt are of potential interest as an ionic conducting materials for batteries. It is readily found experimentally, however, that $LiAlCl_4$ is not in equilibrium with elemental lithium.

If the maximum applicable lithium activity is only slightly reduced by the presence of Al this would not matter very much because a two-phase mixture such as LiAl/Al could be acceptable as an anode material. Since the phase diagram and thermodynamic properties of the ternary system Li-Al-Cl have not been known, it has not been possible to make predictions on the stability range at the time of discovery of fast ionic conduction in $LiAlCl_4$. The stability of $LiAlCl_4$ has therefore been determined using two different electrochemical methods. A coulometric titration process has been employed to vary the composition of lithium with the help of an auxiliary solid electrolyte, $Li_4SiO_4 + Li_3PO_4$ (solid solution) [6]. The result is shown in Fig. 1. The two plateaus correspond to 3-phase equilibria in the direction toward and opposite of the lithium corner. The right hand plateau is influenced by a decreased chlorine partial pressure and is approximately 1 V higher in the case of the presence of pure chlorine gas. It is found that $LiAlCl_4$ is stable over a large voltage range, but only at extremely low lithium activities. It would be necessary to find as well an anode as a cathode material which exist within these

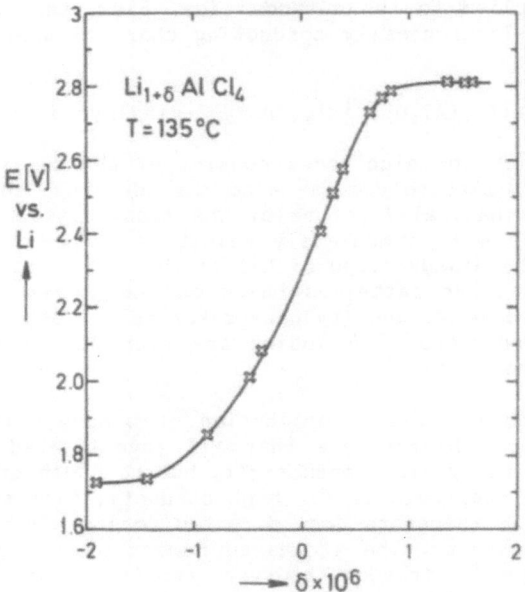

Fig. 1   Coulometric titration curve for the phase $Li_{1+\delta}AlCl_4$. Plateaus correspond to the activities of the 3-phase regions in the direction toward and opposite of the lithium corner under the existing chlorine partial gas pressures.

limits. This appears to be difficult for batteries which require a large variation of the composition for deep discharge. The situation is much simpler in the case of other devices with a small variation of the composition and the application of two electrodes with similar activities.

The knowledge of the Gibbs energies of formation allows one to calculate the thermodynamically most stable configuration, i.e., of the phase equilibria which are shown in Fig. 2. The observed plateaus in the coulometric titration curve correspond to the 3-phase equilibria LiCl, Al, $LiAlCl_4$ and Cl_2, $LiAlCl_4$, $AlCl_3$. The dependence of the cell voltage for the 3-phase region Cl_2, $LiAlCl_4$, $AlCl_3$ from the chlorine partial pressure has been used to construct chemical chlorine gas sensors [7-9].

The phase equilibria as shown in Fig. 2 have been confirmed by studying a galvanic cell using $LiAlCl_4$ as the electrolyte which is in equilibrium with LiCl and Al at one hand side and $AlCl_3$ and Al at the other hand side. This is another method to measure the Gibbs energy of formation of $LiAlCl_4$.

The applicable activity range of the electroactive component (lithium) is directly observed from the coulometric titration curve (Fig. 1) by application of Nernst's equation. The activities of the 3-phase regions may be related to the three Gibbs energies of formation. To that purpose we consider the addition of δ electroactive ions A to the three phases $A_{x_i} B_{y_i} C_{z_i}$ (i = 1,2,3). The corresponding Gibbs energy of the corresponding reaction ΔG is then given by [10]

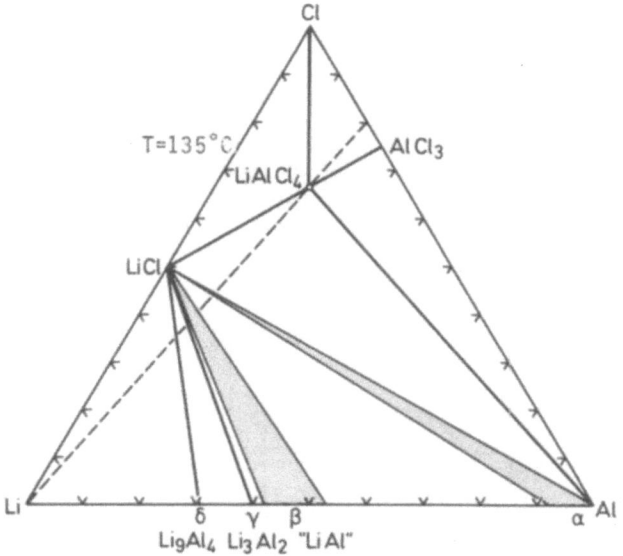

Fig. 2   Phase diagram (Gibbs triangle) of the ternary system Li–Al–Cl
          with the fast ionically conducting ternary lithium compound
          LiAlCl$_4$. The direction of the variation of the stoichiometry by
          coulometric titration is shown by the broken line.

$$\Delta G = - \frac{\delta}{d} \sum_{i=1}^{} (-1)^i d_{i1} \, \Delta G_f^0 (A_{x_i} B_{y_i} C_{z_i}) \qquad (9)$$

where d is the determinant formed by the stoichiometric numbers of the
three compounds

$$d = \begin{vmatrix} x_1 & y_1 & z_1 \\ x_2 & y_2 & z_2 \\ x_3 & y_3 & z_3 \end{vmatrix} \qquad (10)$$

and d_{i1} is the minor formed by eliminating the i-th row and first column
(stoichiometric numbers of the electroactive component). The cell voltage
with reference to an electrode which consists of elemental A, the activity of A and the Gibbs energy of reaction ΔG are then related by

$$E = - \frac{kT}{zq} \ln a_A = - \frac{\Delta G}{zq\delta} \qquad (11)$$

according to Nernst's law and the equivalence of electrical and chemical
energy. z is the charge number of the mobile ions. The working electrode
with the 3-phase equilibrium is assumed to be located at the right hand
side of the electrolyte. From Eqs. (9) and (11) the following expression
is obtained for the activity of A for the equilibrium of the three phases
$A_{x_i} B_{y_i} C_{z_i}$ (i = 1,2,3)

$$a_A = \exp(- \frac{zqE}{kT}) = \exp \{ - \frac{1}{kTd} \sum_{i=1}^{3} (-1)^i d_{i1} \Delta G_f^0 (A_{x_i} B_{y_i} C_{z_i}) \} \qquad (12)$$

In addition to the consideration of the electroactive component, the stability has to be also concerned with the activity range of the other components. $LiAlCl_4$ may only exist under a certain chlorine partial pressure which therefore needs to be controlled. Also, aluminum may show a reaction with the electrodes depending on its activity, similar to the reaction of zirconium with platinum. Knowledge of the phase diagram and the Gibbs energies of formation of all compounds allows readily to determine the activity range of all components. Eq. 12 may be generalized by assuming (fictively) any other component to be mobile. Following the same steps as for the derivation of Eq. (12), the activities of B and C are in any 3-phase region as follows

$$a_B = \exp \{- \frac{1}{k\bar{T}d} \sum_{i=1}^{3} (-1)^i d_{i2} \, \Delta G_f^0 \, (A_{x_i} B_{y_i} C_{z_i}) \} \tag{13}$$

$$a_C = \exp \{- \frac{1}{k\bar{T}d} \sum_{i=1}^{3} (-1)^i d_{i3} \, \Delta G_f^0 \, (A_{x_i} B_{y_i} C_{z_i}) \} \tag{14}$$

d_{i2} and d_{i3} are the minors which are formed by eliminating the i-th row and second and third column of d, respectively. $LiAlCl_4$ is on one side stable in equilibrium with chlorine gas of 1 atm. The minimum chlorine partial pressure is given by the 3-phase equilibrium $LiAlCl_4$, $LiCl$, Al. The corresponding chlorine partial pressure is calculated to be 10^{-74} atm at 25°C. Such a small number is of no practical relevance with regard to the direct exchange of chlorine between the gas phase and the solid. It becomes important, however, if an equilibrium by the mixture of other gases, e.g., H_2 and HCl, is used. The number of gaseous species is in this case large enough to exchange and equilibrate with the sample.

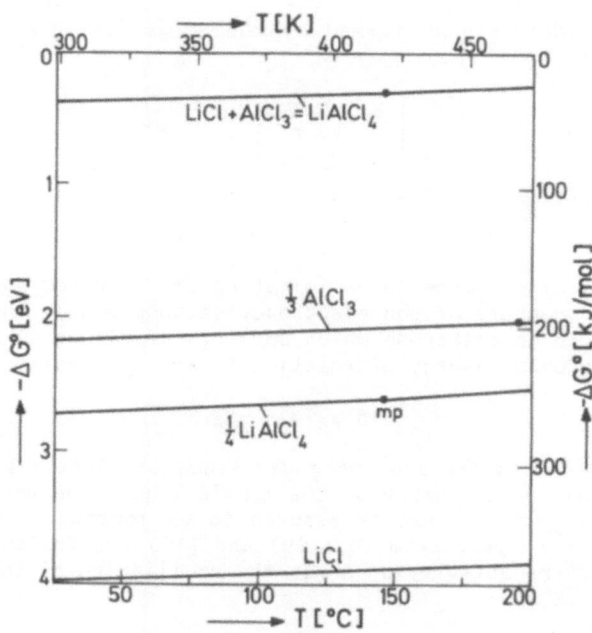

Fig. 3   Standard Gibbs energies of formation of the binary and ternary compounds of the system Li–Al–Cl as a function of temperature.

234

LiAlCl$_4$ is stable in equilibrium with elemental aluminum but not with elemental lithium. This behavior is caused by the fact that LiCl is a very stable compound and AlCl$_3$ is a very unstable compound (Fig. 3). Adding lithium to the ternary compound tends to displace aluminum under the formation of LiCl and Al. This behavior is in contrast to the situation in ternary silver salts. AgI is thermodynamically less stable than RbI. Silver has therefore no tendency to replace rubidium and does therefore not react with the fast ion conductor Ag$_4$RbI$_5$.

According to this knowledge, ternary lithium ion conductors should be searched among those compounds which include a second cation which forms a more stable binary salt than lithium (analogous to Ag$_4$RbI$_5$). This requirement is, however, only fulfilled for a very small number of elements. Lithium salts are among the most stable compounds in contrast to silver salts which are generally rather unstable. In fact, the high stability of lithium compounds is the major reason for looking into lithium systems for batteries rather than looking into silver systems since the electrical energy is equivalent to the chemical energy of reaction.

Among the ternary lithium salts with a chemically more reactive second cation, none of the materials studied so far have shown high conductivity with applicability in practical devices. Another route has therefore been taken in order to obtain lithium ion conductors which are stable against reaction with lithium and also show high decomposition voltage. Two or more binary lithium salts which are all stable with lithium are combined in this approach to form multinary compounds. Lithium is the only cation but several anions are present. Since all binary lithium salts are stable against reaction with lithium, the multinary compound will be also stable with lithium since we may assume that no other compound will exist which has a higher lithium content. This concept has been applied to the systems Li$_3$N-LiCl, Li$_3$N-LiBr, Li$_3$N-LiI and others. Several new solid electrolytes with high ionic conductivity and thermodynamic stability have been found [11-13]. All ternary compounds exist along quasibinary sections lithium nitride - lithium halide.

A comparatively simple technique allowed the determination of ternary phase diagrams. The pellet-shaped samples were contacted at one side with an elemental lithium electrode and at the other side with an inert molybdenum electrode. The shape of the current-voltage curve with the negative polarity at the lithium electrode allowed determination of the decomposition voltage from the extrapolation of the linear increase of the current to the voltage axis. The decomposition voltage corresponds to the cell voltage of the 3-phase region in the opposite direction to the lithium corner. The lithium activity and the Gibbs energies of formation of the involved compounds may be calculated as indicated by Eq. 12. Knowledge of the Gibbs energies of formation of the binary compounds allows one to calculate a value for the Gibbs energy of formation of the ternary compounds for any possible 3-phase equilibrium and to search for the most stable configuration. The results are shown in Figs. 4 - 6 which also include the value of the decomposition voltage. The most interesting compound, Li$_9$N$_2$Cl$_3$, is stable with lithium and allows the presence or application of a voltage of 2.21 V at 650 K. This compound also shows fast conductivity. It has an antifluorite type structure ("Li$_{1.8}$N$_{0.4}$Cl$_{0.6}$") and 10% of the lithium sites are vacant. It may be assumed that these structural defects allow high ionic conductivity for the lithium ions. The presence of vacancies in the lithium partial lattice is apparently necessary for the formation of the antifluorite structure.

The addition of LiCl to Li$_3$N increases drastically the decomposition voltage from 0.270 V to 2.21 V at 605 K. The stability of the ternary

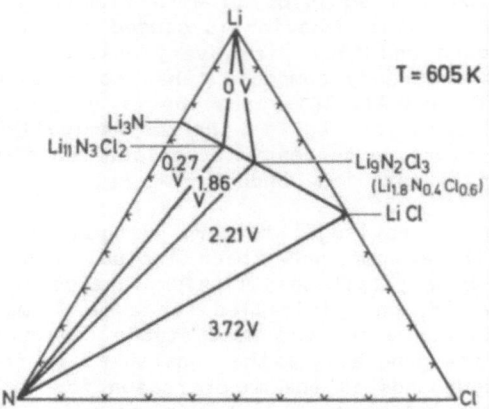

Fig. 4   Gibbs triangle of the ternary system Li-N-Cl. The indicated voltages are the emfs which are observed for the corresponding 3-phase regions using a lithium ion conductor with reference to elemental lithium. These numbers are related to the lithium activity according to Nernst's equation and have been determined by decomposition of the solid lithium ion conductors along the quasi-binary section Li_3N-$LiCl$.

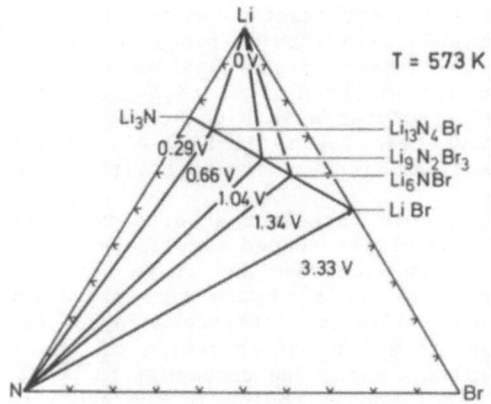

Fig. 5   Gibbs triangle of the ternary system Li-N-Br. The voltages correspond to the lithium activities of the corresponding 3-phase regions (see Fig. 3).

phase is in between those of the two binary compounds, i.e., between 0.270 and 3.72 V. This is the reason why on the one hand side a material with a high ionic conductivity but low stability (Li_3N) has been combined with a very stable compound ($LiCl$) in order to form an ionic conducting material that possibly assumes high stability and conductivity. Because of the high stability of the lithium halides compared to lithium nitride, all ternary phases are in equilibrium with the nitrogen corner. Nitrogen has no tendency to displace the chlorine in the compound or to show a reaction since other ternary compounds have not been found. The thermodynamic stability is, however, limited to a certain minimum nitrogen partial pressure.

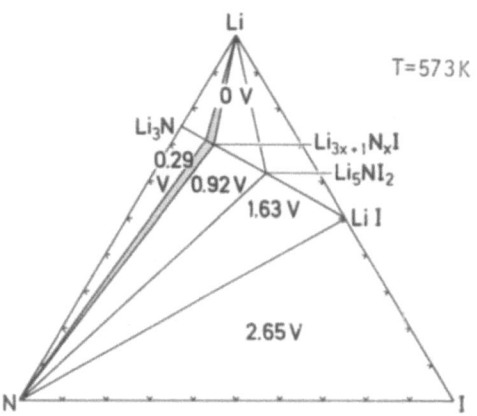

Fig. 6   Gibbs triangle of the ternary system Li-N-I. The voltages correspond to the lithium activities of the corresponding 3-phase
regions (see Fig. 3) and were determined from decomposition
measurements.

None of the ternary compounds is in equilibrium with the halide
corner of the Gibbs triangle. The compounds will react under high chlorine partial pressures and displace the nitrogen to form the more stable
lithium halide. The value for the halide partial pressure range may be
easily calculated from the information provided by the decomposition
measurements. Knowledge of the Gibbs energies of formation of all compounds of the ternary phase diagrams allows one to determine the activities of all components in any 3-phase region by making use of Eqs. (12) –
(14). The result of such an evaluation for all phases along the quasibinary section $Li_3N-LiCl$ is shown in Fig. 7. The ranges of lithium activity
and the partial pressures of nitrogen and chlorine are plotted for which
the phases are stable. The ternary compounds exist under any practical
nitrogen partial pressure whereas reduced chlorine pressures are required.

The treatment given so far has nearly exclusively been concerned
with the thermodynamic stability without looking at the kinetic properties of the materials. It might appear that thermodynamic and kinetic
properties are independent of each other. In fact, these properties are
closely related to each other as will be shown in the next section.

THE STABILITY-KINETICS DILEMMA

In addition to thermodynamic stability, practically useful solid ion
conductors should have high ionic conductivity, preferably at ambient
temperature. A look at the list of presently known solid electrolytes
shows that these two requirements are difficult to match, at least for
simple crystalline structures with small crystallographic unit cells.
Silver and copper compounds often show high ionic conductivities but
generally very small decomposition voltages. In contrast, most alkali ion
conductors are thermodynamically very stable but show very low conductivities. An exception is the fast ionic conduction in Li_3N which shows,
however, a small decomposition voltage of about 0.3 V at ambient tempera-

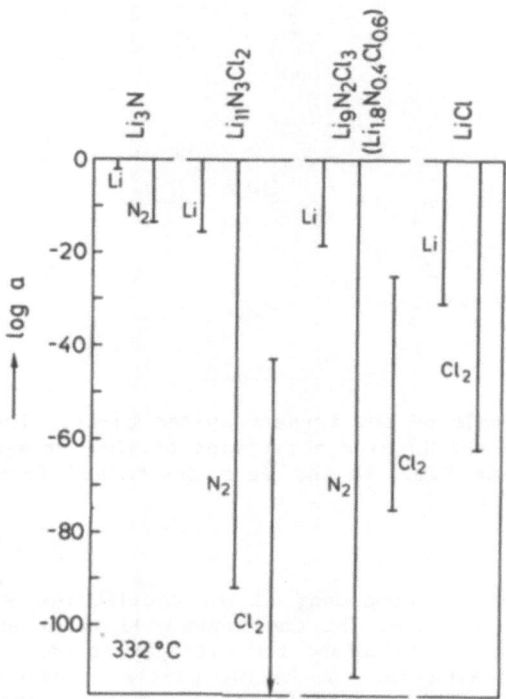

Fig. 7   Activity and partial pressure ranges of all components of all
          compounds of the ternary system Li-N-Cl. All compounds are
          stable against reaction with elemental lithium and nitrogen of
          1 atm pressure (except LiCl). The phases are only stable under
          reduced chlorine partial pressures.

ture. This relation between the ionic conductivity and thermodynamic
stability appears to be very general. The activation enthalpy of the
conductivity is taken as an indicator of the mobility of the ions. This
quantity is plotted against the decomposition voltage. A proportionality
is found within an astonishingly small band independent of the structure
of the compounds. This empirical relationship may be qualitatively understood by the fact that the decomposition voltage or Gibbs energy of formation are an indication of the binding energy of the components of the
compound. The more the anions and cations are bound to each other the
less they become mobile and vice versa.

     The simple relationship is no longer valid if more complex struc-
tures, generally multinary systems, are taken into consideration. It
occurs that specific structures are formed in which a high structural
stability results from the partial lattice of the immobile components but
a large disorder and high mobility occurs in the partial lattice of the
mobile ions. Such a situation is found in the case of the β-aluminas
which form stable spinel blocks with separate fast conducting planes in
between. Voltages higher than 2 V are applicable in spite of the fact
that the ions are very loosely bound in the crystal. The search for thermodynamically stable fast ionic conducting solids should consider especially materials with such structural pecularities.

The materials properties change generally very drastically over the stability range of a compound. The activities of the components may vary by several orders of magnitude whereas the composition is varied only slightly. It will be shown in the next section that these thermodynamic parameters are closely related to the electronic materials properties.

THERMODYNAMIC-ELECTRONIC RELATIONSHIPS

The ionic conductor should accommodate a certain difference in the chemical potential of the neutral electroactive component A. This may be split into a variation of the ions A^{z+} and the electrons e^-

$$d \ \mu_A = d \ \mu_{A^{z+}} + z \ d \ \mu_{e^-} \tag{15}$$

The variation of the chemical potential of A is approximately of the same order of magnitude for all compounds. The variation of the chemical potential of electrons depends, however, strongly on the electronic concentration of the material

$$d \ \mu_{e^-} = kT \ d \ \ln c_{e^-} + kT \ d \ \ln \gamma_{e^-} \tag{16}$$

where γ_{e^-} is the activity coefficient of the electrons which is assumed to be practically constant in the present qualitative consideration. The concentration of electrons varies generally proportionally to the stoichiometric change. The variation of small concentrations of electrons produces a large change of their chemical potential and, according to Eq. (15), also of the chemical potential of A. Solid ionic conductors which require small electronic concentrations have therefore only very narrow stoichiometric stability ranges ('line phases'). This is in agreement with the findings in semiconductors. The mobility of the electronic species is very slow, however, in fast ionic conductors in contrast to the situation in semiconductors. Eqs. (15) and (16) show also that the high electronic concentration in metals tends to form very wide stoichiometric ranges of the existing phases. Narrow stability ranges may be therefore considered to be a good indication for possibly fast ionic transport of the compound.

SUMMARY

Stability of fast ion conductors is required against reaction with any of the electrodes at any degree of discharge of the galvanic cell. Thermodynamic stability is preferred over kinetic stability since the fast motion of ions and the continuous cycling of charges (in secondary cells) enhances the tendency to approach thermodynamic equilibrium. Even very thin reaction layers may cause a drop in the cell voltage or increased resistance.

Fast silver and copper ion conductors could be easily modified by adding other cations to form multinary compounds. The stability with silver or copper was not affected because of the higher stability of the salt of the additional cation as compared to the silver or copper compound. The ionic conductivity is often very high but the decomposition voltage is very low and makes these materials unattractive for practical applications.

In contrast, the conductivity is generally low in alkali ion conductors whereas the decomposition voltage is high. Ternary compounds with two different types of cations are for the vast majority of materials unstable with the elemental alkali metal. This is caused by the high stability of the alkali salt as compared to the salt of the additional cation. The alkali metal tends to displace the additional cation. A solution in this case is to form multinary compounds with the alkali ion as the only cation but several types of anions. The multinary compound remains stable with the elemental alkali metal and shows a decomposition voltage which lies between the values of the binary compounds. Thermodynamic and kinetic properties are generally related in an unfavourable manner. The ionic conductor may be optimized for the type of application. The electronic concentration plays an important part for the stability range of compounds. Small concentrations of electrons with low mobilities are required for fast ion conductors.

REFERENCES

[1]   W. Weppner and H. Schulz, Proceedings 6th Intl. Conference on
        Solid State Ionics, Garmisch-Partenkirchen, Sept. 6-11, 1987,
        North Holland Publ. Comp., in print
[2]   W. Weppner, Electrochemical Measurement Techniques, this volume
[3]   W. Weppner and R.A. Huggins, in: Fast Ion Transport in Solids,
        Electrodes and Electrolytes (P. Vashishta, J.N. Mundy, and
        G.N. Shenoy, Eds.) North-Holland, New York, Amsterdam, Oxford
        (1979), p. 53
[4]   W. Weppner, J. Electroanal. Chem. Interfac. Electrochem. $\underline{84}$, 339
        (1977)
[5]   Q.G. Liu and W.L. Worrell, Sensors and Actuators $\underline{2}$, 385 (1982);
        W.L. Worrell, J. Electroanal. Chem. $\underline{168}$, 355C (1984)
[6]   W. Weppner and R.A. Huggins, Solid State Ionics $\underline{1}$, 3 (1980)
[7]   W. Weppner and R.A. Huggins, in: Fast Ion Transport in Solids;
        Electrodes and Electrolytes (P. Vashishta, J.N. Mundy, and
        G.K. Shenoy, Eds.) Elsevier North Holland Inc., New York, N.Y.
        1979, p. 475
[8]   W. Weppner, Sensors and Actuators $\underline{12}$, 107 (1987)
[9]   W. Weppner, Novel Solid State Galvanic Cell Gas Sensors, this vo-
        lume
[10] W. Weppner, Chen Li-chuan, and W. Piekarczyk, Z. Naturforsch. $\underline{35a}$,
        381 (1980)
[11] P. Hartwig, W. Weppner, and W. Wichelhaus, Mat. Res. Bull. $\underline{14}$, 493
        (1979)
[12] P. Hartwig, W. Weppner, and W. Wichelhaus, and A. Rabenau,
        Solid State Communications $\underline{30}$, 601 (1979)
[13] P. Hartwig, W. Weppner, W. Wichelhaus, and A. Rabenau, Angew. Chem.
        $\underline{92}$, 72 (1980); Angew. Chem. Int. Ed. Engl. $\underline{19}$, 74 (1980)

DEGRADATION OF CERAMICS IN ALKALI-METAL

ENVIRONMENTS

Michel Barsoum

Department of Materials Engineering
Drexel University
Philadelphia, Pa. 19104

I. INTRODUCTION

The high reactivity of alkali metals and their containment is an important issue that impacts several technologies such as high energy batteries, thermonuclear reactors, high pressure Na vapor lamps and alkali metal thermoelectric convertors. This paper addresses the various mechanisms by which degradation of ceramics occurs when placed in an alkali metal environment with special emphasis on the problems associated with corrosion and electrolytic degradation of solid electrolytes. The term degradation in this paper will be taken to mean an increase in the average concentration of the alkali metal in the material in which it is in contact, whether by simple ambipolar diffusion as in the case of material reduction and coloration, or by chemical reaction to form a higher alkali oxide containing phase, or by dendritic formation and electrolytic degradation of solid electrolytes.

The thermodynamic criteria for the selection of binary and ternary oxide systems which are thermodynamically stable against alkali metals, and that could possibly be used as alkali-ion conducting solid electrolytes, are reviewed. The importance of kinetics in determining whether a solid will react and form a reaction layer or will simply darken is discussed. Higher driving forces for reaction and higher temperatures favor reaction layer formation whereas at low temperatures the system responds to the chemical driving force by coloring.

In addition to chemical stability, a solid electrolyte must be electrolytically stable, i.e. it must exhibit a range of activity of the electroactive species over which the electronic contribution to the total conductivity remains negligible. Reduction of the electrolyte is accompanied by a discoloration and an increase in the electronic conductivity which is a contributing factor in the rapid degradation of the electrolyte. The mechanisms by which coloration occurs and their ramifications will be addressed. This phenomena is not restricted to solid electrolytes but is responsible for the degradation of other ceramics and glasses such as glass feedthroughs in lithium cells and ceramic envelopes in Na vapor lamps.

Corrosion and degradation of solid electrolytes is further complicated by the fact that the passage of current can rapidly increase the rate of corrosion and in some instances cause rapid degradation and shorting of the electrolyte. For example, it has long been recognized that the limiting factor for the practical application of β-alumina is its rather short lifetime. The breakdown of β-alumina often appears as rapid (Na filled) crack propagation through the electrolyte. The mechanisms and causes of this type of rapid deterioration will be discussed.

II. IMPORTANCE OF CHEMICAL AND ELECTROLYTIC STABILITY

The stability of a material in an alkali environment is critical for the sucessful application of solid electrolytes. If a solid electrolyte were to react with the electrode to form either a reaction layer or a colored zone with electrical properties that are different from the original electrolyte serious problems could arise, the most damaging being the shorting of the cell.

The problem is summarized in Fig.1 where a reaction layer or color zone is assumed to grow as a result of the chemical or the electrochemical instability of the electrolyte in its environment. This section will deal with the growth of such a layer either in the presence or absence of an externally applied constant current, **I**, as shown schematically in Figs.1 a and b. In Fig.1a, a constant current is passed through the solid electrolyte and the growing reaction layer whereas in Fig. 1b the reaction layer grows as a result of the free energy change associated with the reaction. The driving force is the electrochemical potential gradient across the layer. In all cases the frame of reference is assumed to be the immobile anion sublattice.

For the sake of illustration we shall assume that the defect incorporation reaction in the reaction layer, RL, or color zone, formed is:

$$M^* \iff M^{+z} + z\, e^{-1} \tag{1}$$

and that the growth is rate limited by the diffusion of the slower of the two species. For the alkali metals $z = 1$ and will henceforth be omitted.

The current density of each species under the combined effect of the applied electric field and the chemical potential gradient is given as:

$$I_e = \sigma_e E + e D_e \operatorname{grad} \mu_e \tag{2}$$

$$I_{M+} = \sigma_{M+} E - e D_{M+} \operatorname{grad} \mu_{M+} \tag{3}$$

where e is the electronic charge and I_e and I_{M+} are, respectively, the electronic and cationic current densities. μ_e and μ_{M+} are the chemical potentials of the electrons and cations respectively. D_j is the diffusion coefficient of the species j. σ_j is the partial conductivity of species j defined as,

$$\sigma_{M+} = e\, u_{M+}\, (c^+) \tag{4}$$

and

$$\sigma_e = e\, u_e\, n \tag{5}$$

$$\sigma_{tot} = \sigma_e + \sigma_{M+} \tag{6}$$

$$t_j = \sigma_j / \sigma_{tot} \tag{7}$$

where σ_{tot} is the total conductivity, t_j and u_j the transference numbers and mobilities of species j respectively and n and c^+ the number of electronic and cationic mobile carriers/cc. The mobility of species j is related to its conductivity by the Einstein relation:

$$D_j = [kT/ \,|\, z_j \,|\, e]\, u_j = kT \sigma_j / c_j z^2 e^2 \tag{8}$$

The total current density, assuming only two mobile species is:

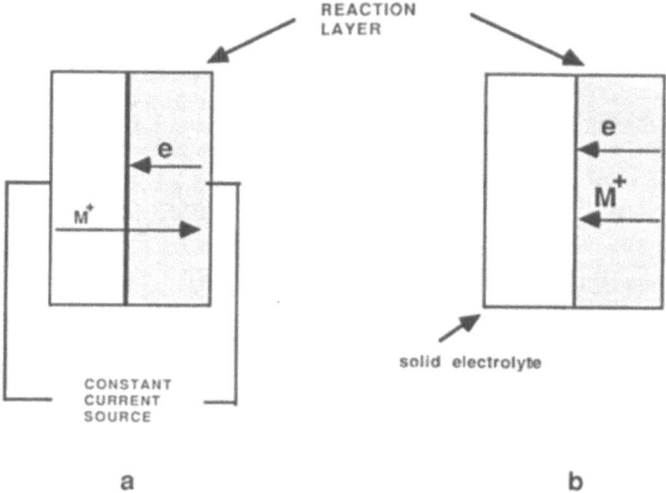

solid electrolyte

CONSTANT
CURRENT
SOURCE

a   b

FIGURE 1. Reaction layer or color zone formation. a) in presence of a constant
externally applied current. Depending on the charge on the rate limiting
species and the direction of the current the reaction layer kinetics will
either be enhanced or retarded. b) Growth of the layer by ambipolar
Wagner type oxidation in the absence of an externally applied electric field.

$$I = I_e + I_{M+} \tag{9}$$

where I is the externally applied constant current density [A / cm^2]. Eliminating E from Eqs. 2
and 3, using Eqs. 6-9 and assuming local thermodynamic equilibrium, results in:

$$I_e = t_e I + [t_{M+} n D_e e / kT] \, grad \, [\mu_{M+} + \mu_e] = t_e I + [t_{M+} t_e \sigma_t / e] \, grad \, [\mu_{M+} + \mu_e] \tag{10}$$

and

$$I_{M+} = t_{M+} I - t_e \, c^+ D_{M+} e / kT \, grad \, [\mu_{M+} + \mu_e] =$$

$$t_{M+} I - [t_{M+} t_e \sigma_t / e] \, grad \, [\mu_{M+} + \mu_e] \tag{11}$$

Eqs. 10 and 11 have general validity and describe the behavior of charged particles
simultaneously subjected to a chemical potential gradient and an electric field. For example, if the
reaction layer shown in Fig.1a is electronic ($t_e = 1$ and $t_{M+} = 0$) while the solid it is growing into
is ionic, then the current in the reaction layer will be electronic, since according to Eq. 10, $I = I_e$.
In the electrolyte, however, where $t_e = 0$ and $t_{M+} = 1$,the current will remain ionic since $I = I_{M+}$
and consequently the second term in Eq. 11 has to be << 1.

If, on the other hand, the reaction layer that forms is ionic in nature, then the total corrosion
current, I_e will be determined by the sum of the electric and diffusive components of the
electrons. It can be easilly be shown, however, that if the condition:

$$\sigma_e \ll I \, e / grad \, [\mu_{M+} + \mu_e] \ll 1$$

is fullfilled then the application of an electric field to the reaction layer would have no effect on the
growth of that layer. The corrosion current would thus be solely determined by the diffusive
component of the electrons.

In the special case where the externally applied current is zero, $I = 0$, Eq. 10 reduces to:

$$I_e = [t_{M+} nD_e e / kT] \quad grad [\mu_{M+} + z\mu_e] \tag{12}$$

or, equivalently, by substituting conductivites for diffusivities using Eq. 8 and assuming local thermodynamic equilibrium:

$$I_e = [1 / e] t_e t_{M+} \sigma_t \quad grad \mu^*_M \tag{13}$$

where $grad \mu^*_M$ is the gradient in the chemical potential of the neutral species. Starting with the defect incorporation reaction, Eq.1, and assuming electron ideality, it can be shown that Eq. 12 is equivalent to the expression:

$$I_e = t_{M+} D_e e [1 + 1 / z (d\mu_{M+} / d\mu_e]) grad c^* \tag{14}$$

derived by Heyne (1). Equations 13 and 14 are also identical to the expressions derived by Wagner (2) for the corrosion or oxidation flux of a material when there is no externally applied electric field imposed on the layer, i.e. when $I = 0$.

The worst possible situation occurs when the reaction layer formed is electronic; the reaction layer growth kinetics will no longer be limited by a slow diffusional process, but will grow as fast as the applied current will allow it. The reaction layer/electrolyte interface will be supplied with M^+ from the left and electrons from the right in Fig. 1a. Up to this point the reaction layer or color front that grows into the electrolyte from right to left was assumed to be uniform. The situation can become even more damaging if the reaction layer is electronic and nonuniform. For example, during cathodic deposition, metal filled cracks will focus the current at the crack tip and growth of that crack will occur very rapidly and eventually short out the electrolyte.

The foregoing analysis thus underscores the importance of chemical stability, especially in the case of solid electrolytes. The following section addresses the thermodynamic criteria required for chemical stability.

III. THERMODYNAMIC STABILITY

a) FREE ENERGY CONSIDERATIONS

For the sake of clarity the following sections will address the thermodynamic stability of various materials against Li. The discussion, however, is quite general and is applicable to all other alkali metal/ceramic interactions.

For a binary oxide, M_x0, to be stable against Li, the free energy change for the reaction, ΔG_{rx}:

$$2 Li + M_x0 \Leftrightarrow Li_20 + xM \tag{I}$$

has to be positive. In other words:

$$\Delta G_{rx} = \Delta G^\circ_f (Li_20) - \Delta G^\circ_f (M_x0) \tag{15}$$

has to be positive, where ΔG°_f (j) is the standard free energy of formation of phase j <u>per mole of</u>

oxygen from the elements. x is the number of M ions per 0 in M_x0. The temperature dependencies of $\Delta G°_f$ of various oxides, normalized to a per oxygen atom basis, are shown in Fig.2, and it follows directly from Eq. 15 that all the oxides that lie above the Li_20 line would not be stable with Li while those below would be. This, as is discussed shortly, is only true if there are no ternary intermediate oxides between the end members, i.e. Li_20 and M_x0.

The situation is more complicated, however, if ternary oxides exist. In that case the oxide will be stable against Li if the free energy change for the reaction:

$$(2 - \alpha) \, Li + Li_\alpha M_\beta 0 \Leftrightarrow Li_20 + \beta \, M \qquad\qquad (\text{II})$$

is positive. That is:

$$\Delta G_{rx} = \Delta G°_f \, (Li_20) - \Delta G°_f \, (Li_\alpha M_\beta 0) > 0 \qquad\qquad (16)$$

On a free energy/composition diagram this condition translates to a situation where the ternary oxide would be stable against Li only if its free energy/mole of O is more negative than that of Li_20. Figure 3 summarizes the various possible free energy/composition diagrams at a given temperature.In Fig.3a, where it is assumed that no ternary compounds form, M_x0 would

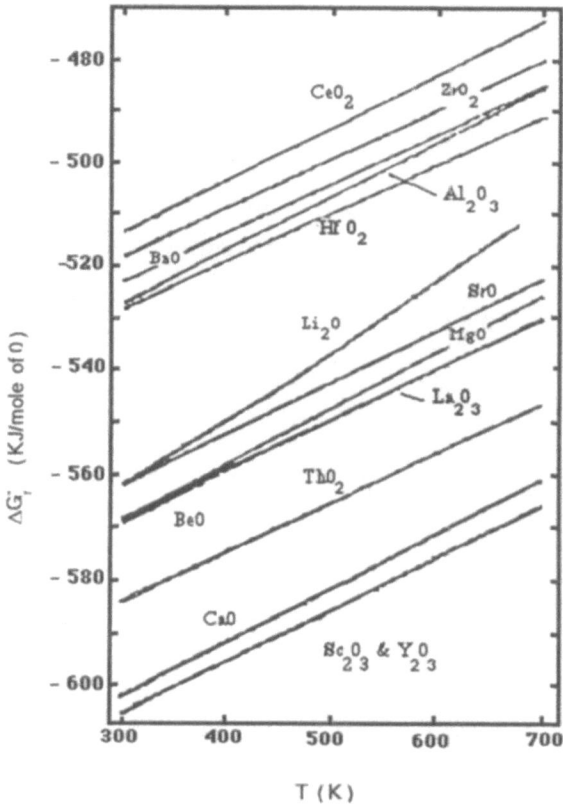

FIGURE 2. Gibbs free energy of formation per mole of oxygen as a function of temperature. Oxides below the Li_20 line will be stable with respect to Li provided no ternary compounds form in the system. After Ref. (3).

be stable with respect to Li but not $M^*_x O$. On such a plot the oxides that are stable with respect to Li would be the same as those that lie above the Li_2O line in Fig. 2, namely: SrO, MgO, La_2O_3, BeO, CaO, and Y_2O_3 etc.

Figs. 3b & c depict the situation where a ternary compound, $Li_\alpha M_\beta O$, exists between the end members Li_2O and M_xO. Two possibilities exist: the free energy of formation/mole of O of the ternary compound is either more or less negative than Li_2O. If it is less negative, the compound is unstable and the terminus of the reaction would be Li_2O (Fig. 3b). If it is more negative, then the ternary compound is stable against Li even though the parent oxide M_xO is not (Fig. 3c).

It should be noted that the free energy of formation of a binary compound being more negative than that of the alkali metal oxide is a necessary but insufficient condition for its stability. The presence of a ternary compound that is more stable than the binary would render tbe latter unstable. This situation is quite common and is discussed below. (see Fig.8 for example.)

b) THERMODYNAMIC STABILITY DOMAIN

Representing the data in the format shown in Fig. 3 has the advantage that it is immediately obvious which compound is the most stable in a given system and hence would be stable when exposed to the pure alkali metal: it is the compound with the lowest Gibbs free energy of formation per mole of oxygen. All other oxides will only be stable at Li activities that are different from one. The purpose of this section is to calculate the activity domain over which a given compound is chemically stable. Again, Li is chosen for the sake of illustration.

Assuming the reaction between a compound $Li_\alpha M_\beta O$ and Li is:

$$(\beta - \alpha)\, Li \;+\; Li_\alpha M_x O \;\Leftrightarrow\; Li_\beta M_y O \;+\; z\, M \qquad\qquad (\,III\,)$$

such that $\beta > \alpha$ and $x = y + z$, if follows that the free energy change for this reaction is given as:

$$\Delta G^{rxn} \;=\; \Delta G^\circ_f\,(Li_\beta M_y O) \;-\; \Delta G^\circ_f\,(Li_\alpha M_x O) + RT \ln K \qquad (17)$$

where K is the equilibrium constant for the reaction, and R and T have their usual meaning. At equilibrium $\Delta G^{rxn} = 0$ and

$$\Delta G^\circ_f\,(Li_\beta M_y O) \;-\; \Delta G^\circ_f\,(Li_\alpha M_x O) \;=\; -\,RT \ln K \qquad (18)$$

If the reactants and products are in their standard states then:

$$K = 1/\,[a_{Li}]^{\,(\beta - \alpha)} \qquad\qquad (19)$$

Substituting in Eq. 18:

$$\Delta G^\circ_f\,(Li_\beta M_y O) \;-\; \Delta G^\circ_f\,(Li_\alpha M_x O) \;=\; (\beta - \alpha)\, RT \ln a_{Li} \qquad (20)$$

Hence if the ΔG°_f's are known then the equilibrium activity of Li for reaction III can be calculated. This equilbrium activity can be converted to an EMF with respect to pure Li since:

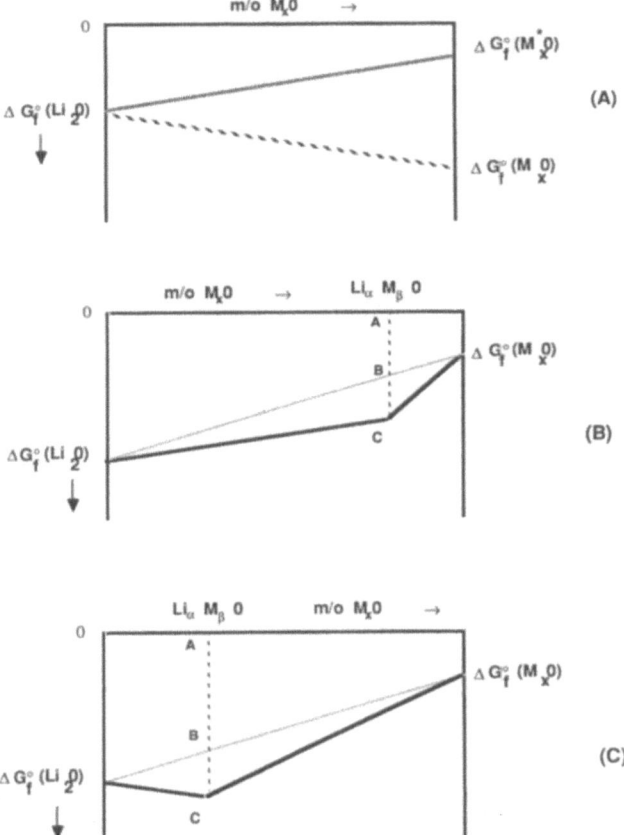

FIGURE 3. Isothermal compositional dependence of the Gibbs free energy of formation per mole of 0 for: a) system in which no ternaries form. M_x0 is stable with respect to Li but M^*_x0 is not; b) Ternary compound forms but is unstable with respect to Li. c) Ternary compound forms that is stable with respect to Li.

$$EMF = RT/F \ln a_{Li} \qquad (21)$$

Up to this point it was assumed that no intermetallic compounds, Li_rM, formed in the Li-M system. Howver, if such phases exist the conditions for stability have to be modified to take into account the free energy of formation of the intermetallic. The reaction now reads:

$$(\beta + zr - \alpha) Li + Li_\alpha M_x O \Leftrightarrow Li_\beta M_y O + z Li_r M \qquad (IV)$$

such that $z + y = x$. The criteria for stability ($\Delta G^{rxn} > 0$) does not change, however. The net effect of the formation of the intermetallic compound is that it places more demands on the stability of the compound than if the intermetallic did not form: the reaction is being shifted to the right. The equilibrium Li activity can be calculated in the same manner as before.

c) COMPATABILITY TRIANGLES

According to the phase rule, mixing three phases together in a ternary system at constant temperature and pressure will completely define the system and fix the activities of all the constituents involved. For example, mixing together the three phases $Li_\beta M_y O$, $Li_\alpha M_x O$ and M would fix the Li activity at the value calculated from Eq. 20.

To illustrate the point, the Li-Si-O system is shown in Fig.4 where each triangle delineates the phases that coexist at equilibrium and form a compatability triangle. For example the phases Si, $Li_2Si_2O_5$ and SiO_2 will, if mixed together and equilibrium is achieved, fix the Li activity at the voltage shown within that triangle. The equilibrium reactions are of type III.

The reaction for the compatability triangles that have oxygen as a phase are of the form:

$$Li_\beta M_y O \Leftrightarrow Li_\alpha M_x O + (\beta - \alpha) Li + 1/2 O_2 \qquad (V)$$

where $\beta > \alpha$. The free energy change for this reaction is:

$$\Delta G^{rxn} = \Delta G^\circ_f (Li_\alpha M_x O) - \Delta G^\circ_f (Li_\beta M_y O) + RT \ln K \qquad (22)$$

where K is the equilibrium constant for the reaction. At equilibrium $\Delta G^{rxn} = 0$ and :

$$\Delta G^\circ_f (Li_\alpha M_x O) - \Delta G^\circ_f (Li_\beta M_y O) = K = -RT \ln [a_{Li}]^{\beta - \alpha} \cdot \sqrt{P_{O2}} \qquad (23)$$

The Li activity can be calculated but in this case it will depend on the oxygen partial pressure, P_{O2}, imposed on the system.

Expressing the data in the form shown in Fig.4 has several advantages. In addition to showing the final phases that would be stable with the alkali metal (the compatability triangle with 0 volts), the range of activities over which each phase is stable and the corresponding reactions are immediately obvious. For example Fig.4 predicts that the first reaction between SiO_2 and Li to form $Li_2Si_2O_5$ will occur when the Li activity corresponds to 1.353 V with respect to pure Li ($a_{Li} = 1.22 \times 10^{-10}$). The next reaction would occur at 1.308 V with $Li_2Si_2O_5$ reacting to form Li_2SiO_3 and more Si according to reaction III. The reactions between SiO_2 and Li thus follow a line joining the points representing SiO_2 and Li in Fig.4.

The terminus of the reaction would occur when the SiO_2 was transformed to Li_2O and $Li_{22}Si_5$ ($Li_{4.4}Si$); the reaction being:

$$42 Li + 5SiO_2 \Rightarrow Li_{22}Si_5 + 10 Li_2O \qquad\qquad (VI)$$

In other words if sufficient Li were to react with SiO_2, the final phases present at equilibrium would be the Li, Li_2O, and $Li_{4.4}Si$.

Once the activities of the metal constituents are calculated for each reaction the thermodynamic stability window, or the range of activites over which the compound is stable, is easily determined. This window is given by the activities or voltages that exist in the compatability triangles adjacent to the compound in question. The phase $Li_2Si_2O_5$ is stable between 1.353 and 1.308 V, Li_2SiO_3 between 1.308 and 0.863 V and Li_4SiO_4 between 0.863 and 0.144 V etc.

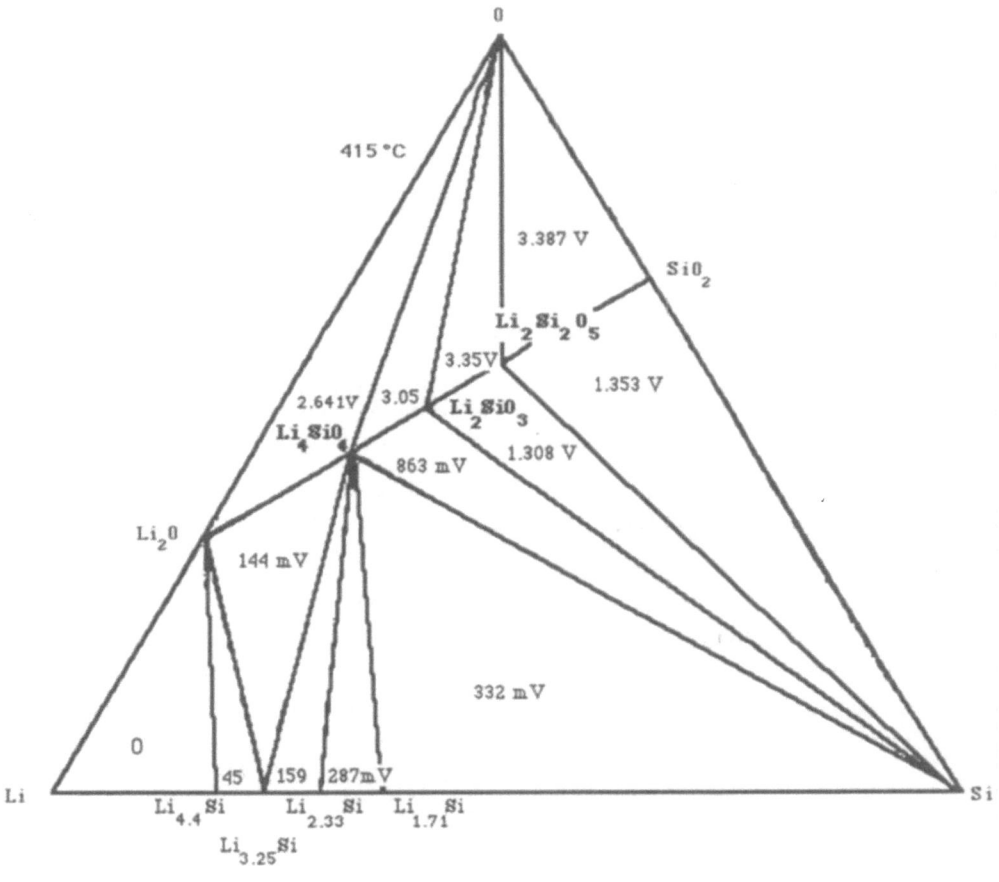

FIGURE 4. Compatability triangles in the Li-Si-0 ternary at 415 °C. Voltages correspond to Li activitites defined by the equilibrium between the phases that make up each triangle. After Ref. (4).

d) ESTIMATION OF UNKNOWN THERMODYNAMIC PARAMETERS

The situation is well defined when the free energies of formation of all the intermediate compounds are known and if the temperature is high enough to ensure that equilibrium is achieved. In many cases, however, the requisite thermochemical data is lacking and it becomes necessary to estimate the unknown free energy data values from what has been reported.

Referring to Fig.3b, and assuming that the ternary phase, $Li_\alpha M\beta O$ is known to exist but its free energy of formation is unknown, then the procedure for estimating the latter is as follows. Given the reaction:

$$(\gamma - \beta)/\gamma \, Li_2O + \beta/\gamma \, M_xO \quad \Leftrightarrow \quad Li_\alpha M\beta O \qquad (VII)$$

the free energy of formation of $Li_\alpha M\beta O$ is given as:

$$\Delta G^\circ_f \, (Li_\alpha M\beta O) = (x-b)/x \, \Delta G^\circ_f \, (Li_2O) + b/x \, \Delta G^\circ_f \, (M_xO) + \Delta G^\circ_f \, (excess) \qquad (24)$$

Reaction VII is written in the form shown to emphasize the fact that the number of moles of oxygen remains constant during the reaction and that the free energy of formation of the ternary phase is made up of two components. The first component is the linear combination of the Gibbs energy of formation of the binary oxides for the composition of the ternary; i.e. by the sum of the first two terms in Eq. 24, which is represented by the distance AB in Fig. 3b & c. The excess free energy, ΔG°_f (excess), constitutes the second component, and results in the formation of the ternary from the end members (the binary oxides in this case) and is represented by the distance BC in Fig. 3b & c. ΔG°_f (excess) is thus the free energy of formation of the ternary from the parent oxides. Estimating ΔG°_f (excess) is not easy but it is usually between 2-10% of the Gibbs free energy of the ternary oxide. In other words, BC is usually between 2 and 10 % of AB. Note that had the Gibbs energy of formation of $Li_\alpha M\beta O$ not been more negative than the sum of its components, i.e. AB, it would not have formed.

In more complex situations where there is more than one ternary compound the same procedure as outlined above can be used. In general the more information that is available about the system the more accurate will be the final estimates for the unknown values.

To illustrate this point actual data for the Li_2O -$B_{2/3}O$ system are plotted in Fig. 5. The free energies of formation of the 1:4, 1:2 and 1:1 phases are known (5), but those of 3:2 and 3:1 have not been determined to date. In order to estimate these values the first step is to draw a straight line between the values that are closest to the unkown compounds: Li_2O and 1:1 in this case. The free energy values on that line (points B and D) are the highest (least negative) possible values of ΔG°_f for the unknowns. The lowest (most negative) estimate for the free energy of formation for the phase 3:2 is given by the point A in Fig. 5, which is the value of ΔG°_f at the intersection of the 3:2 composition and the extrapolation of the line joining the points 1:2 and 1:1. In other words the free energy of formation of 3:2 cannot be more negative than point A, otherwise the system could lower its free energy by 1:1 disproportionating into 3:2 and 1:2. It follows from the preceding argument that the free energy of formation of the unknown compound 3:2 has to lie between points A and B, while that for 3:1 lies between C and D. Repeating the interpolation procedure the best values for 3:2 and 3:1 at 700 K are shown in Fig.5

It should be noted that the procedure outlined above is a very powerful tool for determining whether new thermodynamic data generated in the literature for ternary compounds is consistent with existing data or not.

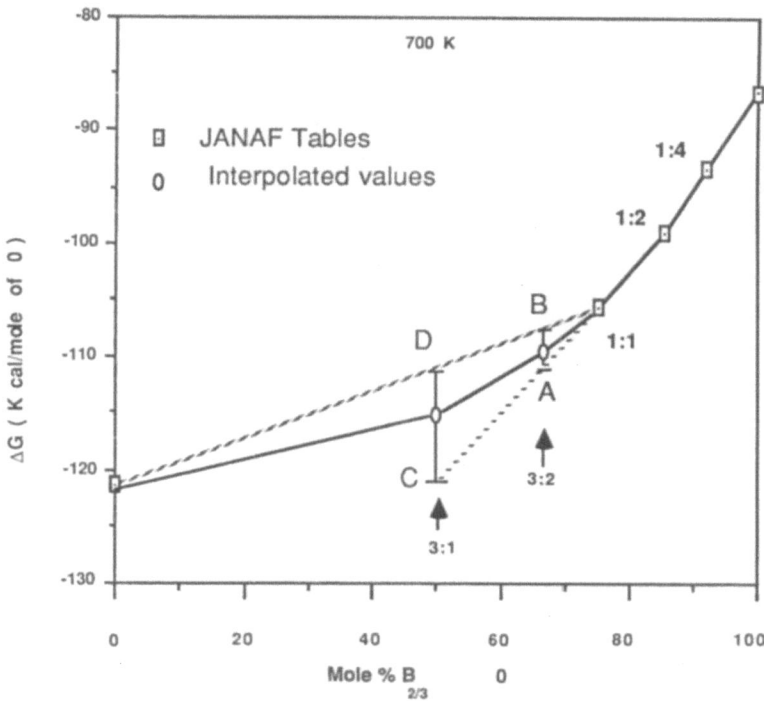

Figure 5. Compositional dependence at 700 K of the Gibbs Free Energy of formation per mole of O for the lithium borate system. Points for 3:2 and 3:1 are interpolated results.

e) EXPERIMENTAL RESULTS

The published experimental results on the stability of ceramics in alkali metal environments substantiate the thermodynamic predictions discussed in the preceding section. In general when the free energy of formation between the alkali metal oxide and the binary metal oxide is large direct reduction of the metal oxide to the metal and Li_2O takes place. However, when the free energy of formation is small, ternary compounds usually form.

Barker et. al. (6) stirred various oxides into an excess of purified liquid Li at 600 °C and found that the reactions followed thermodynamic prediction: TiO_2, ZrO_2 reacted with Li to form Li_2O and the appropriate metal; ThO_2 did not react with Li as expected from Fig. 2. HfO_2, on the other hand, was found to react with Li to form a ternary oxide, believed to be $LiHfO_2$.

We used an electrochemical technique to study the reaction between Al_2O_3 and Li as a function of the activity of the latter (8,9). The lithium activity was varied by coulometric titration. The equilibrium Li activity for the reaction was manifested as a plateau in the EMF versus time curves. The working electrode consisted of Sn into which Al_2O_3 powder was pressed to form a composite electrode. A constant current was applied for a predetermined time and then interrupted. Typical relaxation curves are shown in Fig. 6a where the EMF was found to increase with time as the Li activity in the working electrode dropped. The reaction proceeded as long as the Li activity in the working electrode was greater than the equilibrium activity at the reaction layer/Al_2O_3 interface. The temperature dependence of the plateau EMFs are plotted in Fig.6b and compared'with the theoretical EMFs that would be expected based on the reaction:

$$2\ Al_2O_3 + 3\ Li \Leftrightarrow 3\ Li\ AlO_2 + Al$$

which was calculated using existing thermodynamic data (5). The agreement between theory and experiment is excellent and proves that the reaction between Li and Al_2O_3 is as written. Furthermore, we identified the reaction product to be $Li_2\ 0.Al_2O_3$ by X-ray analysis. The possibility of LiAl formation is precluded in this case because the activity of Li was kept very low at all times.

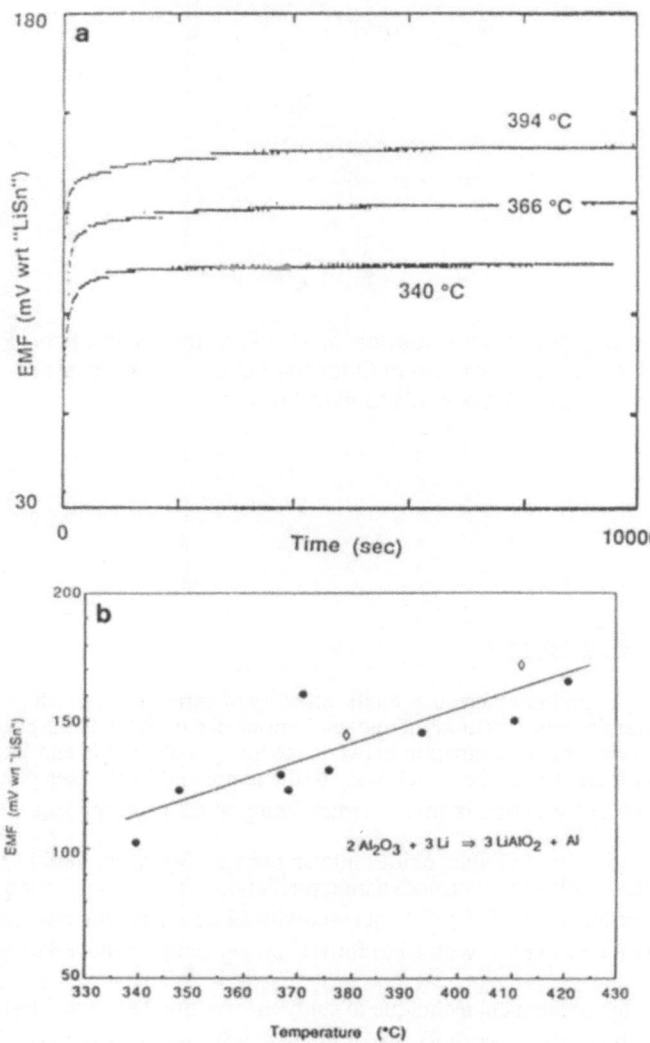

FIGURE 6. a) EMF relaxation curves as a function of time and temperature after the interruption of the applied current for the Li/alumina system.(b) Temperature dependence of plateau EMF's for two different runs. Solid line is theoretical EMF for the reaction shown as calculated using the JANAF data (5).

Reactions between lithium borate glasses and molten Li have also been investigated (10) as a function of temperature (200 and 250 °C) and composition. The final reaction products consisted principally of $3Li_2O.B_2O_3$, $Li_2O.2B_2O_3$ with some $3Li_2O.2B_2O_3$. As the lithia content of the glass was increased the kinetics of layer formation slowed down, clearly indicating that the most stable oxide in that system was shifted to the lithia side as expected from Fig. 5. Since the free energy of formation of $3Li_2O.B_2O_3$ is not known it is still not clear whether the final reaction product would be $3Li_2O.B_2O_3$ or Li_2O. However, when the reaction layers were grown at activities that were less than unity the final reaction product was unambiguously identified as $3Li_2O.B_2O_3$ (11).

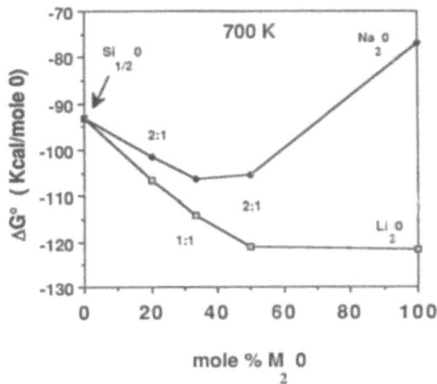

FIGURE 7.   Compositional dependence at 700 k of the Gibbs free energy of formation per mole of O for the Na_2O - $Si_{1/2}O$ and Li_2O - $Si_{1/2}O$ systems. The most stable phases according to this diagram are 1:1 in the Na_2O - $Si_{1/2}O$ and Li_2O in the Li_2O - $Si_{1/2}O$. The data for the 2:1 compositions (i.e. 2 M_2O - $Si_{1/2}O$) were obtained from Barin et.al.(6) the rest of the data are from the Janaf Tables (5).

The thermodynamic data at 700 K for the Li_2O -$Si_{1/2}O$ and Na_2O -$Si_{1/2}O$ systems are summarized in Fig.7 where it is clear that the most stable oxide in the Na_2O -$Si_{1/2}O$ system is the metasilicate, $Na_2O.SiO_2$ or 1:1 composition. This has been confirmed in several studies where silicate glasses and silica have been exposed to Na; the final reaction product formed was always found to be the 1:1 composition.The situation in the Li_2O -$Si_{1/2}O$ system is not as clear cut and cannot be determined from Fig. 7 since Li and Si form intermetallic compounds. It appears, however, that the reaction between Li and SiO_2 is given by reaction VI rather than the reaction:

$$SiO_2 + 4Li \Rightarrow 2Li_2O + Si \qquad\qquad (VIII)$$

suggested by Bunker et.al.(12). The uncertainty lies in the fact that the final reaction products have not been unequivocally identified. The most compelling evidence that the reaction between

Li and silica is that given by reaction VI is the fact that when the reaction products are exposed to water a vigorous reaction occurs with the evolution of hydrogen gas (13, 14). Furthermore, the reaction layer was found to be conductive and chemical analysis showed it to contain high levels of Li. Both observations are inconsistent with reaction VIII but consistent with reaction VI and, more importantly, the thermodynamic predictions.

In confirmation of the results of Skarstad et.al.(12), Maschhoff and his co-workers (15) have examined the reactions of evaporated lithium films with silica surfaces using in situ x-ray Photoelectron Spectroscopy (XPS). They concluded that the major products of the reaction are composed of oxidized Li, a metasilicate anion (SiO_3^{2-}) and possibly a lithium-rich Si alloy. The XPS results also seem to indicate that at low Li coverage the formation of non-bridging oxygens is favored over the formation of the Li rich Si alloy. At higher Li activities, on the other hand, Si reduction occurs. This is consistent with the predictions of Fig.4 where at low Li activities no Li-Si intermetallics compounds are expected and only form at higher Li activities.

The importance of this reaction lies in the use of silicate based glasses as insulating feedthroughs in Li cells. It has been observed that, even at ambient temperatures underpotential deposition of metallic Li from solution at the interface between the insulating material and the metal current collector resulted in the attack of the glass and the formation of a conductive layer that eventually led to cracking of the glass and short circuiting of the cell (15).

Despite the great interest in beta alumina solid electrolytes in recent years the thermodynamics of the Na_2O -$Al_{3/2}O$ system are still not well established. A critical review of the available data will be published in the near future (16) but a summary of that analysis is shown in Fig.8 a & b for 600 K. The results indicate that the most stable phase in this system is neither β-alumina ($NaAl_7O_{11}$) nor β"-alumina ($NaAl_5O_8$) but $Na_2O.Al_2O_3$ or the 1:1 phase. The differences in free energy, however, are not large and kinetic considerations may explain why β–alumina does not react with liquid sodium for extended periods of time. However, β-alumina and other ceramics will darken and color when exposed to liquid Na or Li as discussed in the next section.

In the preceding analysis the role of nonmetallic impurities dissolved in the alkali metal have not been addressed. It is well established that nonmetallic impurities, such as nitrogen, carbon and oxygen play an important role in the reactivities and reactions between metals and ceramics and alkali metals (19, 20, 21). For example, Barker et. al.(6) noted that dissolved nitrogen in Li was extremely reactive towards the metal produced by the reduction of the metal oxide and both binary and ternary nitrides have been observed as reaction products. Furthermore, it should be mentioned that grain boundary attack is also an important consideration. For example Fig.1, predicts that MgO should be stable against molten Li; however, experiments have shown that while single crystals are immune to Li, MgO polycrystals will react, most probably by grain boundary attack of a silica rich layer (22).

IMPLICATIONS FOR SOLID ELECTROLYTES

As discussed in other chapters in these proceedings, the addition of various "salts" such as chlorides, sulphates, iodides, etc., to borate and silicate based glasses, usually enhances the ionic conductivity of these glasses. Most of these glasses are thermodynamically unstable with respect to Li to begin with, and the addition of these salts further weakens the glass network and are thus detrimental to the their chemical stability. For example, the addition of LiCl to a lithium borate glass has been shown to substantially increase the rate of reaction of that glass in molten Li (10).

Based on the preceding discussion there are several strategies that can be employed to enhance the stability of a material towards alkali attack. The most obvious is to determine the stable phase in a given system and use it. In many cases, however, such phases are poor ion conductors. An alternate strategy is to maximize the alkali oxide concentration, which would result in moving the free energy of the system towards the stable, i.e. the alkali metal oxide, end. A third strategy is to add oxides that have a lower free energy of formation than the alkali metal oxide, such as CaO, Y_2O_3 and ThO_2. This will necessarily lower the free energy of the system since the geometric mean of the free energy of the resulting mixture must be lower than the original composition. This strategy would be especially effective for β"-alumina where the difference in free energy between it and the stable 1:1 composition is very small to begin with.

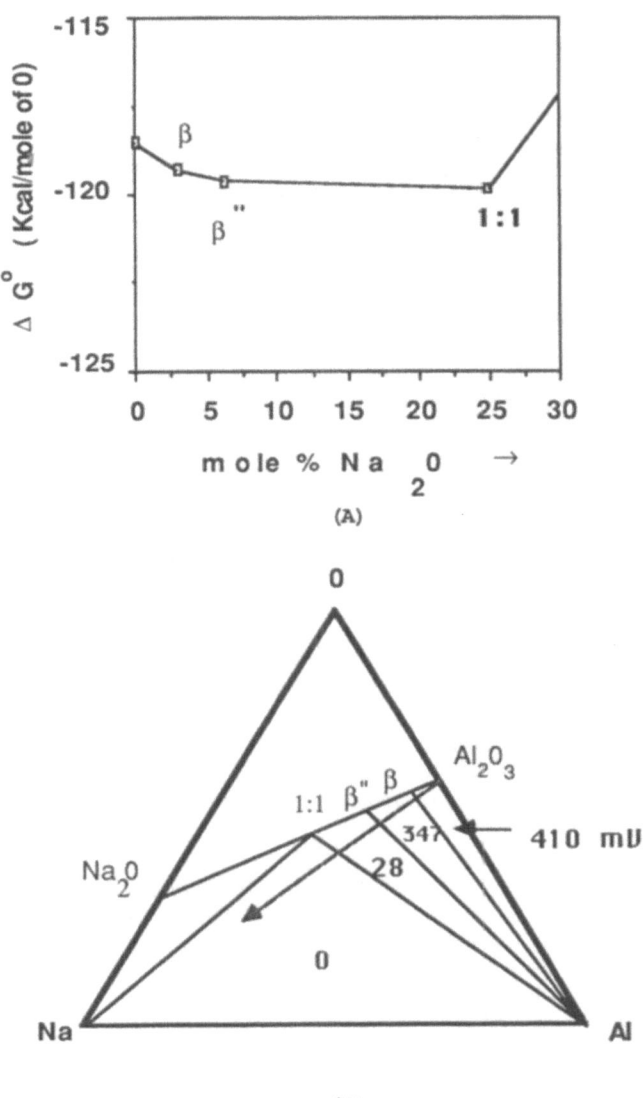

FIGURE 8. Compositional dependence, at 600 K, of the Gibbs free energy of formation per mole of 0 in the $Na_2O-Al_{2/3}O$ system. (A) Data for the 1:1 and alumina were

obtained from the Janaf Tables. The date for β-alumina ($Na_2O.11Al_2O_3$) and

β"-alumina ($Na_2O.5Al_2O_3$) were obtained from Refs. 17 & 18 respectively.

(B) Corresponding ternary diagram. The results indicate that β"-alumina is not stable with respect to Na.

IV. CHEMICAL AND ELECTROCHEMICAL COLORATION

It is well established that exposure of ceramics to alkali metals can result in their coloration and darkening. The importance of coloration lies in the fact that it is a form of degradation that usually results in an increase in the electronic conductivity of the ceramic which is detrimental to the operation of most devices are in contact with alkali metals. For the most part ceramics that are used in contact with Li or Na are chosen either for their insulating properties, as in the case of sealant glasses, or for their ionic conductivity. Electronic conductance poses an obvious problem in either case and limits the applicability of the material.

Whether a material will degrade by coloration or chemically react seems to be related to the free energy change for reaction discussed in the previous section. The smaller the driving force for the reaction the larger the kinetic barrier looms and the more likely coloration will occur and vice versa. However, this does not imply that in the absence of a driving force for reaction, coloration would not occur since in many systems coloration occurs within the thermodynamic stability domain of the material (30). It should be pointed out that the basic atomic process in both coloration and reaction layer formation during alkali metal attack are usually the same; viz an increase in the average concentration of the alkali metal in the ceramic relative to the other constituents by an ambipolar diffusion process.

For most materials that come in contact with alkali metals the defect incorporation reaction that results in coloration has been found, as is discussed in detail below, to be:

$$M \Leftrightarrow M^+ + e^{-1}$$

a) COLORATION OF SILICATES

Lau and McMillan (23-24) investigated the coloration of silica and silicates exposed to Na vapor and their results are summarized in Table I. At temperatures below about 400 °C, the glasses darkened. Associated with the darkening was the simultaneous appearance of non-bridging oxygens as determined by ESCA and Auger analysis of the colored layers. As the temperature was increased, however, a reaction layer formed that was identified to be Na_2SiO_3 in all cases, as expected from Fig. 7. The temperature at which the reaction layer formed increased as the soda and alumina contents of the glass increased. The addition of either of these oxides to SiO_2 has the effect of lowering the free energy of the glass to a value closer to that of Na_2SiO_3, reducing the driving force for reaction and thus increasing the temperature at which the reaction will occur.

Lau and McMillan argued that the propensity for discoloration, indicated in column 3 of Table I, bears no correlation to the temperature at which the reaction layer forms but instead was correlated to the ratio of non-bridging to bridging oxygens, R, in the glass; the higher R the more resistant the glass was to coloration. With the notable exception of quartz that correlation does hold (compare columns 2 with 3 in Table I). Lau et. al. further argued that because of this correlation the nonbridging Si-O⁻ was more stable towards chemical attack than the Si-0 bond. The reason quartz behaved so differently from fused silica is not clear but it could be related to the bond angle distribution in the glass that would render some bonds weaker than others and thus more susceptible to attack.

Brinker and Klein's results (25-26) for coloration of silicates in the temperature range 300-400 °C, showed that if the SiO_2 content of the glasses were greater than 70 mole % a reaction layer formed, whereas for glasses with a silica content < 70% coloration occurred. These results are consistent with those of Lau and McMillan and again emphasise the importance of the driving force for the reaction and its relation to kinetics. The closer the glass composition is to the 1: 1 phase the smaller the driving force for reaction and coloration occurs instead.
One of the more difficult problems associated with coloration of ceramics is identifying the nature of the defect causing the coloration. For example, Brinker and Klein maintain that

TABLE I. Reaction products (identified using X-ray powder diffraction) of glasses and quartz after having been exposed to sodium for 3 h. Third column rates the glasses for their propensity for discoloration (1 being highest) With the exception of quartz the propensity for coloration correlates with the R value of the glasses. (Ref. 24)

GLASS	R	COLOR.	Temperature of exposure to sodium (°C)						
			300	350	400	450	500	550	600
VITROUS SILICA	0	1				←	Na_2SiO_3		→
QUARTZ	0	5				←	Na_2SiO_3		→
20 NaO-80SiO *	0.143	2				←	Na_2SiO_3		→
30 NaO- 70SiO	0.273	6					← Na_2SiO_3		→
30NaO-5AlO-65SiO	0.238	4					← Na_2SiO_3		→
30NaO-10AlO-60SiO	0.20	3						← Na_2SiO_3	→

* NaO, AlO and SiO refer to soda, alumina and silica contents in mole %.

the coloration occurred as a result of atomic Na diffusion in the glass and that the color centers are small crystals of atomic sodium. They cite as evidence the fact that some colored glasses when heated in flowing dry air at 400 °C became colorless indicating that the Na was not chemically bound to the glass. Furthermore, electron diffraction patterns of some of the colored samples showed diffraction rings that were ascribed to small (< 100 A) randomly oriented crystals with d spacing that closely matched those of atomic sodium.

On the other hand, Lau et.al., have shown that coloration is associated with non-bridging oxygen formation and that sodium was present in the glass as ions. Their model for discoloration, shown in Fig.9a, involves the breaking of an 0-Si bond and the formation of a nonbridging oxygen. The extra electron is assumed to localize on the Si atoms to form an E' type color center. The authors further proposed, based on the optical absorption spectra, that Si-Si bonds could exist. The formation of such bonds, where two E' sites combine to form a Si-Si bond, are shown in Fig. 9b.

Thus evidence exists for both types of defects and most likely both are present. This is not surprising since coloration can be viewed as being a solid solution between the metal and the ceramic. Hence the same parameters that determine whether precipitation or supersaturation will occur also apply to coloration; i.e. concentration, quench rate, etc.. In that regard it is instructive to note that Brinker and Klein's samples were exposed to Na for much longer times (days to several month) than Lau et. al. (12 -24 h). It would thus be reasonable to assume that

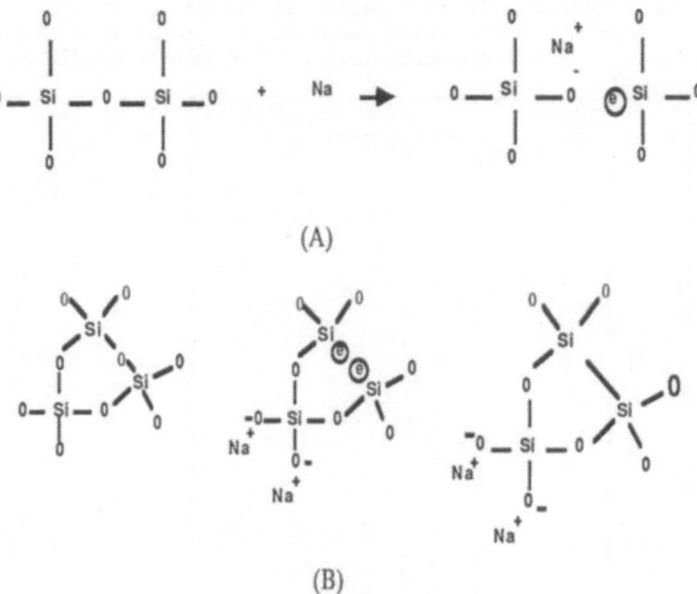

Figure 9. Complementary models for the discoloration of silica by Na. (a) Localization of electron at a non-bridging site to form an E' center. (b) Formation of a Si-Si bond from the joining together of two E' centers resulting in the formation, on an atomic scale, of sodium silicate and non-stoichiometric silicon oxide.

the longer exposure times resulted in an increased concentration of Na in the glasses which resulted in the precipitation of Na.

b) COLORATION OF BETA ALUMINA

Due to its technological importance numerous studies have addressed the problem of the chemical and electrochemical coloration of β-alumina in Na . It is well established that immersion of β-alumina in liquid sodium sometimes results in its darkening. Transmission electron microscopy of the discolored regions fail to reveal any features that could be attributed to discoloration. The chemical darkening is thus believed to be due to point defects rather than second-phase precipitates (27). The influence of the chemical coloration on electrolyte lifetime was investigated by D. Gourier et. al. (28) who measured the lifetime of the electrolyte [measured as total charge transfer (Ah) before the detection of a short circuit] as a function of immersion time in molten Na at 330 °C before the passage of the current. Their results, shown in Fig. 10, clearly indicate that the lifetime strongly decreases as a function of the time the tubes were placed in Na. Subsequent E.S.R. studies on the blackened material revealed the presence of a weak line superimposed on the Mn (II) spectrum (the Mn was present as an impurity). This line was attributed to the conduction electrons associated with metallic sodium. The width of the signal suggested that the sodium particles size was less than 0.5 μm. Based on their work on the coloration of single and polycrystals of Na β-alumina De Jonghe et.al. (29) proposed the following reaction to explain the phenomena:

$$(O_o)\beta \implies V_O^{\bullet\bullet} + O_E (Na) + 2 e^{-1} \qquad\qquad IX$$

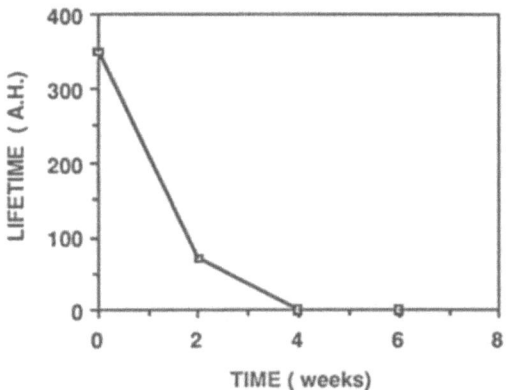

FIGURE 10.   The effect of exposure of β"-alumina to Na at 330° C prior to electrolysis. The lifetime was found to be a strong function of the coloration. After Ref. (28).

where E refers to the Na electrode into which the oxygen dissolves and β refers to the electrolyte. In other words, the oxygen is postulated to diffuse out of the electrolyte into the Na electrode leaving behind oxygen vacancies and electrons which combine to form a color center. The fact that the coloration was bleached in the presence of air at 300 °C but not in an evacuated and sealed ampule was taken to be compelling evidence that the coloration involves the removal of oxygen from the β-alumina. The bleaching occurred in well defined layers and only in the direction of the conduction planes. In the absence of air the heating did not produce bleaching, implying that the defects responsible for the coloration were oxygen vacancies. Furthermore since the activation energy of the process was found to be about half the activation energy for electronic conduction in β-alumina, it was concluded that the rate limiting step in the process was the diffusion of oxygen vacancies into the electrolyte.

Choudhury (18) measured the free energy of formation of β"-alumina (his results were used in Fig. 8) and pointed out that since the activity of Na_2O in the Na electrode saturated with oxygen was unity while the activity in the electrolyte was less than unity, diffusion of oxygen from the electrolyte into the electrode was a thermodynamic impossibility. Choudhury estimated that for oxygen to dissolve out of the electrolyte the concentration of oxygen in the electrode would have to be less than $\approx 10^{-20}$ ppm..

An alternate reaction scheme that is thermodynamically favored, is the simpler incorporation reaction:

$$2\,Na\ (liq)\ \Rightarrow\ 2\,Na^{\bullet}_i + 2\,e^{-1} \qquad\qquad X$$

followed by the trapping of the electrons by the oxygen vacancies to form a color center:

$$V_O^{\bullet\bullet} + 2\,e^{-1}\ \Rightarrow\ V^x_O \qquad\qquad XI$$

It is well established (30) that the displacement of the color zone results from the displacement of the free electron and its subsequent entrapment to form a color center further on (V^x_O in this case). It follows that the progression rate of the color zone is determined by the mobility of the electrons and their lifetime. Stsikov et. al. (31) observed that the color zone was influenced by the applied field as shown in Fig. 11. What they failed to note, however, was that the color zone seemed to move towards the anode, indicating that the rate limiting step in the

259

color producing defects were effectively negatively charged, consistent with reaction X. Had the rate limiting step in the coloration process been positively charged oxygen vacancies as postulated by De Jonghe et. al., the color zone would have migrated to the left as a result of the applied field, contrary to the results shown in Fig.11.

The fact that the color zone was influenced by the electric field at all is significant and is evidence that the color zone has some partial electronic conductivity since according to Eq. 10 the applied field affects the migration of the minority species only if the layer into which it is growing is partially electronically conducting. This is consistent with the results of Weber (32) who showed that the blackened layer was found to have increased electronic conductivity and produced a broad optical absorption.

The concentrations of oxygen vacancies and Na interstitials are interrelated through the mass action expressions of the various defect reactions. Increasing the concentrations of Na interstitials would result in an increase in the concentration of oxygen vacancies. The oxidation of the ceramic according to:

$$\frac{1}{2} O_2 + V_o^{\infty} + 2 e^{-1} \Rightarrow O_0^x$$

would thus eliminate the color centers near the surface and cause bleaching and could possibly explain the observations of DeJonghe et.al. (29).

Chemical coloration also occurs when ceramics are exposed to lithium. However, because the free energy of formation of the Li_2O is so negative, in many cases reaction layer formation is favored over coloration. As mentioned earlier, the exposure of lithium borate glasses to Li resulted in the formation of a reaction layer that was identified as $3Li_2O.B_2O_3$ (11). However, when $3Li_2O.B_2O_3$ was immersed in molten Li it turned black quite rapidly at 330 °C . The blackening was reversible and bleaching occurred upon heating in air.

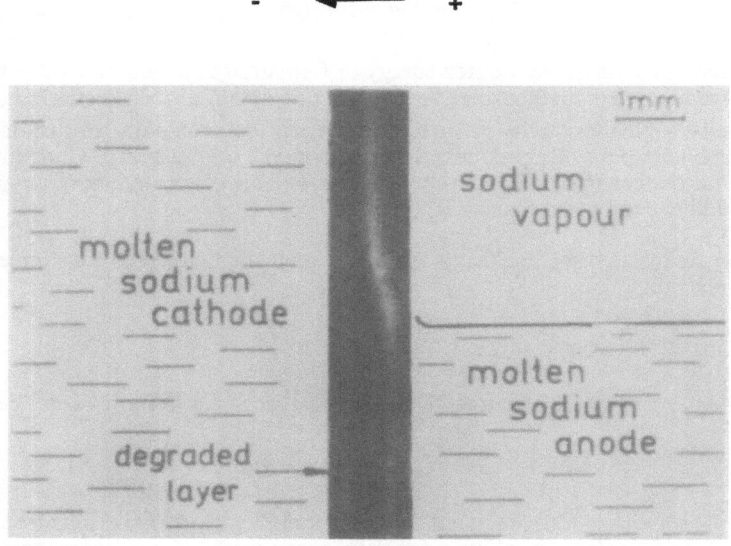

FIGURE 11. Micrograph of a thin section of β"-alumina membrane after passing about 1500 A/cm^2 at 1 A/cm^2 at 350 °C. Current was passed from right to left.After Ref.(31).

V. ELECTROLYTIC DEGRADATION

The degradation processes discussed up to this point were thermodynamic in origin and are related to the inherent chemical and electrolytic instabilities of the materials involved. There are other forms of degradation mechanisms, however, that come into play as a result of the passage of an ionic current through a solid electrolyte. In the following sections these electrolytic degradation mechanisms are considered.

The importance of these degradation mechanisms can be best illustrated by noting that the most severe problem that has hindered the wide scale use of β-alumina solid electrolytes in Na/S cells is electrolytic in nature. The problem was identified early on and involves the rapid propagation of Na filled cracks which eventually lead to electronic short circuit. The problem occurs only above a critical current density, i_{crit}; below that threshold no degradation is observed for long time periods.

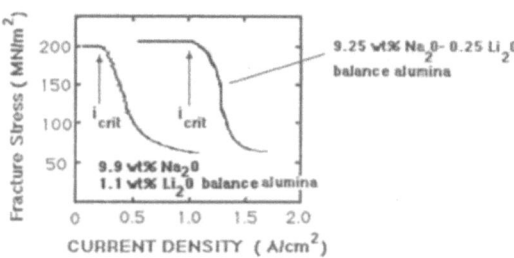

FIGURE 12. Dependence of mechanical degradation on current density in 10 min fused salt electrolysis test on two different solid electrolytes. For current $<i_{crit}$ little or no degradation is observed. Fracture stress was measured in 3 point bending. After Ref. (33).

Evidence for the rapid degradation in strength of two different β-alumina compositions beyond i_{crit} is shown in Fig. 12 where the fracture stress in three point bending is plotted as a function of current density. The degradation in strength is due to the extension of metal filled cracks in the electrolyte as a result of the ionic current. Because of the nonuniform potential distribution around the cracks, the current is focused to the crack tip, as shown in Fig. 13, where the ions are neutralized. If the ionic flux at the crack tip is greater than the flux that exits the crack, a pressure buildup will ensue that results in the extension of the crack and the ultimate failure of the ceramic.

The evidence in support of this mechanism is:

1) Simple immersion of the electrolytes in Na causes no deterioration in strength (33).

2) Na $\rightarrow$ Na$^+$ electrolysis (discharging) causes no deterioration, whereas Na$^+$ $\rightarrow$ Na (charging) causes a deterioration that is a function of current density.

3) The lifetime is related to the surface finish; the rougher the finish the shorter the lifetime. (34)

4) Above a critical current density, the fracture strength decreases as the time of electrolysis increases (33) indicating that the average flaw size increases as a result of the applied current.

5) Na streaks from the tip of machining cracks were observed upon room temperature electrolysis of the β-alumina demonstrating the concept of current focusing (35).

There are three models that have been suggested to explain this phenomena: a Critical Fracture Model (CFM) originally suggested by Armstrong, Dickinson and Turner (34); a stress corrosion or subcritical crack growth model advanced by Richman and Tennenhouse (33); and a supersaturation model proposed by Nicholson (36).

The experimental evidence to date points to the CFM model as being the most accurate. There is little evidence that stress corrosion cracking occurs in these electrolytes. Shetty el. al. (37) measured the stress corrosion exponent, n, from the slopes of log velocity -log stress intensity plots for tests carried out under molten Na. The slopes were quite large (n= 562 and 355) indicating that stress corrosion is not the primary mechanism for electrolytic degradation.

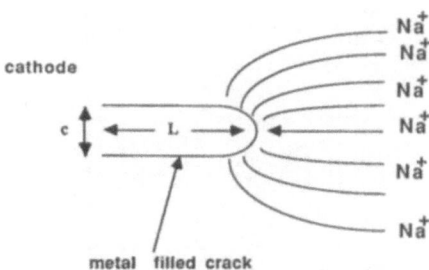

FIGURE 13.   Current focusing at crack tip. L is length of crack and c its width.

Nicholson (36) proposed a model in which sodium precipitation occurs in the grain boundaries adjacent to current focusing flaws by the oxidation of local oxygen ions. A process that leads to the formation of color centers and sodium atoms in the grains. The latter coalesce and form Na colloids, microcracking the ceramic. These microcracks later join the originating flaw and promote its extension. Nicholson further proposed that due to current focusing more Na ions arrive at certain sites within the electrolyte than are discharged which presumably leads to the precipitation of Na and collapse of the lattice. The major flaw with this model, as Virkar (38) pointed out, is that the electrochemical potential of the ions was not taken into account; the ions diffuse down an electrochemical gradient rather than a chemical gradient as suggested by Nicholson. The rapid increase in the electrostatic potential would very rapidly stop further influx of Na ions to a given region. Furthermore, the Nicholson model cannot explain many experimental observations such as the rapid increase, by orders of magnitude, in the critical current density as the melting point of Na is exceeded. Furthermore, Hunt et. al. (39) have shown that at high temperatures very high current densities (15 A cm^{-2}) can be passed through Na/Na cells as long as the Na on the egress side was in the gas phase. This observation strongly suggests that the presence of a solid or liquid dendrite in the electrolyte is required to cause the degradation.

The following discussion addresses the critical fracture model.

If the applied current and the crack geometry are such that the Na flux at the crack tip is greater than the flux exiting the crack at the electrode/electrolyte interface, a Poiseuille pressure will build up. This pressure is given by (34) :

$$\Delta P = 8 \, \eta \, i \, V_m \, L^2 / zFc^3$$

where L is the length of the crack, c its radius (see Fig. 13). η, i, V_m, z, F are, respectively, the viscosity of the fluid, the current density, the molar volume, number of equivalents per mole (z=1 in this case), and Faraday's constant. This hydrostatic pressure buildup will cause the stress intensity, K_1, at the crack tip to increase. When K_1 reaches the critical stress intensity, K_{1C}, of the ceramic, fracture occurs. The fracture is not catastrophic, however, because as the crack extends, its volume increases, relieving the pressure. The process then repeats itself till failure occurs.

The pressure buildup in the metal filled crack depends on the crack geometry and the rate of accumulation of material in the crack and, as Shetty et. al (37) pointed out, is a nontrivial problem because of the interdependence of the crack shape and the pressure distribution: one determines the other. The various models put forward to explain this phenomena (34,37,40) differ in their sophistication and assumptions as to crack geometry and response to the pressure buildup. The details of the various models will not be discussed in this paper. The interested reader is referred to a recent review article by Ansell (41) that compares and summarizes the various models and their assumptions.

Theoretical calculations have shown that the critical current density, i_{crit}, at which this mode of failure, classified as mode I failure, will occur is:

$$i_{crit} = \alpha K^4_{1C} . c^2 . L^{-n} . \eta^{-1} \tag{25}$$

α is a constant containing elastic constants, and n is a parameter that ranges between 1 and 3 depending on the way in which the elastic relaxation is taken into account and on the details of the crack geometry.

The evidence for this model comes from various sources, the most convincing being the effect of increasing the fracture toughness of the ceramic on i_{crit} (38). The enhancement in i_{crit}, was found to be about 3 times that of untoughened β"-alumina, but was lower than predicted. This model also explains the observations of Kuribayashi and Nicholson (42) who measured the fracture stress of notched β—alumina bars in point bend tests as a function of current density and applied stress and found that degradation was enhanced under both compressive and tensile stresses. Kuribayashi and Nicholson argued that their results prove the invalidity of the CFM model. Virkar (38), however, counter argued that these observations were in complete accord with the expectations of the CFM model: in tension the applied stress would enhance the stress intensity factor at the crack tip and enhance degradation, while in compression the pinching off of the crack opening at the surface would reduce the flux of Na out of the crack, enhancing the Poiseuille pressure buildup and reducing the critical current density.

The major problem with the various crack fracture models put forward to explain the degradation is that the predicted critical current densities are orders of magnitude higher than the observed ones. The models predict values of the order of 1500 Acm^{-2} , whereas the experimentally determined values are about two orders of magnitude lower. Several possibilities have been proposed to explain the discrepancy and include:

1) Anisotropy of fracture toughness. Hitchcock and De Jonghe (43) measured the fracture toughness anisotropy for β-Alumina single crystals using a hardness indent method. They found that the fracture toughness normal to the 00.1 planes to be $\approx$ 2MPa. $m^{1/2}$ while parallel to the 00.1 planes it was $\approx$ 0.16 MPa. $m^{1/2}$. When the latter value is inserted in Eq.25, values for the critical current densities agree reasonably well with experimental results.
2) The tortuosity of the crack path for liquid Na is likely to enhance the Poiseuille pressure that is generated at the crack tip (37).
3) Non-complete wetting of electrolyte and current intensification at the edges of the nonwetted areas. Virkar et. al. (35) have shown that significant current focusing can occur around non-wetted areas at room temperature.
4) Localized heating at the crack tip and failure due to thermal stresses. Kawamoto and

Kobayashi (45) calculated the temperature distribution subject to the condition of current focusing at a crack tip and demonstrated that the thermal stresses would be sufficient to cause fracture even in the absence of cracks.

Buechele et. al. (46) investigated the effect of microstructure on Mode I degradation using acoustic emission to detect the onset of degradation. A fine grained and a coarse grained ceramic were used and the results are shown in Fig. 14. Na/Na cells were operated at 350 °C and most of the area of the electrolyte was covered with a sealing glass except for a small area of about 0.2 cm^2 left unexposed for current flow.The current density was increased linearly until failure was detected. The criteria for failure was significant acoustic activity above background (point A in Fig. 14a) and the onset of sustained acoustic activity involving flaw extension (point B in Fig. 14a).

The authors used Weibull statistics to interpret the failure data on the premise that if failure depended on the statistical distribution of flaws, then the critical current density would also be determined statistically. The survival probability would thus be related to the critical current density by:

$$P_s = \exp \left[-\int_A \{ i - i_u / i_o \}^m \ dA \right] \qquad (26)$$

where P_s is the survival probability, A the area of the sample, i the current density at failure, i_u the current density below which failure does not occur and i_o the curent at which the survival probablity is 1/e and m the Weibull modulus. Implicit in Eq. 26 is that the surface flaws are responsible for failure. It was pointed out by the authors that because of current focusing at the edges of the unmasked area, failure occurred at an average current density that is lower than if the electrode current was uniform. i_u was assumed to be zero.

The acoustic activity is shown in Fig. 14a and the results for both grain size materials are shown in Fig.14b where ln i for the onset of sustained acoustic emmision are plotted versus ln ln P_s. The results clearly show that the average critical current density was inversely proportional to the grain size. Assuming that the flaw size is of the order of the grain size, this constitutes strong evidence that this mode of degradation is related to the flaw size distribution in the electrolyte. The experimental results also seem to indicate that $i_{crit} \propto 1/L$, in other words n in Eq. 25 should be 1

Current focusing and electrolytic degradation is not limited to β-alumina and Na but has been observed in other electrolytes such as $Na_{3.2}Zr_2P_{0.8}Si_{2.2}0$ (42) and with other electrodes such as Li. Barsoum et. al. (10) observed electrolytic degradation in Li/glass/Li cells where the electrolyte was a lithium borate glass. The lifetime, as determined by the time it took to short out the cells at 225 °C, was found to be inversely proportional to the applied current density. Scratching of the glass surface was found to have a deleterious effect on the lifetime as expected. The cells shorted out as a result of the extension across the glass electrolyte of very thin Li filled cracks. In a subsequent study (45), cells of the type: "Li-Sn"/ glass/ "Sn" were used to titrate Li into Sn electrodes. "Li-Sn" refers to a 2 phase electrode (Li saturated Sn and the LiSn intermetallic). The lifetime of these cells was found to be very long (> 100 hrs) at comparable current densities as were used in the Li-Li study. These result strongly suggests that the presence of a molten Li cathode was a necessary condition to short the cells.

A different mode of degradation was observed, however, in the "LiSn" cells. Above a certain current density, that depended on the electrolyte composition, Li metal dendrites were observed to form at the nodes of the molybdenum mesh that was used to contain the molten Sn as shown schematically in Fig. 15. These dendrites were localized and imbedded below the glass surface and did not appear to grow into the electrolyte. None of the cells that exhibited this behavior shorted out, once again demonstrating the importance of the liquid penetration of the electrolyte. Given the location of the Li dendrites and the fact that the Li grew into the glass seem to indicate that the most likely process for the formation of these Li dendrites is one of electron injection into the glass that results in the neutralization of the Li ions within the glass.

264

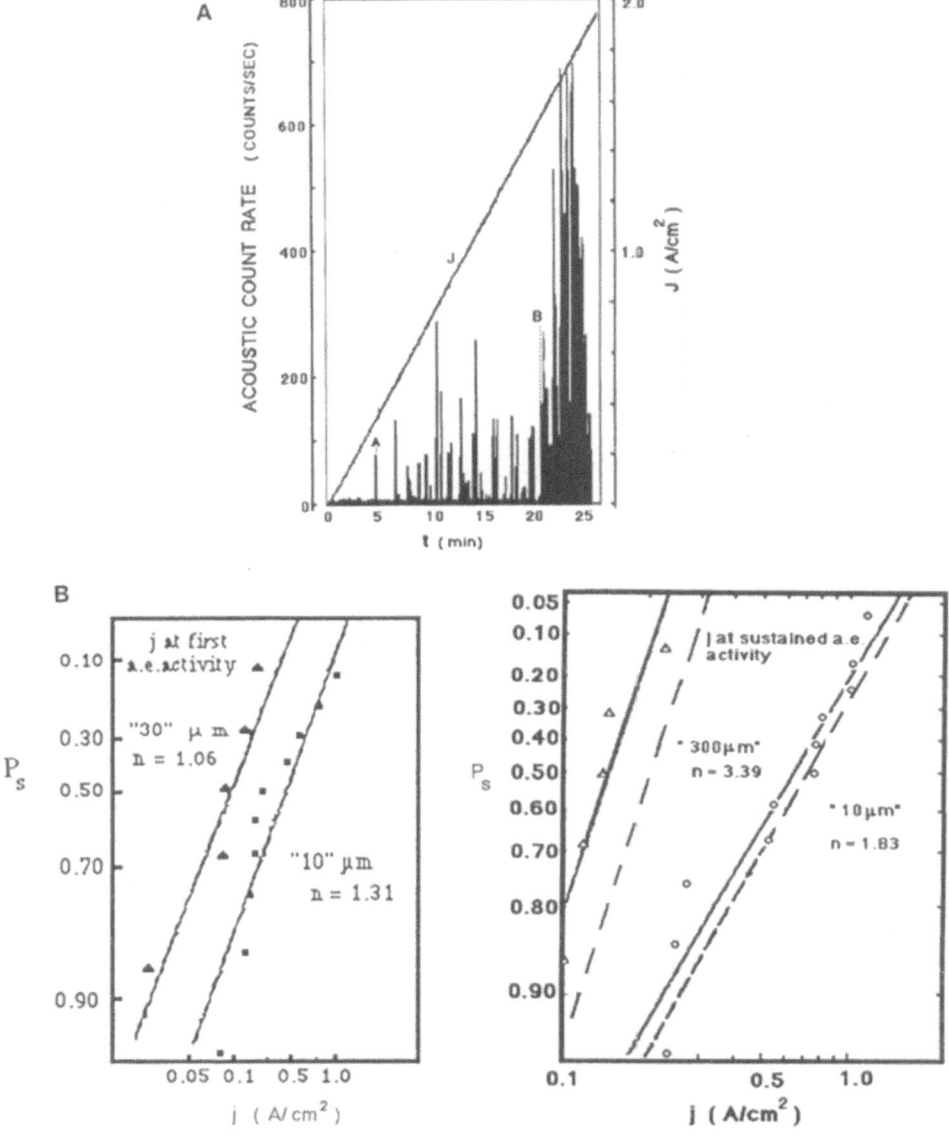

FIGURE 14. (a) Typical acousic activity as a function of time at 350 °C as the current is
increasing at a consant rate. The first significant acoustic events occur at a
current density A, that is considerably lower than that for sustained emission,B.
(b) Grain size dependence of theWeibull plots for current densities at which
sustained emission was observed. The solid lines are for j=i/A; dashed lines are
calculated for uniform current densities. After Ref. (46).

VI. ASYMMETRIC POLARIZATION

Asymmetric polarization is defined as a situation where the discharge resistance, i.e. when Na ions move from the liquid Na electrode through the electrolyte to the sulphur electrode, is much higher than the reverse reaction. Electrolyte compositions that display asymmetric polarization have been shown to suffer from short cell life caused by premature electrolyte failure probably due to high local current densities. Cell life seldom exceeds 100-200 Ah/cm^2 when the condition of asymmetric polarization exists. Life is usually terminated by cracking of the tube which, upon examination, appears darkened and crumbly.

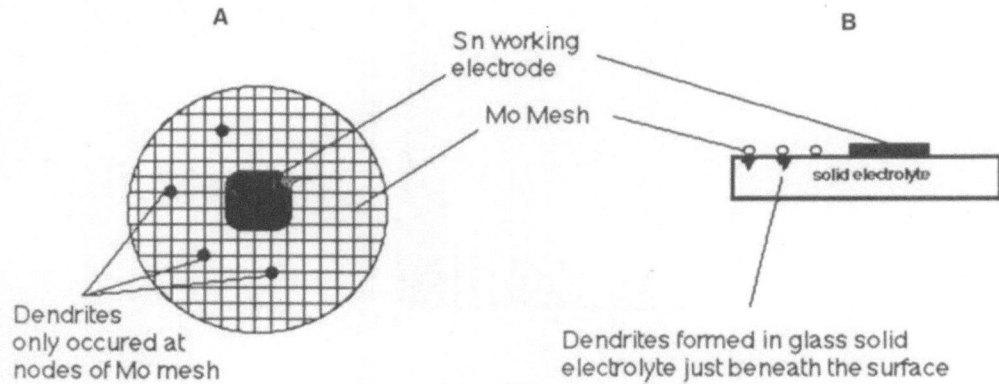

FIGURE 15.   Schematic of dendritic formation in glass electrolytes. (A) Top view and (B) Side view. The dendrites formed exclusively at the nodes of the Mo mesh that was used to contain the Sn working electrode. No dendrites formed below the Sn electrode.

The experimental arrangement and the equivalent circuit used to measure asymmetric polarization is depicted schematically in Fig. 16. The preferred technique is cyclic voltametry where a current that flows through the circuit is measured as a function of a sawtooth voltage that is applied between two Na electrodes (or a fused NaNO$_3$ bath and a Na electrode). The current is usually measured as a potential drop between one of the Na electrodes and a third reference electrode (between electrodes 2 and 3 in Fig.16a). The corresponding equivalent circuit is shown in Fig. 16b. In the absence of asymmetric polarization, the resistance is independent of the polarity of the applied voltage as shown in Fig. 16c (curve 1). The response shown in curve 2 in Fig.16c, however, is typical of cells that exhibit asymmetric polarization. The slope of the curve, which is inversely proportional to the resistance, is lower when the Na is moving from the electrode into the electrolyte than during the reverse cycle. Comparing the slopes for curves 1 and 2 it is clear that the overall resistance of cells that exhibit asymmetric polarization are higher than those that do not.

Three factors seem to affect the extent of asymmetric polarization: the presence of a thin passivating layer on the surface of the as manufactured electrolyte (47-49), the activity of oxygen in the Na electrode (50) and temperature. It is important to note that the presence of a thin passivation layer *per se* is not sufficient to cause asymmetric polarization since the resistance of that layer, if nothing else happened to it, would contribute to the potential drop for both directions of the current. A layer has to grow or be modified as a result of the passage of the current to result in polarization in one direction but not the other.

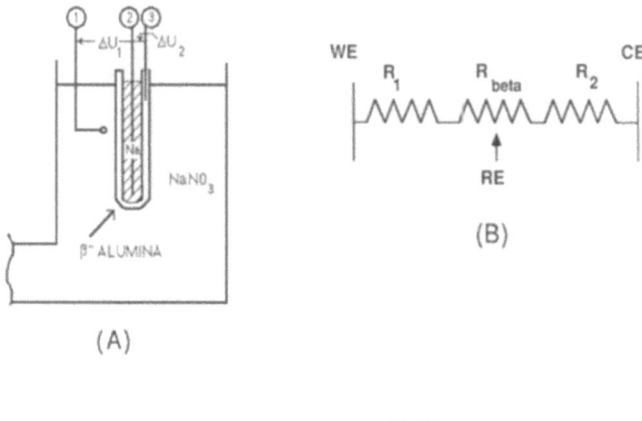

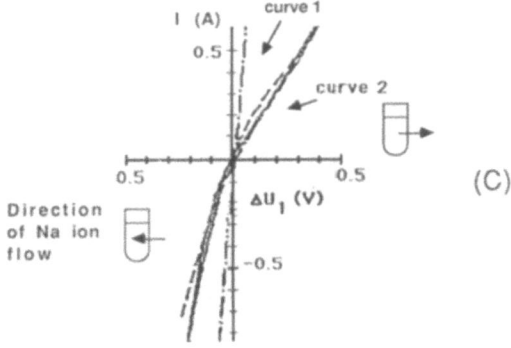

FIGURE 16. (A)Typical arrangement for measuring asymmetric polarization. electrode 1 is a platinum reference electrode, electrode 2 is the inner sodium electrode and 3 is a small reversible sodium electrode. (B) Simplified equivalent circuit; (C) typical voltammetric circuit for two compositions. Curve 1, symmetric polarization and low overall cell resistance. Curve 2 exhibits asymmetric polarization where the resistance during discharging of the Na is greater than during charging. After Ref. (49).

If a layer is initially present on the surface of the electrolyte, the asymmetric polarization appears to be due to the anodic modification of that film rendering it more resistive (49). Breiter and Dunn (49) measured the total charge transfer upon charging and discharging of polarized cells and found that at all sweep rates the charge passed during anodic passivation was greater than the cathodic charge for converting to the initial film. They concluded that the excess charge was used in modifying the layer. The nature of that modification is not clear at this stage. In support of the model for the presence of a passivating layer, Stsikov et. al. (31) pointed out that the AC impedance spectra of cells that showed asymmetric polarization were similar to those frequently observed in metal passivation processes in electrolytic solutions.

The effect of electrolyte composition on asymmetric polarization was investigated by Singh (47,48) who has shown that asymmetric polarization was displayed whenever the electrolyte composition contained more than about 80 vol% of the β"-alumina phase regardless of the exact starting composition of the electrolyte. He also showed that a chemical treatment of the tubes, either with phosphoric acid or by washing with deionized water, was effective in circumventing the asymmetric polarization by removal of the resistive surface film. The resistive film was found to consist mainly of 38 to 43 % Na_2O and 45 to 57 % Al_2O_3 which was significantly higher than the bulk ceramic. Singh proposed a mechanism by which this film would form during the rapid zone-sintering process used in the production of the tubes, based on evaporation of the soda during the heating cycle and its subsequent condensation during the cool down cycle.

The effect of oxygen concentration in the Na electrode on asymmetric polarization was studied by Mailhe et. al.(50). They have shown that depending on the amount of oxygen present in the electrode a passivating film can form in situ as a result of the passage of a current. Five oxygen concentrations were investigated. The highest was Na saturated with oxygen and the lowest was Na that was filtered through stainless steel wool and then reacted with titanium sponge at 350 °C for 24 hrs. The degree of polarization was correlated to the oxygen content: the higher the oxygen content the more severe the polarization. The removal of Na at the electrode/electrolyte interface results in a local increase in the concentration of oxygen at that interface. If the increase in concentration exceeds the solubility limit of oxygen, Na_2O would tend to form on the surface of the electrolyte and result in an increased interfacial resistance.

Mailhe et. al. (50) also report that at 350 °C, polarization was absent in most cases, regardless of the oxygen content, presumably due to the high solubility of oxygen at the higher temperatures where precipitation of Na_2O was unlikely.This result is consistent with the finding of Breiter and Dunn (48) who observed a declining polarization effect with time at 350 °C in tubes that started with a thin passivation layer. They argued that possibly the current distribution became more uniform or that the conduction planes became unblocked due to the relatively high fields as a function of time. In light of the results of Mailhe et. al. a more likely explanation would be the slow dissolution of the prexisting passivation layer in the Na with time at 350 °C.

The reason why asymmetric polarization would result in premature failure of the electrolyte is not entirely clear but is believed to be related to current focusing. A nonuniform layer could channel the current through flaws in the passive film, resulting current focusing and electrolytic degradation. Hence, the best strategy for long life electrolytes should include choice of compositions that do not form a thin passivation layer, operation of cells at temperatures at or above 350 °C and starting with ultraclean Na that would prevent the in situ formation of such layers.

VII. CONCLUSIONS

Alkali metals, particularly Li, because of their high reactivity are very challenging elements to contain and work with. As outlined in this paper the best strategy is to design and choose ceramics that are thermodynamically stable. Coloration, a sign of chemical instability, and the concomitant changes in the electrical properties is another factor to contend with. If used as electrolytes, the combined effects of electrical and chemical driving forces place even more stringent demands on the materials that come in contact with alkali metals. Finally, the passage of current can result in degradation mechanisms that are more related to the mechanical properties of the ceramic, such as fracture toughness, than to the chemical or electrochemical instabilities.

BIBLIOGRAPHY

1. L. Heyne in 'Solid Electrolytes', Ed. S. Geller, Springer Verlag, Berlin, 1977, p.169.
2. C. Wagner, "Beitrag zur Theorie des Anlaufvorgangs", Z. Phys. Chem., **B21** (1933) p. 26.
3. E.E. Hellstrom and W. Van Gool, "Constraints for the Selection of Lithium Solid Electrolytes", Revue de Chimie Minerale, **17**, (1980) p.263.
4. Aspandiar, R.F., PhD. Dissertation, Stanford University, 1984.
5. JANAF Thermochemical Tables, 2nd Ed. 1972.
6. I. Barin and O. Knacke, "Thermochemical Properties of Inorganic Substances", Springer-Verlag, New York-Heidelberg, (1973).
7. M.G. Barker, I.C. Alexander and J. Bentham, " The Reactions of Liquid Lithium with TiO_2, ZrO_2, HfO_2 and ThO_2" J. of Less-Common Metals, **42** (1975) p. 241.
8. M.W. Barsoum and K.Pytlewski, " Thermodynamics and Kinetics of Li/Ceramic Interactions", Presented at Electrochemical Soc. Spring Meeting, Abstract # 46, Philadelpha, PA (1987).
9. K. Pytlewski, Masters Thesis, Drexel University, August 1987.
10. M.W. Barsoum, M. Velez, H.L. Tuller and D.R. Uhlmann, "Reactions at Alkali Metal-Glass Interfaces", in Surfaces and Interfaces in Ceramic and Ceramic-Metal Systems. Eds. J.Pask and A. Evans, Plenum Publishing Co. (1981) p. 567.
11. M.W. Barsoum and H.L. Tuller, "In-Situ Determination of the Kinetics of Reaction between Lithium and Fast-Ion Conducting Lithium Borate Glasses". Solid State Ionics, **18 & 19**, (1986) p. 388.
12. B.C. Bunker, C.J. Levy, C.C. Crafts in Power Sources, Vol.**8**, Ed. J. Thompson, Academic Press, London 1981 pp. 57-62.
13. P. Skarstad, D. Merritt and N. Istephanous, "Corrosion of Glass Feedthrough Insulators by Lithium". To appear in Proceedings of 15th International Sources Power Symposium, Brighton, England, Sept. 1986.
14. N.S. Istephanous , K. Fester, D. Merritt, P. Skarstad and D. F. Untereker, " Glass Seal Corrosion in Liquid Li Electrolyte Batteries", Electrochemcial Soc. Extended Abstracts 84-2, Abstract # 146.
15. B.L. Maschhoff, K.R. Zavadil and N.R. Armstrong, " Reactions of Evaporated Lithium Films With Silica Surfaces" Appl. Surf. Sci. **27**, (1986) pp.285-298.
16. M. Barsoum, Submitted to J. Mat. Science.
17. F.A. Elrefaie and W.W. Smeltzer, "The Stability of β-alumina ($Na_2O.11Al_2O_3$) in Oxygen Atmospheres", J. Electrochem. Soc., **128**, (1981) p.1443.
18. N.S. Choudhury, "Thermochemical Stability of β- and β"-Alumina Electrolytes in Na/S Cells: Parts I & II", J. Electrochem. Soc., **133**, (1986) pp.425-431.
19. O.M. Sheedharan and J.B. Gnanamoorthy; " Oxygen Potentials in Alkali Metals and Oxygen Distribution Coefficients between Alkali and Structural Metals-An Assesment", J. of Nuc. Mat., **89**, (1980) pp.113-128.
20. Singh, R.N., "Compatability of Ceramics with Liquid Na and Li", J. Am. Ceram. Soc., **59** (1976) p.112.
21. W.H. Cook, "Corrosion Resistance of Various Ceramics and Cermets to Liquid Metals", Oak Ridge National Lab. Rept ONRL-2391 (1960).
22. W.D. Thonig, J.T.A.Roberts and R.N. Singh, "Materials Studies in Support of Liquid Metal-MHD Systems", CONF-740414-I, Argonne National Lab, Illinois, 1974.
23. J. Lau and P.W. McMillan, "Interaction of Na with Simple Glasses: Part I. Vitreous Silica", J. of Mat Sci. **17**, (1982), pp.2715-2726.
24. J. Lau and P.W. McMillan, "Interaction of Na with Simple Glasses: Part II" *ibid*. **19**, (1984), pp.881-889.
25. C.J. Brinker and L.C. Klein, " Behaviour of Silicate and Borosilicate Glasses in Contact with Metallic Na, Part I ". Phys. Chem. Glass., **21**, (1980) pp. 141-145.
26. C.J. Brinker and L.C. Klein, " Behaviour of Silicate and Borosilicate Glasses in Contact with Metallic Na, Part.II", *ibid.*, **22**, (1981) p. 23.
27. D. Gourier, A. Wicker and D. Vivien, "E.S.R. Study of Chemical Coloration of β and β"-Aluminates by Metallic Na". Mat. Res. Bull, **17**, (1982) pp.363-368.
28. L. C. DeJonghe, L. Feldman, A.Beuchele, " Slow Degradation and Electron Conduction in Na/β-Aluminas." J. of Mat .Sci. **16**, (1981), pp.780-786.

29. L. C. DeJonghe and A.Beuchele, " Chemical Colouration of Na/β—Aluminas". J. of Mat. Sci. **17**, (1982), pp.885-892.

30. P.Fabry and M. Kleitz, " Electrochemical Coloration and Redox Reactions in Solid Ionic Conductors" in Electrode Processes in Solid State Ionics, Eds. M. Kleitz and J. Dupuy, Proceedings of NATO Advanced Study Institute, Sept. 1975. D. Reidel Pub. Co.1976. pp.331-365

31. G. Stsikov, P. Yankulov and E. Budevski, " Nonstandard Behavior of Polycrystalline β/β"-Alumina Membranes in Na Environment" Solid State Ionics, **18&19,** (1986) pp. 631-635)

32. N. Weber, Energy Conversion, **14,** (1974) p.1.

33. R.H. Richman and G. J. Tennenhouse, " A Model for the Degradation of Ceramic Electrolytes in Na-S Batteries", J. Amer. Cer. Soc., **58,** (1975) p.63.

34. R.D. Armstrong, T. Dickinson and J. Turner," The Breakdown of β—Alumina Ceramic Electrolyte", Electrochem. Acta, **19,** (1974) p.187.

35. A. Virkar, L. Viswanathan & D. Biswas, "On the Deterioration of β"-Alumina Ceramics Under Electrolytic Conditions", J. Mat. Sci, **15,** (1980) pp.302-308.

36. P.S. Nicholson, "A Supersaturation Model for the Degradation of Na β-aluminas" J. Mat. Sci, **18,** (1983) pp.1597-1603.

37. D.K. Shetty, A.V. Virkar and R. S. Gordon, " Electrolytic Degradation of Lithia-Stabilized Polycrystalline β-alumina" in Proceedings of Internat. Symp. on Fracture Mechanics of Ceramics entitled " Fracture Mechanics of Ceramics" Vol. 4. Ed. R.C. Bradt, Plenum Press, N.Y. (1978) p.651

38. A. Virkar, " The Role of Superimposed Stresses on the Degradation of Solid Electrolytes", J. Mat. Sci, **21,** (1986) p.859.

39. T.K. Hunt, N. Weber & T. Cole, " Na β-alumina at high Current Density", in Fast Ion Transport in Solids Ed. P. Vashishta, J. N. Mundy and G.K. Shenoy , North Holland Press, N.Y. (1979) p.95.

40. L.A. Feldman and L. C. DeJonghe, " Initiation of Mode I Degradation in Na/β-alumina Electrolytes", J. of Mat. Sci. **17,** (1982), pp.517-524.

41. R. Ansell, "Review: The Chemical and Electrochemical Stability of β-Alumina" J. Mat. Sci, **21,** (1986) p.365.

42. K. Kuribayashi & P.S. Nicholson, "The Deterioration of Na ion Conductors under Applied Stress" J. Mat. Sci, **18,** (1983) pp.1597-1603.

43. D. Hitchcock and L. C. DeJonghe, " Fracture Toughness Anisotropy of Na β-Alumina", J. Amer. Soc., **66,** (1983), C204.

44. H. Kawamoto and H. Kobayashi, " Investigation of Temperature Distribution and Thermal Stress of β-Alumina under Conditions of Na Ion Transport". Extended Abstracts of Electrochem. Soc.Fall Meeting 1986. Abstract # 80, p.114.

45. M. Barsoum, PhD. Thesis, Massachusttes Institute of Technology, June 1985.

46. A. Buechele, L. DeJonghe, and D. Hitchcock, " Degradation of Na β"-Alumina: Effect of Microstructure", J. Electrochem. Soc., **130,** (1983) pp. 1042-1049.

47. R.N. Singh and N. Lewis, " Role of Electrolyte-Na Interface Behaviour to the Degradation of a Na-S Cell", Solid State Ionics, **9 &10,** (1983).

48. R.N. Singh, " Surface Characterization of Na β-alumina Electrolyte", J. Amer. Cer. Soc., **67,** (1984) p.637.

49. M.W. Breiter and B. Dunn, "Time Dependence of the Asymmetric Resistance of Polycrystalline β"-Alumina", Electrochim. Acta **26** (1981) pp.1247-1251.

50. C. Mailhe, S. Visco and L. De Jonghe, "Oxygen Activity and Asymmetric Polarization at the Sodium/ β"-Alumina Interface", J. of Electrochem Soc., (1987) p.1121.

270

MIXED IONIC-ELECTRONIC CONDUCTION IN FAST ION CONDUCTORS AND CERTAIN SEMICONDUCTORS

I. Riess

Physics Department
Technion IIT
Haifa 32000, Israel

ABSTRACT

Ionic and electronic currents can coexist in the so-called mixed ionic electronic conductors (MIEC). When these conductors interact with non-uniform chemical surroundings, electrical currents are induced in the MIEC. Applied electric fields affect not only the ionic and electronic currents, but also induce chemical changes in the MIEC. Under certain conditions a quasi p-n or p-i-n junction is formed, the structure of which depends on the applied voltage.

This lecture deals with the I-V characteristics and four point conductivity measurements performed on mixed ionic electronic conductors as well as induced chemical changes. The possible use of a MIEC in a fuel cell is also discussed.

INTRODUCTION

Mixed Conductivity

"Electrolytes" are commonly defined as materials that conduct ionic species only. Many ionic solids have this property. Due to the tight binding of the electrons, they have a wide energy gap and are electronic insulators. As discussed in other chapters, certain crystallographic and glass structures allow for rather easy motion of ions. Usually only one kind of ion is mobile. A typical example is α-AgI, in which the Ag^+ ions "flow" within the matrix formed by the rigid I^- sublattice. By changing slightly the chemical composition of the solid, electronic conduction may be introduced. This can be induced either by doping with foreign atoms or by changing the stoichiometry, i.e. the ratio of the various chemical components of the solid. The material then becomes a mixed ionic electronic conductor, MIEC. CeO_2 + 10 mol% Gd_2O_3 (D-CeO_2), for example, is a solid electrolyte at 800°C and atmospheric pressure, i.e. its ionic transference number, t_i, is unity. It is a fast O^{2-} conductor with an ionic conductivity $\sigma_i \sim 0.1 \ \Omega^{-1} cm^{-1}$ at 800°C. The oxide is, however, reduced at this temperature under a low oxygen partial pressure. The oxygen vacancies formed in the oxide are donors, which are ionized at elevated temperatures, contributing quasi-free electrons to the conduction band. The oxide becomes a mixed ionic electronic conductor.[1] In this way t_i can be "tuned" in the range $0 < t_i \leq 1$.

Mixed conduction has also been observed in predominantly electronic conductors. α-$Ag_{2+\delta}S$ is a semiconductor with a small gap of 0.4 eV.[2,3] In contact with sulphur, $\delta \leq 10^{-6}$, the solid is practically an intrinsic semiconductor. For larger values of δ ($\delta \leq 3 \times 10^{-4}$) it is an n-type semiconductor. The excess silver ions act as donors. However, in addition to the electronic conduction high ionic conduction is also observed, because of the high concentration of interstitial silver ions, $Ag_i\cdot$, and the rather high ion mobility in the α phase ($T \geq 180$°C). The solid however remains a predominantly electronic conductor, since the electron mobility is much higher than that of the silver ions.

Mixed ionic-electronic conduction has also been observed in less exotic semiconductors. Ge doped with Li is such an example. This material is of interest as a γ-radiation solid state detector.[4] Li in Ge is a donor. The dopant, Li, is mobile near room temperature. When an electric field is applied, both electrons and Li ions migrate. Since the electrodes block Li transport, a change in Li distribution is built up under an applied voltage. This redistribution is taken advantage of in detecting the γ rays.[4]

There are important processes that occur because of mixed ionic-electronic conduction in solids. Tarnishing, i.e. the oxidation of metals depends on the simultaneous diffusion of ions and electrons or holes through the tarnishing layers.[5] The tarnishing layer grows because of ion migration of the oxidant or metal through the layer. If not for the electronic compensating current, polarization would occur and the process would stop.

Layers formed on metals during the passivation process[6] or during fabrication of capacitors, e.g. Ta_2O_5 placed between Ta and MnO_2[7], exhibit current rectification. The oxide layer is n-type on the metal side and p-type on the the oxidant side. This p-n junction differs from conventional p-n junctions given that the defect ions are slightly mobile under operating conditions resulting in a time dependence of the electric properties.[8] This can also result in the development of I-V relations different from that of common p-n junctions.

The photographic process also depends on the simultaneous diffusion of ions and electons or holes. An important step in the photographic process in silver halides is the migration of photo-excited electrons and interstitial silver ions, $Ag_i\cdot$, to form aggregates of neutral silver atoms at certain defects (internally, at jogs and dislocations and externally, on the crystal surface).[9]

Mixed ionic-electronic conductors can replace the purely electronically conducting electrodes,[10] whether metals[11] or semiconductors[12] that are usually applied to solid electrolytes. This should have the advantage that both electronic charge and matter can be transported across the whole area of the electrode resulting in enhanced kinetics.

Theoretical Background

Transport in MIEC was previously analyzed theoretically for five limiting cases:[13]

MIEC of uniform composition. Interacting with a uniform chemical surrounding, with reversible electrodes applied to it. The current-voltage relation is then linear and the coefficient of proportionality yields the total conductivity σ_t.

For MIEC interacting with different chemical surroundings, four configurations were considered for the steady state:

Polarization. Under this condition the ionic current vanishes and the current is electronic only. If the experiment is done with one electrode blocking for ions and the other reversible, this is the Hebb-Wagner polarization method.[13,14] In this case, the chemical potentials within the sample are fixed by the sample's environment. If both electrodes are blocking to ions or the environment does not contain the ionic species, the chemical potential depends on sample history.[15] The polarization method can be used to determine conductivities and other thermodynamic properties.[16,17,18]

Open circuit. Under these conditions the total current (ionic plus electronic) vanishes. The coupled motion of ions and electronic species is denoted as ambipolar. It occurs, for instance, in tarnishing as mentioned before and in fuel cells as a leakage current which consumes fuel without yielding power.[19]

Zero electronic current. This is achieved either by short circuiting the electrodes or by using electrodes blocking to electrons.

Chemical diffusion. When the internal electric field vanishes (and the temperature is uniform), charges diffuse due to concentration gradients alone. As we show later, this happens in certain cases under polarization conditions. It occurs in predominantly electronic conductors with a high concentration of electrons, in which the electronic current is limited or blocked by electrodes.[20]

Scope

We present a theoretical evaluation of the I-V characteristics and defect distributions in MIECs. This includes the topics:

1. Four point conductivity measurements on a MIEC.[21]

2. Fuel cell based on a mixed ionic-electronic conductor.[19]

3. MIEC with a graded n-type or p-type charge distribution.[22]

4. MIEC in which the charge distribution forms a p-n or p-i-n quasi junction.[23]

The solution of the transport equations will be for arbitrary voltages and nonblocking electrodes thus eliminating many of the restrictions existing in previous theories.

ASSUMPTIONS AND DEFINING EQUATIONS

Notation

The mixed ionic-electronic conductor, MIEC, will for simplicity of notation be presented by XY. It may conduct electrons in the conduction band, holes in the valence band and one kind of ionic defect, say, X'. The latter is either a real ionic species, e.g. an interstitial silver ion, $Ag_i{}^{\cdot}$, in α-$Ag_{2+\delta}S$ or a quasi ionic species, e.g. a copper vacancy, V'_{cu}, in Cu_2O or an oxygen vacancy, $V_o{}^{\cdot\cdot}$, in ceria. $X^{\cdot}$ will represent both native defects and foreign ions introduced by doping, e.g. $Li_i{}^{\cdot}$ in Ge. Y represents either a rigid sublattice of the other ions (when the defects are native ones) or the rigid host material (when the defects are foreign atoms). We use here the Vink-Kröger defect notation.[24]

The environments of the MIEC at each electrode may be different. The surroundings are either the electrode material, e.g., a silver electrode on α-$Ag_{2+\delta}S$, or when the electrode is inert but not blocking, the material adjacent to the electrode, e.g., oxygen gas near a porous Pt electrode on D-CeO_2.

Interactions

Quasi free electrons interact with ions to form neutral species

$$X^{\cdot} + e' \leftrightarrow X^{x} \tag{1}$$

Due to reactions at the electrodes, material can be transported through the MIEC in the form of ions. When local equilibrium prevails, Eq. (1) yields

$$\mu_{X+} + \mu_e = \tilde{\mu}_{X+} + \tilde{\mu}_e = \mu_X \tag{2}$$

where μ and $\tilde{\mu}$ are the chemical and electrochemical potentials respectively. An n-type semiconductor is formed by excess of X species (high μ_x), while a p-type semiconductor is formed by a deficiency in X (low μ_x).

If both conduction electrons, e', and holes, h', are present in comparable concentrations, we must also pay attention to their recombination reaction:

When local equilibrium prevails,

$$\mu_e + \mu_h = \tilde{\mu}_e + \tilde{\mu}_h = 0 \tag{4}$$

Reversible Electrodes

We assume that the electrodes are reversible unless stated otherwise. At a reversible electrode, E, we have unimpeded passage of electrons between the MIEC and the electrode as well as local chemical equilibrium between the MIEC just under the electrode and the surroundings at the electrode, even when current is flowing. Therefore

$$\tilde{\mu}_e(E) = \tilde{\mu}_e \text{ (in electrode E)} \tag{5}$$

where $\tilde{\mu}_e$ (E) is the electrochemical potential of electrons in the MIEC just under the electrode E.

$$\tilde{\mu}_e(E) + \tilde{\mu}_{X^+}(E) = \mu_X \text{ (at electrode E)} \tag{6}$$

where $\tilde{\mu}_{x^+}$ (E) is the electrochemical potential of X ions in the MIEC just under the electrode and μ_x (at electrode E) is the chemical potential of X in the surroundings near the interface electrode/MIEC.

One can define $\tilde{\mu}_{x^+}$ at the electrode by

$$\tilde{\mu}_{X^+} \text{ (at electrode E)} = \mu_X \text{ (at electrode E)} - \tilde{\mu}_e \text{ (in electrode E)} \tag{7}$$

For a reversible electrode we find from Eqs. (5), (6), and (7)

$$\tilde{\mu}_{X^+}(E) = \tilde{\mu}_{X^+} \text{ (at electrode E)} \tag{8}$$

Constraints

1. The temperature throughout the system is uniform.

2. We allow for a non-uniform surrounding to interact with the MIEC. In addition, a voltage is applied to the MIEC. This results in an electric current. The MIEC is therefore not in thermodynamic equilibrium. However, it will be assumed that:

    a. the system has reached steady state, and

    b. that local equilibrium prevails with the various thermodynamic parameters having well defined local values.

3. Interactions between the ions, electrons, and holes are formally represented by Eqs. (2) and (4), which when applied to the electrode regions will define boundary conditions for the transport equations. Further coupling between those charges is an indirect one through the electric field within the MIEC. Direct coupling between fluxes is neglected.[25]

The four point van-der-Pauw conductivity measurement and a general solution of the transport equations for a one dimensional configuration will be given within the framework of these constraints. The discussion will then concentrate on more specific models which fulfill further constraints:

4. Local neutrality or quasi-local neutrality (i.e. where the difference between positive and negative charge concentrations is much smaller than the concentration of either component).

5. The concentration of mobile ions, $N = [X^+]$, electrons, n, and holes, p, are related by

$$N >> n, p \qquad (9a)$$

in one model, or by

$$N \sim n \text{ and } n >> p. \qquad (9b)$$

in a second model.

6. "Ideal gas" statistical behaviour of the electrons and holes. This holds also for ions for the case $N \sim n$.

7. The mobility of the various charges is uniform, i.e. independent of variation in concentation within the MIEC.

Transport Equations

The equations governing charge transport are:

$$j_i = -(\sigma_i/q) \nabla \tilde{\mu}_{X^+}, \quad \sigma_i = q\nu_i[X^+] \qquad (10a)$$

$$j_e = (\sigma_e/q) \nabla \tilde{\mu}_e, \quad \sigma_e = q\nu_e n \qquad (10b)$$

$$j_h = -(\sigma_h/q) \nabla \tilde{\mu}_h, \quad \sigma_h = q\nu_h p \qquad (10c)$$

where indices e, h, and i refer to electrons, holes, and ions respectively, and j - current density, σ - conductivity, q - absolute value of electron charge, and ν - mobility. Cross coefficients representing direct coupling between fluxes are omitted.[25,26]

To solve these differential equations, boundary
conditions must be specified. These are given by the value of
the chemical potentials of X (μ_x) in the (non-uniform)
surroundings interacting with the MIEC and by the applied
voltage.

In addition to the boundary conditions one has to be more
specific about the geometry, the nature of disorder (Cf. Eqs.
(9a), (9b)), the statistics of the various charge carriers,
and the kind of electrodes used. The system discussed will
follow the constraints mentioned above.

APPLICATION OF THE VAN-DER-PAUW METHOD TO CONDUCTIVITY
MEASUREMENTS ON MIEC[2][1]

In a four point conductivity measurement the current
flows through two of the electrodes, while a voltage is
measured on the two other electrodes, which we denote here as
voltage probes.

A high impedance voltmeter is used to minimize the
current through the probes with the aim of eliminating any
voltage drops on the probes due to contact resistances. This
does not work when the sample is a MIEC. The high impedance
voltmeter can eliminate the total current, i.e., the sum of
ionic and electronic currents, but not each component
separately. If the probes are not reversible, a voltage drop
will exist which will distort the readings.

We shall follow van-der-Pauw[2][7] and look for a solution
for a planar singly connected sample (see Fig. 1).

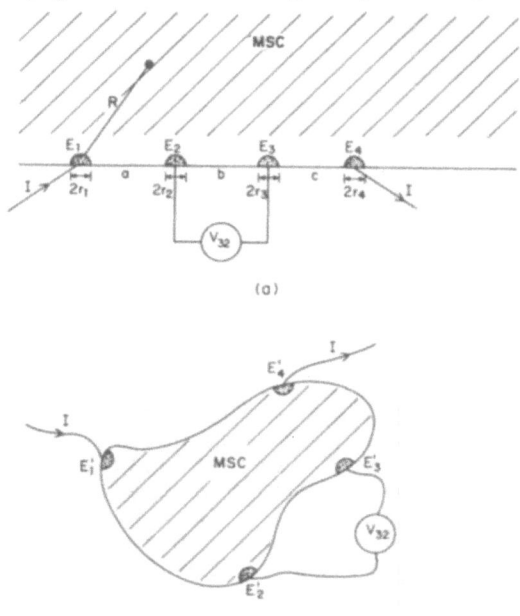

Fig. 1. Planar singly connected sample with four electrodes:
a) infinite half xy plane; b) finite sample. (From
Ref. 21, Fig. 1).

Reversible probes

If the probes can be made approximately reversible, one can solve Eqs. (10a), (10b) for a planar n-type MIEC with roughly uniform composition. The latter requires that the voltage applied to the current carrying electrodes, V_{app}, and any non-uniformities in the chemical composition in the surrounding be small. Quantitatively this means

$$qV_{app} << k_B T \qquad (11a)$$

to avoid excessive changes in the MIEC due to polarization, and

$$\Delta\mu_{ij} \equiv \mu_X(E_i) - \mu_X(E_j) << k_B T, \quad i,j = 1,2,3,4 \qquad (11b)$$

where k_B is the Boltzmann constant, T the temperature, and indices i, j = 1, 2, 3, 4 refer to the electrode number. In Fig. 1, the current I flows through electrodes E_1 and E_4. We denote it here by I_{14}. The voltage difference on probes E_3, E_2 is denoted by V_{32}. By changing the electrical contacts, the current becomes I_{12} and the corresponding voltage difference V_{34}. These four measured quantities are related by

$$exp \ \ (-\pi d\sigma_t \mid V_{32} + (\sigma_i/\sigma_t)(\Delta\mu_{32}/q) \mid / \mid I_{14} \mid) \qquad (12)$$

$$+ \ \ exp \ \ (-\pi d\sigma_t \mid V_{34} + (\sigma_i/\sigma_t)(\Delta\mu_{34}/q) \mid / \mid I_{12} \mid) = 1$$

where d is the sample thickness and σ_t is the sum of the ionic and electronic conductivities. If $\Delta\mu_{32} = \Delta\mu_{34} = 0$. i.e. the surrounding is uniform, Eq. (12) reduces to an implicit equation for σ_t. Once σ_t is determined, one can change the surroundings to induce a known difference of, say, $\Delta\mu_{32}$ and repeat the two I, V ,measurements. Using Eq. (12) σ_i can now be determined. If the sample is made circular with equally spaced electrodes and the surrounding is uniform, σ_t can be written explicitly as

$$\sigma_t = (\ell n2/2\pi d)(\mid I_{14}/V_{32} \mid + \mid I_{12}/V_{34} \mid) \qquad (13)$$

Blocking Probes

If the voltage probes are far from being reversible, one can still resort to blocking electrodes. To measure the electronic conconductivity, ions are blocked. Three electrodes should be blocking, say, the two voltage probes and one current carrying electrode. This enables one to repeat the experiment for two sets of electrical connections I_{14}-V_{32}, I_{12}-V_{34} with blocking voltage probes. The fourth electrode should be made as reversible as possible, so that the sample can come into equilibrium with a known surrounding. This is important in order to be able to refer the measured conductivity to a well defined state of the sample. The measured currents and voltages are related by:

$$\exp(-\pi d\sigma_e \mid V_{32}/I_{14} \mid) + \exp(-\pi d\sigma_e \mid V_{34}/I_{12} \mid) = 1 \qquad (14)$$

where electrodes E_2, E_3, E_4 are blocking to ions. This in an implicit equation for σ_e. For a circular sample with equally spaced electrodes, an explicit equation for σ_e can be obtained from Eq. (14), which looks formally like Eq. (13).

To determine σ_i one can measure σ_t and σ_e and calculate $\sigma_i = \sigma_t - \sigma_e$. Alternatively, one can measure σ_i directly using three electron blocking electrodes in a way analogous to the determination of σ_e. The blocking of the electrons will be achieved by using solid electrolytes. This is shown schematically in Fig. 2. The voltmeter connected to the outer sides of the two solid electrolytes measures the difference in the electrochemical potential of electrons. The latter is related to that of the ions and μ_x in the surroundings. Choosing uniform surroundings, i.e. $\Delta\mu_{ij} = 0$, enables a simple solution. The equation that relates σ_i and the measured I, V parameters looks formally as Eq. (14), with σ_i substituted for σ_e.

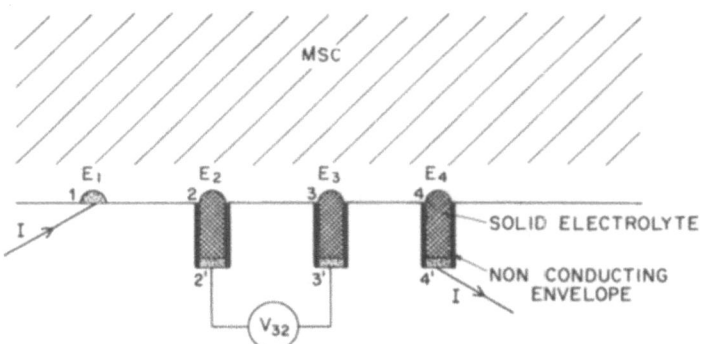

Fig. 2. Planar sample with three electron blocking
electrodes in a uniform surrounding of X. The
contacts are 1, 2´, 3´, 4´ are non-blocking.
(From Ref. 21, Fig. 2).

To summarize: the transport equations were solved for a particlular geometry (planar, singly connected) and for a rather uniform composition of the MIEC. This allowed for small applied "driving forces" (Eqs. (11a) and (11b)) only. The limitation on the applied voltage is required only for non-reversible electrodes. No specific assumptions were made regarding ionic disorder and whether the electrons are degenerate or not.

We shall discuss next some general properties of a one-dimensional configuration.

GENERAL SOLUTION OF THE TRANSPORT EQUATIONS FOR A ONE-DIMENSIONAL CONFIGURATION[2 2]

The configuration under consideration ·is shown in Fig. 3. Electrodes E_o, E_L are reversible. The sample is subject to two "driving forces" an applied voltage V and a chemical potential difference $\mu_x{}^L$ - $\mu_x{}^o$. These are also the boundary conditions for integration. In the steady state the total current density is uniform. The ionic current density, j_i, is also uniform throughout the MIEC, since no material is accumulated or depleted inside the MIEC. Hence, the total electronic current density $j_{el} = j_e + j_h$ is also uniform. This enables integration of the transport equations. The integration yields the following I-V relations: (from Eqs. (2), (10a)-(10c)) for the ionic current,

$$I_i = (V_{th} - V)/R_i, \quad I_i = Sj_i \tag{15a}$$

where

$$R_i \equiv \int_0^L dx/S\sigma_i \tag{15b}$$

and

$$-qV_{th} \equiv \mu_X^L - \mu_X^o \tag{15c}$$

S is the cross section of the MIEC, perpendicular to the current direction x. For the total electronic current

$$I_{el} = -V/R_{el}, \quad I_{el} = Sj_{el} = S(j_e + j_h) \tag{16a}$$

where

$$R_{el} \equiv \int_o^L dx/S(\sigma_e + \sigma_h) \tag{16b}$$

If one electronic specie, say the electron, is predominant ($\sigma_e \gg \sigma_h$, $j_e \gg j_h$) then,

$$I_e = -V/R_e, \quad I_e = Sj_e. \tag{17a}$$

where

$$R_e = \int_o^L dx/S\sigma_e \tag{17b}$$

These relations may look, misleadingly, simple, as if the system could be represented by an equivalent circuit with an Emf of V_{th} having an internal resistance R_i (in series) and an electronic shunt R_{el} in parallel to both the Emf source and

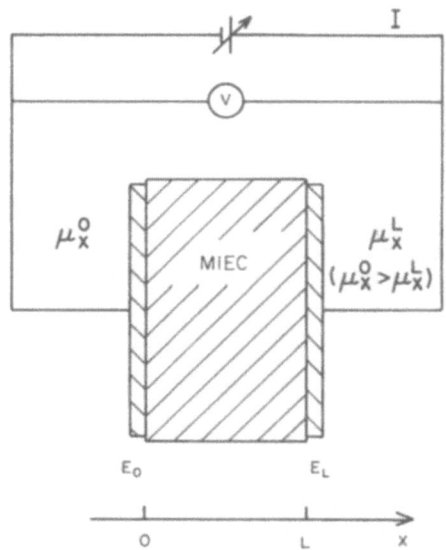

Fig. 3. Schematic presentation of a MIEC connected to an
external voltage V. E_o, E_L are reversible
electrodes. $\mu_x{}^L$, $\mu_x{}^o$ are the chemical potentials of
X in the surrounding near E_o, E_1 respectively.
(From Ref. 22, Fig. 1).

R_i. However, R_e, R_{e1}, and R_i are not simple resistors. They
are functions of the applied voltage and therefore do not
follow Ohm's law.

The total current, I, through the MIEC is the one sensed
in the external circuit. From Eqs. (15a), (16a),

$$I = I_{el} + I_i = (V_{th} - V)/R_i - V/R_{el} \tag{18}$$

Under open circuit (oc) conditions the total current I
vanishes, but

$$I_{el} = -I_i \neq 0 \ (I = 0) \tag{19a}$$

i.e., there is a leakage current of material through the MIEC
in the open circuit conditions. The open circuit voltage is

$$V_{oc} = R_{el}V_{th}/(R_i + R_{el}) \equiv \bar{t}_i V_{th} \leq V_{th} \tag{19b}$$

where R_{e1}, R_i are evaluated at $V = V_{th}$. The average ionic
transference number is t_i. The open circuit voltage is close
to the maximum possible value V_{th} for $R_i \ll R_{e1}$, but is much
smaller than V_{th} for $R_i \gg R_{e1}$.

When the chemical "driving force" vanishes, $\mu_x{}^L = \mu_x{}^o$, V_{th} vanishes (Eq. (15c)) and the I-V relation reduces (from Eq. (18)) to

$$I = -(1/R_i + 1/R_{el})V, \quad V_{th} = 0 \tag{20}$$

In order to be more specific on the dependence of R_i, R_e, R_{e1} on V and other parameters, V_{th}, T, and to obtain a detailed I-V relation, one has to discuss specific models. This will be done in the next sections.

MIEC WITH A HIGH AND UNIFORM CONCENTRATION OF MOBILE IONS; APPLICATION TO FUEL CELLS BASED ON MIEC[19,22]

A fuel cell converts the chemical energy of the oxidation reaction of the fuel directly into electrical energy. A solid state (high temperature) fuel cell is based on an oxygen conducting solid electrolyte, e.g. stabilized $(S)ZrO_2$, which at elevated temperatures is a fast anion conductor. The configuration is shown schematically in Fig. 3 with the understanding that $\mu_x{}^o$, $\mu_x{}^L$ are the chemical potentials of neutral oxygen vacancies, $V_o{}^x$, on the fuel side (o) and on the air side (L) respectively. By the interaction $V_o{}^x + 1/2O_2 = O_o$ the difference $-2(\mu_x{}^L - \mu_x{}^o)$ equals the difference in the oxygen chemical potential between the two sides of the fuel cell.

Energy waste is determined in part by heat loss due to the internal resistance of the electrolyte. It is therefore desirable to minimize R_i. There are not many fast oxygen conductors that have a lower resistivity and can replace S-ZrO_2. One candidate is CeO_2 doped e.g. with 10 mol% Gd_2O_3. The ionic conductivity of this oxide is three times higher than that of yttria S-ZrO_2 at $800°C$.[28,29] However, D-CeO_2 has the disadvantage that it is reduced at high temperatures and low oxygen partial pressures $(P(O_2) < 10^{-16}$ atm) that prevail on the fuel side of the cell. In this section we shall discuss the I-V characteristic, power output and efficiency of a fuel cell based on a MIEC reduced on the fuel side with quantitative examples calculated for D-CeO_2. We shall show that this cell will operate with less energy loss than the one based on S-ZrO_2 with equal geometry and temperature.

The mobile ionic species in D-CeO_2 are oxygen vacancies $V_o{}^{\cdot\cdot}$ introduced by lower valent dopants. Their concentration is high, on the order of a few mol%. Under reducing conditions, additional oxygen vacancies (and conduction electrons) are formed. Under the conditions of interest, this additional concentration of vacancies can be neglected to a first approximation. The concentration of $V_o{}^{\cdot\cdot}$ defects is thus uniform under all relevant conditions. Hence, the gradient of the ion chemical potential vanishes.

$$\partial\mu_{X+}/\partial x = 0 \tag{21}$$

The conduction electrons, on the other hand, may have a dramatic effect on the conductivity due to their high mobility. For D-CeO$_2$, the electron mobility is about 20 times larger than the ionic one.[30] Hence, the local transference number in D-CeO$_2$ on the fuel side may be considerably less than unity, $t_i < 0.9$, where

$$t_i = \sigma_i/(\sigma_e + \sigma_i) \tag{22}$$

To proceed we rewrite Eqs. (10a), (10b) for the one-dimensional configuration and for "ideal electron gas" statistics. The electron chemical potential is

$$\mu_e = \mu_e^o + k_B T \ell n \ n. \tag{23}$$

We also use the relations

$$\tilde{\mu}_{X^+} = \mu_{X^+} + q\phi \tag{24a}$$

$$\tilde{\mu}_e = \mu_e - q\phi \tag{24b}$$

where ϕ is the electric potential in the MIEC. This yields

$$j_i = -q\nu_i[X^+]\partial\phi/\partial x \tag{25a}$$

$$j_e = k_B T \nu_e \partial n/\partial x - q\nu_e n \partial \phi/\partial x \tag{25b}$$

where we have also used Eq. (21). j_i and j_e are uniform in the steady state. Hence, from Eq. (25a) the electric field $\varepsilon = -\partial\phi/\partial x$ is a constant. This is an important result of the model as it enables the integration of the transport equation (25b), in which only n is a function of location x. A boundary condition is determined from $\mu_x L - \mu_x o$ and Eqs. (1), (21), (23). This yields

$$n_o/n_L = \exp(q\beta V_{th}), \quad \beta = 1/K_B T \tag{26}$$

where n_o, n_L are the concentrations of electrons at the boundaries o and L, respectively. The integration of Eq. (25b) yields the distribution of electrons, n(x)

$$n(x) = \quad n_o[1 - (1 - exp(-\beta q V_{th})) (1 - exp(-\beta q(V_{th} - V)x/L))/ \tag{27}$$

$$(1 - exp(-\beta q(V_{th} - V)))]$$

We must bear in mind that the electron concentration reflects the concentration of oxygen vacancies created by reduction, and therefore the local deviation from stoichiometry due to reduction. We find that n(x) depends on the applied voltage V

(Eq. (27)). This is demonstrated in Fig. 4. The change of n(x) with V means that the applied voltage affects the local and overall deviation from stoichiometry. For a forward bias (V > 0) the width of the reduced layer increases, while for a reversed bias the width decreases. For V = V_{th}, n is linearily graded.

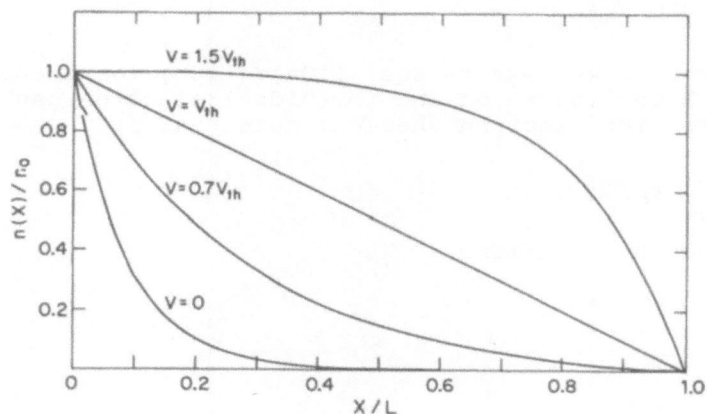

Fig. 4. n(x) vs. x for different values of V for V_{th} = 1 V, T = 1000K (n_L/n_o = 10^{-5}), t_i^L ~ 1 (From Ref. 22, Fig. 2).

The overall deviation from stoichiometry, δ, is an integral over n(x),

$$\delta = \quad (V_m n_o / N_A) \left[1 + (1 - exp(-\beta q V_{th})) \right] * \tag{28}$$

$$[1/(\beta q(V_{th} - V)) - 1/(1 - exp(-\beta q(V_{th} - V)))]$$

where V_m is the molar volume and N_A is Avogadro's number. The dependence of δ on V is shown in Fig. 5. δ is seen to be most sensitive to changes in V for V V_{th}.

For a large forward bias (V > $2V_{th}$) the reduction occurs throughout most of the MIEC and reaches a maximum value for which n(x) = n_o. The asymptotic values of the electronic conductivity and current density are, therefore,

$$\sigma_e^o \approx q n_o \nu_e, \quad V > 2V_{th}, \quad \beta q V >> 1 \tag{29a}$$

and

$$j_e \approx \sigma_e^o (V_{th} - V)/L, \quad V > 2V_{th}, \quad \beta q V >> 1 \tag{29b}$$

For large reverse bias V < -V_{th}, the MIEC deviates slightly from stoichiometry and n(x) reaches its lowest possible value n_L almost throughout the whole MIEC. Therefore

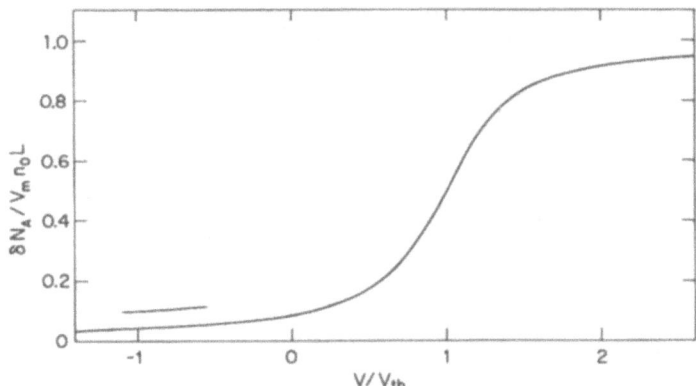

Fig. 5.   δ vs. V (for V_{th} = 1V, T=1000K, $t_i \sim 1$).   (From
Ref. 22, Fig. 3).

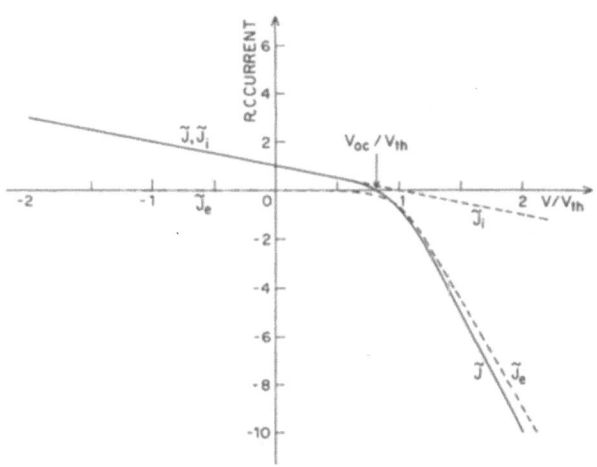

Fig. 6.   Electronic, ionic, and total current vs. voltage
(for V_{th} = 1V, T=1000K, $t_i \sim 1$).   (From Ref. 22,
Fig. 6).

$$\sigma_e^L \approx qn_L\nu_e, \quad -V > V_{th}, \quad -\beta qV \gg 1 \tag{30a}$$

and

$$j_e \approx \sigma_e^L(V_{th} - V)/L, \quad -V > V_{th}, \quad -\beta qV \gg 1 \tag{30b}$$

The asymptotic I-V relations are linear, but the coefficient of proportionality may differ by a few orders of magnitude. This is shown in Fig. 6. The dependence of the ionic current on the applied voltage is quite different, it is linear,

$$j_i = \sigma_i(V_{th} - V)/L \tag{31}$$

We continue now to examine this system with emphasis on the high temperature fuel cell application, i.e. as a power source. The voltage we are interested in is within the range $0 < V < V_{oc}$. The open circuit voltage V_{oc} is given by

$$V_{oc} = V_{th} - (k_B T/q)\ell n((\sigma_e^o + \sigma_i)/(\sigma_e^L + \sigma_i)) \tag{32}$$

$$= V_{th} - (k_B T/q)\ell n(t_i^L/t_i^o) \equiv \bar{t}_i V_{th}$$

The difference between the open circuit voltage and V_{th} is seen to increase with V_{th} and $1-t_i$. If the MIEC were a predominantly electronic conductor throughout its entire width, so that $\sigma_e \gg \sigma_i$ and $\sigma_o o \exp(-\beta qV_{th}) = \sigma_e L \gg \sigma_i$, then $V_{oc} = 0$, as one would expect. For D-CeO$_2$, $t_i{}^L$ - 1. Allowing $t_i o$ to be as low as 0.1 the average ionic transference number, t_i, is still quite high: $t_i = 0.8$.

The power output is

$$P = IV \tag{33a}$$

where

$$I = S(V_{th} - V)[\sigma_i - \sigma_e^L(exp(\beta qV) - 1)/ \tag{33b}$$

$$(1 - exp(-\beta q(V_{th} - V)))]/L$$

One should operate the fuel cell far from the short circuit conditions where V -> 0, P -> 0 and from open circuit conditions, since there I-0, P-0 and fuel is still being consumed. Maximum power is expected to be delivered for V - 0.5 V_{th}. as can be calculated from Eqs. (33a), (33b).[19] The maximum for D-CeO$_2$ (with $t_i = 0.1$) is found at V = 0.48 V_{th}. The maximum power is $P_{max} = 0.97 \times (0.25\ V_{th}{}^2/R_i)$, where $R_i = L/S\sigma_i$. P_{max} is only 3% lower than one would find, if the electronic conductivity could be neglected and $t_i o = t_i L = 1$. The efficiency of energy conversion at the maximum power output is 0.46, i.e. 8% less than the efficiency of 0.50 that could be obtained, if $t_i o = t_i L = 1$, as in S-ZrO$_2$. One has to bear in mind that R_i in D-CeO$_2$ is much lower than in S-ZrO$_2$.

The conclusions are: a MIEC, which is unstable with respect to reduction on the fuel side can still be considered for use in a fuel cell if its ionic conductivity is sufficiently high.

p-n AND p-i-n QUASI-JUNCTIONS[2][3]

We consider a mixed ionic-electronic conductor subjected as in Fig. 3 to different surroundings, one at $x = 0$ with $\mu_x o$ sufficiently high to turn the MIEC into an n-type material, and the other $\mu_x L$ sufficiently low to turn the MIEC near $x = L$ into a p-type material. This forms a p-n or p-i-n quasi junction. The boundary values of n and p (n_o, n_L and p_o, p_L) depend on the local equilibrium established with the surroundings. However, the charge distributions inside the MIEC depend on the applied voltage within the limits set by n_o, n_L, p_o, p_L. The possibility of observing both n-type and p-type behaviour in the same MIEC is not a hypothetical one and has been observed before, e.g. in S-ZrO$_2$,[31] Cu$_2$O,[32] and Ta$_2$O$_5$.[33]

The p-n or p-i-n junctions formed are different from the conventional p-n junction. In the conventional p-n junction the ionic defects are frozen in and only electrons and holes come into thermal equilibrium. In the MIEC the defects are mobile. Due to the "driving force" $\mu_x L - \mu_x o = 0$, the MIEC is never in equilibrium (also when V = 0) but only in a steady state. No space charge is formed and the mobility of the ions allows local neutrality to be maintained, though one side is n and the other p (while a current is steadily flowing). This may change, if the MIEC is quenched thereby freezing in the ionic defects, yet retaining the electrons and holes in their respective bands. If this is the case, one can "shape" the p, n and ionic defect distributions at elevated temperature by a proper choice of the applied voltage, μxL, μxo, and T followed by quenching.

Quantitative analysis requires the solution of Eqs. (10a)-(10c) taking into consideration electron-hole interactions, Eq. (4).[23] The distribution of n(x), p(x) in the MIEC for different applied voltages is shown in Fig. 7. For large forward bias ($V > 2V_{th}$) the transition region between the n and p regions is narrow. The electronic conductivity of the whole MIEC is high. For large reverse bias ($-V > V_{th}$) the n and p regions are narrow with a wide intrinsic region between them. The electronic conductivity is low, determined by the large intrinsic region. This leads to the following asymptotic I-V relations

$$j_{el} \approx q n_o \nu_{el} (V_{th} - V)/L, \quad V > 2V_{th}, \quad \beta q V >> 1 \tag{34a}$$

and

$$j_{el} \approx 2q C \nu_{el} (V_{th} - V)/L, \quad -V > V_{th}, \quad -\beta q V >> 1 \tag{34b}$$

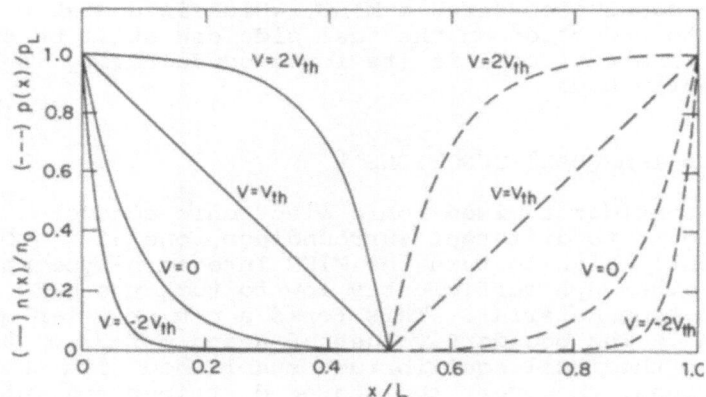

Fig. 7. n(x) and p(x) vs. x for different values of V (for V_{th} = 1V, T = 1000K, $n_L/n_o = 10^{-5}$, $P_o = n_L$, $P_L = n_o$, $V_e = V_h$). (From Ref. 23, Fig. 2).

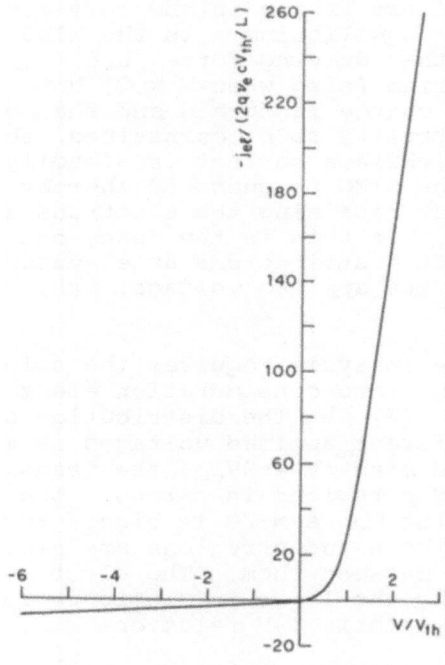

Fig. 8. Electronic current j_{el} vs. V (for conditions as in Fig. 7). (From Ref. 23, Fig. 3).

where $\nu_{e,i}$ is the mobility of electrons and holes (assumed here to be equal) and C is the concentration of electrons and holes in the intrinsic region. The ionic current follows Eq. (31) and is linear in V.

The asymmetry in the I-V relation of the electronic current is shown in Fig. 8 and does not follow conventional p-n junction rectification. For large V, $J_{e,i}$ depends on V linearly, not exponentially. The asymmetry is not due to a space charge, but rather to a difference in composition induced by the applied voltage (see Fig. 7).

EQUAL CONCENTRATION OF MOBILE IONS AND CONDUCTION ELECTRONS[22]

In the previous two sections, we assumed $[X^+] \gg n$, p. As mentioned in the introduction, one is interested also in understanding a model with equal concentrations of mobile ions and, say, electrons, $[X^+] = \alpha n \gg p$, α-1, which is relevant, e.g. to reduced ceria, CeO_{2-x}, and Li doped Ge.

Solving the transport equations (10a), (10b) under the condition

$$[X^+] = n \tag{35}$$

yields:

$$[X^+] = n(x) = n_o - (n_o - N_L)x/L \tag{36}$$

The electron distribution n(x) (and ionic distribution $[X^+]$) changes linearly between the boundary values n_o and n_L. $[X^+]$, n, and hence the deviation from stoichiometry δ are independent of the applied voltage. The I-V relations are:

$$j_e = -2\nu_e k_B T(n_o - n_L)\, V/LV_{th} \tag{37a}$$

and

$$j_i = 2\nu_i k_B T(n_o - n_L)(V_{th} - V)/LV_{th} \tag{37b}$$

i.e. the currents are linear in V. One expects the electron current to be predominant.

The local ionic transference number is uniform

$$t_i = \nu_i/(\nu_e + \nu_i) = \bar{t}_i \tag{38}$$

BLOCKING ELECTRODES, POLARIZATION, AMBIPOLAR DIFFUSION AND CHEMICAL DIFFUSION[22]

We would like to show now that the topics mentioned in the heading are covered by the present theory and constitute special cases of the theory, even though we have assumed that the electrodes are reversible. We consider a MIEC conducting electrons and ions.

Electrodes Blocking to Electrons ($j_e = 0$)

This case is covered by the theory when $V = 0$. To show that this is the relevant value of V, let us consider one electrode blocking to electrons, say E_o. Any applied voltage will appear at the blocking electrode and the voltage, V, seen by the MIEC will vanish. For $V = 0$ the theory yields $j_e = 0$ as required. One should notice that the composition of the MIEC under electrode E_o is not fixed by the surroundings at $X = 0$, since this interaction is partially blocked. Instead, the composition there changes until a steady state is established, in which the flux of ions through the MIEC equals the flux of atoms passing through electrode E_o. This determines n_o and V_{th} that appear in the theory.

Electrodes Blocking to Ions ($j_i = 0$)

If E_o were blocking to ions, j_i would vanish. This is treated by the theory when $V = V_{th}$. The latter equality arises as follows: The composition of the MIEC under electrode E_o is not fixed by interaction with the surroundings, since the interaction is partially blocked. Instead, the composition there is determined by the applied voltage, and no changes until V_{th} equals the applied voltage, V.

When both electrodes block ions, the ionic current must vanish in the steady state, i.e. $V_{th} = V$. The applied voltage must be limited to avoid electroplating. A linear graded distribution of electrons and defects is built-up in the MIEC as is the case for $V = V_{th}$, with the constraint that the overall composition does not change. This fixes also the boundary values of n_o, n_L.

The value of n_o or n_L (and μ_e^o, μ_e^L) determines also the chemical potential of X in the MIEC (Cf. Eq. (2)). When the applied voltage is too high, μ_x^o or μ_x^L may exceed the solubility limit and X will precipitate at the boundary, i.e. it will be electroplated. If μ_x^o or μ_x^L is too low other components of the solid will leave it.

Polarization

By "polarization" we mean zero ionic current under an applied voltage. This corresponds to the case $V = V_{th}$, whether V_{th} is fixed by the surrounding or adjusts itself to the applied voltage under blocking electrodes.

Chemical diffusion alone occurs when the electric field, ε, in the MIEC vanishes. For the two detailed models discussed, this happens when: (I) $N \gg n$, p, $\varepsilon \equiv 0 \longrightarrow V = V_{th}$; (II) $N = n \gg p$, $\varepsilon = 0 \longrightarrow V = V_{th}/2$.

Ambipolar Diffusion

Ambipolar motion occurs under open circuit conditions when the total current vanishes, $I_{el} = -I_i$, $V = V_{oc}$.

REFERENCES

1. T. Kudo and H. Obayashi, "Oxygen Ion Conduction of the Fluorite-type Lanthanoid Oxide $(Ce_{1-x}Ln_xO_{2-x/2})$:I," J. Electrochem. Soc. 122:142-7 (1975).

2. H. Rickert, V. Sattler, and Ch. Wedde, "Thermodynamische Groessen der Elektronen in α-Silbersulfid," Z. Phys. Chem. NF 98:339-50 (1975).

3. J. Sohege and K. Funke, "Composition Dependent Electronic Conductivity and Hall Coefficient of β-$Ag_{2+\delta}S$," Ber. Bunsenges. Phys. Chem. 88:657-63 (1984).

4. G.F. Knoll, "Radiation Detection and Measurement," John Wiley & Sons (1979) pp. 414-70.

5. H. Rickert, "Electrochemistry of Solids, An Introduction," Springer-Verlag (1982) pp. 157-67.

6. F. Huber and M. Rottersman, "Graded p-n Junctions in Thin Anodic Oxide Films of Titanium," J. Appl. Phys. 33:3385 (1962).

7. R.I. Taylor and H.E. Haring, "A Metal-Semiconductor Capacitor," J. Electrochem. Soc. 103:611-13 (1956).

8. A.P. Belova, L.G. Gorskaya and L.N. Zakgeim, "Electrical Properties of Thin Oxide Films on Aluminum, Tantalum and Zirconium," Soviet Physics - Solid State 3:1348-53 (1961).

9. H. Schmalzried, "Solid State Reactions," Verlag Chemie and Academic Press (1974) pp. 190-95.

10. J. Vedel, "Electrode Reactions at Electrode - Solid Electrolyte Interfaces," in Electrode Processes in Solid State Ionics, M. Kleitz and J. Dupuy eds., D. Reidel Pub. Co. (1976) pp 223-59.

11. D. Braunshtein, D.S. Tannhauser, and I. Riess, "Diffusion-Limited Charge Transport at Platinum Electrodes on Doped CeO_2," J. Electrochem. Soc. 128:82-9 (1981).

12. M. Kertesz, I. Riess, D.S. Tannhauser, R. Langpape and F.J. Rohr, "Structure and Electrical Conductivity of $La_{0.84}Sr_{0.16}MnO_3$," J. Solid State Chem. 42:125-29 (1982).

13. L. Heyne, "Electrochemisty of Mixed Ionic-Electronic Conductors" in "Solid Electrolytes," S. Geller ed., Springer-Verlag 1977, pp. 169-221.

14. C. Wagner, "Galvanic Cells with Solid Electrolytes Involving Ionic and Electronic Conduction," in Proc. Int. Comm. Electrochem. Thermodyn. Kinetics (CITCE) 7:361-77 (1957).

15. L.H. Allen and E. Buhks, "Copper Electromigration in Polycrystalline Copper Sulfide," J. Appl. Phys. 56:327-35 (1984).

16. C. Wagner, "Equation for Transport in Solid Oxides and Sulfides of Transition Metals," Prog. Solid State Chem. 10:3-16 (1975).

17. G.J. Dudley nd B.C.H. Steele, "Theory and Practice of a Powerful Technique for Electrochemical Investigation of Solid Solution Electrode Materials," J. Solid State Chem. 31:233-47 (1980).
18. Ref. 5, pp. 101-10.
19. I. Riess, "Theoretical Treatment of the Transport Equations for Electrons and Ions in a Mixed Conductor," J. Electrochem. Soc. 128:2077-81 (1981).
20. J. Berger, I. Riess, and D.S. Tannhauser, "Dynamic Measurement of Oxygen Diffusion in Indium-Tin Oxide," Solid State Ionics 15:225-31 (1985).
21. I. Riess and D.S. Tannhauser, "Application of the van -der-Pauw Method to Conductivity Measurements on Mixed Ionic-Electronic Solid Conductors," Solid State Ionics 7:307-15 (1982).
22. I. Riess, "Current-Voltage Relations and Charge Distribution in Mixed Ionic Electronic Solid Conductors," J. Phys. Chem. Solids 47:129-38 (1986).
23. I. Riess, "Voltage-Controlled Structure of Certain p-n and p-i-n Junctions," Phys. Rev. B35: 5740-43 (1987).
24. F.A. Kroeger, "The Chemistry of Imperfect Crystals," 2nd ed. Vol. II, North-Holland Publ. Comp. (1974) p. 14.
25. S. Miyatani, "Experimental Evidence for the Onsager Reciprocal Relation in Mixed Conduction," Solid State Commun. 38:257-59 (1981).
26. Ref. 5, pp. 87-8, 217-19.
27. L.J. van-der-Pauw, " A Method of Measuring Specific Resistivity and Hall Effect on Disks of Arbitrary Shape," Philips Res. Rept. 13:1-9 (1958).
28. H. Obayashi & T. Kudo, "High Temperature Electrolysis/Fuel Cells: Materials Problems" in "Solid State Chemistry and Energy Conversion," J.B. Goodenough and M.S. Whittingham eds., Am. Chem. Soc. (1977) pp. 316-63.
29. I. Riess, D. Braunshtein and D.S. Tannhauser, "Density and Ionic Conductivity of Sintered $(CeO_2)_{0.82}-(GdO_{1.5})_{0.18}$," J. Am. Ceram. Soc. 64:479-85 (1981).
30. D.S. Tannauser, "The Theoretical Energy Conversion Efficiency of a High Temperature Fuel Cell Based on a Mixed Conductor," J. Electrochem. Soc. 125:1277-82 (1978).
31. Ref. 5, p. 120.
32. K. Hauffe, "Reactionen in und an Festen Stoffen," 2nd ed. Springer-Verlag (1966) pp. 180-7.
33. Y. Sasaki, "p-i-n Junction in the Anodic Oxide Film of Tantalum," J. Phys. Chem. Solids 13:177-86 (1960).

NOVEL SOLID STATE GALVANIC CELL GAS SENSORS

Werner Weppner

Max-Planck-Institut
für Festkörperforschung
Heisenbergstr. 1
D-7000 Stuttgart 80
Fed. Rep. Germany

Solid electrolytes transduce the ratio of the activities of the electroactive component at both electrodes directly into an electrical voltage. Oxygen ion conductors such as zirconia are commercially successfully employed on this basis as sensors for oxygen partial pressures in automobile exhaust systems, in molten steel and for food packing. It is of considerable practical interest to expand this technique to the detection of other gases and to operate the galvanic cells at lower temperatures. There are, however, severe restrictions at the present time with regard to the available materials. Especially, it may be hardly expected that fast solid ionic conductors may ever be found for heteroatomic gas molecules such as CO_x, NO_x or SO_x.

The problem may be partially overcome by employing compounds as solid electrolytes that are composed of the gaseous species as an immobile component. The operation of these 'galvanic cells of second type' is based on the internal equilibration between the various components of the solid electrolyte according to Duhem-Margules' relation at constant temperature and total pressure

$$\sum_i n_i d\mu_i = 0 \qquad\qquad (1)$$

where μ_i and n_i are the chemical potential and stoichiometric number of the i-th component, respectively. Molecular species may be considered in this thermodynamic sense as a component of the solid. This principle is employed in a developmental stage in SO_2/SO_3 sensors based on silver and alkali sulfates [1-3].

This approach still requires elevated temperatures for sufficiently high conductivities and is restricted to few of the presently known fast solid ionic conductors which have suitable immobile components of the gaseous component. On the other hand, a large variety of materials became known in recent decades that show high pure ionic conductivity preferably for Ag, Cu, Na or Li-ions at room temperature. Neither the mobile nor immobile component corresponds to gaseous species that are of practical interest, however. Nevertheless, these fast solid electrolytes may be employed for gas sensors in 'galvanic cells of third type'. The coupling between the partial gas pressure and the activity of the electrochemically mobile component is achieved with the help of gas sensitive additional phases. These are preferrably thin fast mixed (ionic and electronic) conductors in order to obtain fast equilibration and therefore short response times [4-8].

The number of gas sensitive phases has to be as large as required by Gibbs' phase rule to establish a well defined relationship among the activities. If the electrolyte and the gaseous species form an N-component system it is necessary to use N-2 phases of that same system at constant temperature and partial pressure. Including the gas phase and the electrolyte, N phases are in equilibrium with each other and all activities are therefore fixed at constant temperature and total pressure. Components that remain unchanged in their ratio because of extremely slow motion and exchange do not have to be taken into account. Each of the N-1 electrolyte and gas sensitive phases has a standard Gibbs energy of formation of

$$\Delta G^o_{f,p} = kT \sum_{i=1}^{N} n_{ip} \ln a_i \qquad (p=1, \ldots, (N-1)) \qquad (2)$$

(k: Boltzmann's constant, T: absolute temperature, a_i: activity of the component i) In addition, the activity of the mobile component in the electrolyte, a_m, is related to the cell voltage E by Nernst's equation. All activities except the one of the gaseous species, a_g, may be eliminated from the set of N equations which results in the following relationship [9]:

$$E = \frac{kT}{z_m qd} \ln a_g \sum_{p=1}^{N-1} (-1)^{1+m} n_{gp} d_{mp}$$

$$- \frac{1}{z_m qd} \sum_{p=1}^{N-1} (-1)^{1+m} \Delta G^o_{f,p} d_{mp} + \frac{kT}{z_m q} \ln a^{Ref}_m \qquad (3)$$

z_m, d and d_{mp} are the charge number of the mobile ions, the determinant formed by the stoichiometric numbers of all compounds and the minor formed from d by eliminating the m-th row and p-th column, respectively. Knowledge of the stoichiometric numbers and Gibbs energies of formation of all N-1 solid phases allows one to determine the activity of the gaseous component a_g from the measured cell voltage with reference to an activity a^{Ref}_m of the mobile component. If these values are not known, the relationship may be established by calibration. In addition, it is not necessary to know and to prepare exactly the N-2 gas sensitive phases. The equilibrium compounds and stoichiometries will be formed by equilibration of the gas sensitive materials in the presence of the gas phase. A very simple procedure to form the gas sensitive phases is possible by passing an ionic current for some period of time through the electrolyte which is exposed to the gas phase at the measuring electrode. The electroactive component reacts with the gas phase and the sensitive layers are formed in-situ. The correct number of equilibrium phases will thereby be formed. The relation between the cell voltage and the activity of the gaseous species is Nernst type but shifted by an expression which includes the Gibbs energies of formation of all compounds involved in the equilibration.

An example is the determination of chlorine partial pressures by using ambient temperature lithium and silver ion conductors. The fast lithium ion conductor $LiAlCl_4$ is prepared with a thin gas sensitive layer of either $AlCl_3$ or $LiCl$ as shown in Fig. 1. The experimentally observed cell voltages are in agreement with Eq. (3) which reads in this special case

$$E = \frac{kT}{2q} \ln p_{Cl_2} - \frac{1}{q} \Delta G_f^o \text{ (LiCl)} \qquad (4)$$

for an LiCl gas sensitive compound and

$$E = \frac{kT}{2q} \ln p_{Cl_2} - \frac{1}{q} [\Delta G_f^o(\text{LiAlCl}_4) - \Delta G_f^o(\text{AlCl}_3)] \qquad (5)$$

for an AlCl$_3$ gas sensitive compound. The voltages are given in both cases with reference to elemental lithium.

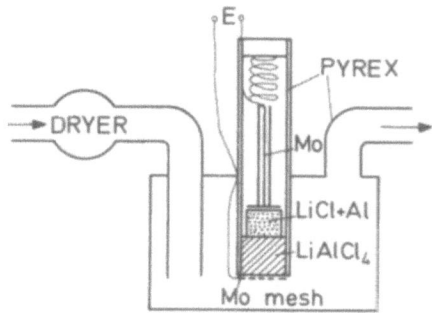

Fig. 1   Experimental arrangement of a chlorine gas sensor using LiAlCl$_4$ as solid electrolyte for lithium ions. The gas sensitive layer may be composed of either AlCl$_3$ or LiCl. The chlorine partial pressure fixes the activities of all components in thermodynamic equilibrium. The lithium activity is measured against a reference electrode which is inserted into the electrolyte.

Also, the fast silver ion conductor Ag-β-alumina may be employed in combination with AgCl as a gas sensitive layer which has been made electronically conducting by suitable additives. This is in principle a four-component system which would require two additional phases. The nature of the β-alumina structure may be considered, however, to have a fixed Al/O ratio which does not get involved in the equilibration. Fig. 2 shows the experimental results which follow closely the theoretical relationship

$$E = \frac{kT}{2q} \ln p_{Cl_2} - \frac{1}{q} \Delta G_f^o(\text{AgCl}) \qquad (6)$$

with reference to a silver electrode

Likewise, other gaseous components may be measured using known fast solid ionic conductors at room temperature. Fig. 3 shows the results of measurements of oxygen partial pressures using Ag-β-Al$_2$O$_3$ as fast ion conductor for silver ions and Ag$_2$O as gas sensitive material. The straight line is limited to the existence range of Ag$_2$O. The applicability to gases such as NO$_x$, CO$_x$ or SO$_x$ with the help of nitrates, carbonates and sulphates has also been demonstrated. The results of the mea-

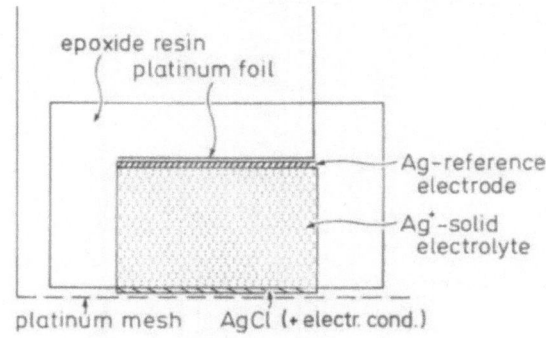

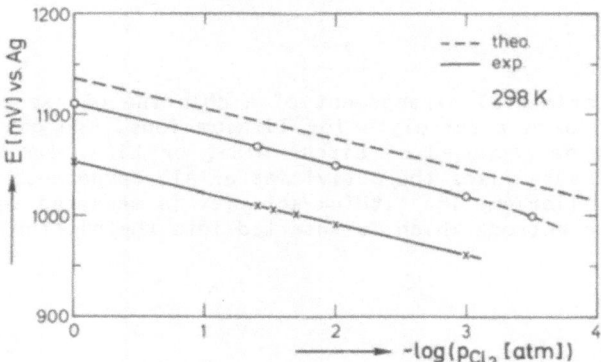

Fig. 2   Experimental rrarrangement and relation between the cell voltage
and the chlorine partial pressure in the case of application of
Ag-β-alumina as solid electrolyte for silver ions and AgCl as
gas sensitive layer with reference to elemental silver. The
theoretically expected behavior based on the Gibbs energy of
formation of AgCl is plotted for reference.

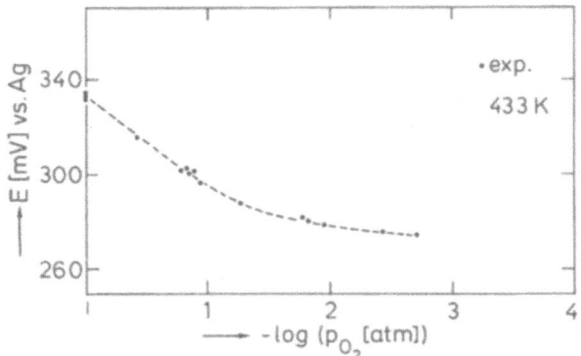

Fig. 3   Relation between the cell voltage and oxygen partial pressure
          for the application of Ag_2O as gas sensitive material and Ag–β
          β'–Al_2O_3 as solid electrolyte for silver ions.

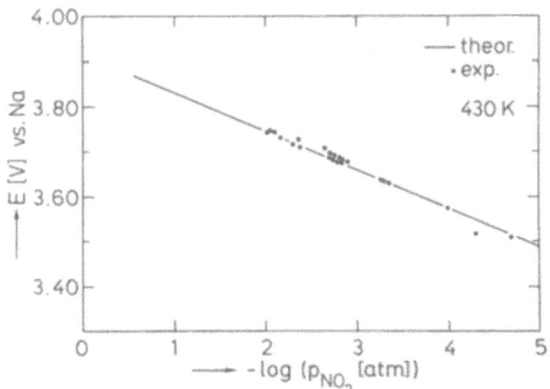

Fig. 4   Relation between the cell voltage and the NO_2 partial pressure
          for the application of sodium nitrate gas sensitive materials
          and Na–β'–Al_2O_3 as solid electrolyte for sodium ions.

surements of NO_2 partial pressures are shown in Fig. 4. Sodium-β''-alumina is employed in this case as solid fast ion conductor for sodium ions. It also appears to be possible to determine partial gas pressures selectively by applying a separate sensor for each component. The voltage readings may be mutually influenced but they may be altogether used to calculate the partial pressures of each gaseous component from the set of n equations with n partial pressures. These aspects and the application to other gaseous species are presently under investigation.

REFERENCES

[1] W.L. Worrell and Q.G. Liu, Anal. Chem. Symp. Ser. 17, 332 (1983)
[2] W.L. Worrell and Q.G. Liu, J. Electroanal. Chem. Interfacial Electrochem. 168, 355 (1984)
[3] Q.G. Liu and W.L. Worrell, in: Phys. Chem. Extr. Metall., Proc. Int. Symp. (V. Kudryk and Y.K. Rao, eds.) Metall. Soc., AIME (1985) p. 387
[4] W. Weppner, German Patent DE 2926172C2 (28 June 1979)
[5] G. Hötzel and W. Weppner, in: Transport-Structure Relations in Fast Ion and Mixed Conductors, Riso Nat. Lab., (1983) p. 401
[6] G. Hötzel and W. Weppner, Solid State Ionics 18,19, 1223 (1986)
[7] G. Hötzel and W. Weppner, Sensors and Actuators, in press
[8] W. Weppner and R.A. Huggins, Solid State Ionics 1, 3 (1980)
[9] W. Weppner, Sensors and Actuators 12, 107 (1987)

REMARKS ON APPLICATION OF FAST ION CONDUCTORS

Joachim Maier

Max-Planck-Institut für Festkörperforschung
Heisenbergstr. 1
7000 Stuttgart-80, FR-Germany

In the following context, in a supplementary, non-systematic manner, different additional remarks on applications of fast solid ion conductors or of ionically disordered semiconductors (mainly with respect to sensors) shall be made.

BULK CONDUCTIVITY SENSORS

At elevated temperatures the well known power law response of the electronic conductivity on the component activity, e.g. σ of an oxide semiconductor on oxygen partial pressure (Po_2) can be used for sensing Po_2. In particular, $SrTiO_3$ and $BaTiO_3$ are appropriate materials[1] and well-investigated in this regard.[2-5] Here at T ~ 1000°C, the response is fast with Pt- or Ag-electrodes. At lower temperatures, both bulk diffusion as well as electrode processes become slow.[5] A remarkable change from a conductive n-type material to a p-type insulator is observed over the available P-range for nominally pure materials. The σ (Po_2) characteristic can be quite precisely changed by intentional doping (e.g. $La_{Sr}^{\cdot}$, Fe_{Ti}').

BOUNDARY CONDUCTIVITY SENSORS

The main disadvantage of bulk conductivity sensors, viz. the necessity of high temperatures, is avoided in the so-called Taguchi-sensor SnO_2 (or analogously in other semiconductive oxides) where the influence of redox-active gases (e.g. H_2 or O_2) is perceived mainly in the space charge regions.[6,7] (O_2: depletion of e^-, $\bar{\sigma}$ decreases; H_2: $\bar{\sigma}$ increases). The main problems are overlap with slow bulk equilibration and appropriate conditioning. The finding that SnO_2 works quite well if it is donor-doped ($Sb_{Sn}^{\cdot}$, $F_O^{\cdot}$) can be explained by the fact, mentioned in my lectures, that these charge carriers which determine the charge density in the space charge regions (most probably the majority charge carriers) determine the space charge profile and thus the concentration profile of all mobile carriers.[8]

An analogous response of ion conductors on Lewis-acid-base-active gases is being studied at present in our laboratory.

EMF SENSORS

The main advantage of activity sensing electrochemical cells is the fact that the signal is theoretically independent of materials constants (cf. the lectures of Weppner). In practice, the main problems are response time and selectivity. The possibility of the performance of such a cell is determined by a sufficiently high rate for the overall electrochemical process (at least adsorption, charge transfer, bulk migration) with respect to the species to be detected, i.e., a low enough resistance for each individual series step must be guaranteed. Many sensor applications are hindered by a low exchange current density (low transfer resistance) being dependent on the defect concentration immediately at the surface. As shown in my lectures, the use of dispersed or multilayered structures can significantly influence this value. By the way, dispersions of e.g. metals in ion conductors can provide a simple way of enhancing the ionic conductivity (below the percolation threshold) or of introducing electronic conduction (above the percolation threshold) without having to have a special interface effect.

The selectivity of an EMF sensor is determined by the requirement that the overall process for interactions with competing species must be low enough. Here, it is sufficient that one series step-e.g., the adsorption step-is blocked. I want to draw your attention to the large experience in the field of surface chemistry concerning selectivity of adsorption processes.

Quite a well-developed know-how on working electrochemical cells can be found in the field of thermochemistry. Many formation cells have been used for measuring thermodynamic values of compounds: simple ones as metal + non-metal (gas) or more complicated ones as described below. When a gas species is involved in the cell reaction, the cell can in principle be used as a gas sensor. If all other components are solids without extended non-stoichiometries, we find immediately from the mass action law

$$\text{cell voltage } \alpha \; \triangle G = \triangle G^\circ \pm RT \ln P_{gas}.$$

The requirements with respect to the electrolyte are very weak (cf. low measuring current densities), i.e. slightest solubilities or other interaction may be sufficient (cf. liquid electrochemistry). One example is the formation cell Au, Na_2CO_3, CO_2/Na-ß-alumina/Na_2ZrO_3, ZrO_2, Au, which we used for determining $\triangle H_f$ and S° of Na_2ZrO_3 rather than for determining the CO_2 partial pressure. Nevertheless, the consistency of the enthalpy and entropy values for different P_{O_2} prove the possibility of such an application. This example shows that the increase of chemical complexity may be quite helpful. Since the oxygen partial pressure in the relevant cell reaction (not too low P_{O_2} values cancels, the applications of oxidic compounds is here very helpful, since no redox interaction appears.

The determination of thermodynamic data itself is an important (indirect) application. With analogous formation cells we recently determined the complete thermochemistry of Nasicon ($Na_{1+x}Zr_2Si_xP_{3-x}O_{12}$) for T=300-1300K and x=0-3. The values unambiguously showed that Nasicon is thermodynamically unstable with respect to elementary sodium. Hence, this ionic conductor is more appropriate as an ion-selective membrane rather than for an application in the Na/S-cell.

ACCELERATION SENSOR

Besides many other applications of fast ion conductors as chemical sensors (cf. ISFETs, pellistors, current limiting sensors, etc.), fast ion conductors may be used as acceleration sensors as shown recently.[11] If a α-AgI crystal is exposed to an acceleration (e.g. if a falling crystal touches the ground) a voltage is induced due to the different transport numbers of cations and anions. The voltage is much higher than in the case of a metal owing to the large effective mass of Ag.

HETEROGENEOUS CATALYSIS

Many other applications, besides sensor applications, such as the use of appropriate materials in batteries or fuel cells, have been mentioned in the other contributions. Lastly, I would like to mention the necessity of bringing experience of defect chemistry, particularly the defect chemistry of boundary layers, into the field of heterogeneous catalysis in order to ensure a better understanding and to support empirical research. In order to do so, we are presently investigating in our laboratory the catalytic influence of heterogeneously doped ion conductors on organic elimination reactions.

REFERENCES

1. K.H. Hardtl, personal communication (1987).
2. J. Daniels, K.H. Hardtl, D. Hennings, and R. Wernicke,
    Philips Res. Repts. 31: 487 (1976).
3. R. Wernicke, Ph.D. thesis, Aachen (1975).
4. Y. Balachandran, N.G. Eror, J. Solid State Chem.
    39:351 (1981).
5. J. Maier, G. Schwitzgebel, H.-J. Hagemann, J. Solid
    State Chem. 58:1 (1985).
6. W. Gopel, Progr. Surf. Sci. 20:1 (1986).
7. J. Maier, W. Gopel, J. Solid State Chem., in press.
8. J. Maier, J. Electrochem. Soc. 134:1524 (1987).
9. J. Maier, U. Warhus, J. Chem. Thermodynamics
    18:309 (1986).
10. J. Maier, U. Warhus, Solid State Ionics 18/19:969
    (1986); U. Warhus, J. Maier, A. Rabenau, J. Solid
    State Chem., in press.
11. M. Betsch, H. Rickert, and R. Wagner, Ber. Bunsenges,
    Phys. Chem,. 89:113 (1985).

SOME FUTURE TRENDS IN SOLID STATE IONICS

Harry L. Tuller

Crystal Physics and Optical Electronics Lab
Dept. Materials Science & Engineering
Massachusetts Institute of Technology
Cambridge, MA   02139

ABSTRACT

The future uses of fast ion conductors as electrochemical sensors and actuators in feed-back control systems are examined.

INTRODUCTION

The recent proliferation of inexpensive microprocessors has now made possible on-line monitoring and control of many complex processes. The extension and application of such techniques to the monitoring and control of chemical processing and combustion represents an important opportunity. The key to the feasibility of such closed-loop systems (see Figure 1) is the existence of suitable electrochemical sensors and actuators to serve the input and output functions. In the following, we discuss some of the opportunities in this field and the requirements this places on solid ionic materials.

SENSORS

There are increasing numbers of applications in which the chemical composition of a liquid, gas, or solid must be determined and monitored. Examples include the oxygen content of auto exhausts, molten steel and glass melts, combustible gases, and humidity and ion concentrations in aqueous solutions. Perhaps the most successful of these applications to date is the stabilized zirconia auto exhaust sensor, illustrated in Figure 2, which is used in millions of automobiles yearly. Here, the air-to-fuel ratio is maintained within close limits to about 14.5, the "stoichiometric ratio" for optimum efficiency of the exhaust catalyst. The output of the sensor (proportional to the oxygen partial pressure in the exhaust gas) is fed to the on-board microprocessor which, in turn, controls the fuel injection system.

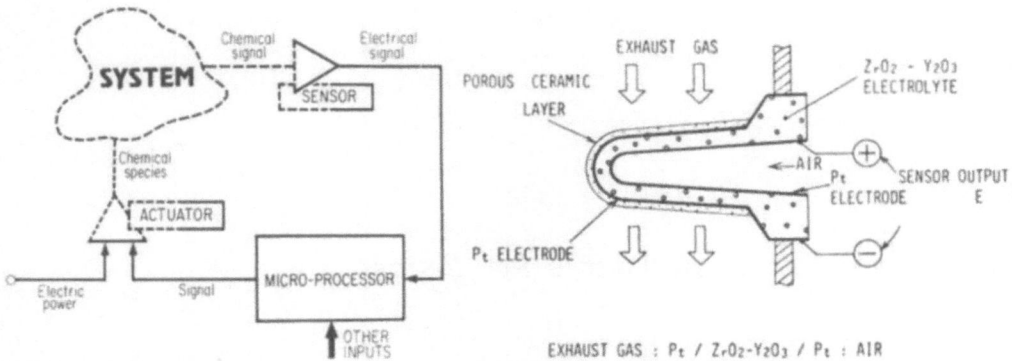

Fig. 1. Schematic diagram of a feedback system for measurement and control of a chemical system using fast ion conductors for both the sensor and the actuator (From Ref. 1).

Fig. 2. Schematic view of a practical ZrO_2 auto exhaust sensor (From Ref. 1).

In general, the following requirements are driving developments in the sensor field. These include: speed, low cost, sensitivity, selectivity, reproducibility, size, and operating life. Some of these factors are discussed in the papers by Weppner[1] and Maier[2]. A major advantage of solid ionic materials is their general robust character enabling operation at elevated temperatures and chemically agressive environments, e.g. auto exhaust gases and molten steel.

Although perhaps not immediately obvious, many of the above requirements of speed, sensitivity, selectivity, and reproducibility, depend more often on the interfacial rather than the bulk properties of the fast ion conductors. In the auto exhaust sensor, for example, platinum plays an essential role in insuring that the gases reach a reproducible equilibrium value at the oxide-platinum interface. For purposes of monitoring specific combustible gases, all of which say are reducing, it is often necessary to acheive selectivity by use of different catalysts or gas-permeable membranes. A number of symposia recently held on chemical sensors[4,5] address these issues in some detail.

Other important criteria relate to the chemical and redox stability of materials of interest. Both these issues have been discussed from a variety of standpoints in the articles by Barsoum,[6] Weppner,[7] and Riess.[8]

ACTUATORS

As mentioned above, closed-loop control systems are playing ever increasingly important roles in process and combustion control, robotics and other manufacturing processes. There is, however, a notable lack of satisfactory

electrochemical sensors and actuators for such applications.
The aforementioned ZrO_2-based auto exhaust sensor is a notable
exception.

Because solid electrolytes satisfy two fundamental and
relevant electrochemical laws, they are, in priciple, ideal
for such applications.[9] The first, Nernst´s law is given by

$$E = RT/nF \ ln(a_m´/a_m´´) \hspace{3cm} (1)$$

where E is the emf generated when an activity gradient
$(a_m´ - a_m´´)$ is imposed across the electrolyte, R the gas
constant, T the temperature in degrees kelvin, F the faraday
constant and n the number of charges passed per atom of
reactant. If $a_m´´$ is fixed as a reference, E may be used to
monitor the activity $a_m´$ in a reaction vessel. This of course
is the basis of operation of the auto exhaust sensor
illustrated in Figure 2 wherein the activities $a_m´$ and $a_m´´$ can
be substituted by their equivalent $Po_2´$s.

The second, Faraday´s law given by

$$M = It/nF \hspace{4cm} (2)$$

which states that M moles of material are transported at the
rate of 1 gram-equivalent per faraday of electricity passed,
where I and t are the current and time respectively. Since
fast ion conductors generally conduct selectively by one type
of ion, the cell acts both as a selective valve and ion pump
in which the flow rate is accurately controlled over a wide
range by controlling the current. The ionic conductivity and
interfacial kinetics ultimately limit the rate of transport.
Figure 3 illustrates a possible apparatus combining the sensor
and actuator function in a solid electrolyte process tube.

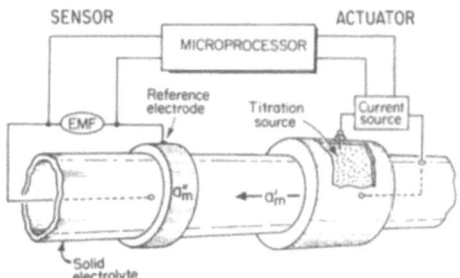

Fig. 3:   Schematic diagram of apparatus which combines
           electrochemical sensor and actuator functions in the
           solid electrolyte process tube. The activity $a_m´$ of
           metal is maintained constant by titration into or
           out from the process tube (From Ref. 9).

Other interesting applications include the use of solid
ionic materials as electrochemical actuators in electrochromic

windows. Such windows would allow for regulation of solar input to building interiors.[10] Here, a key requirement is that the electrolyte be readily formed in thin film form compatible with a number of other constituents including an insertion compound such as WO_3, which colors upon the insertion of H or Li ions, as well as the sources and sinks for the ions. The most attractive choices appear to be polymer or glass electrolytes given that they can be fabricated in very thin layers and operate under near ambient conditions.

In summary, many applications, in addition to energy conversion devices, e.g. batteries and fuel cells, show promise for future utilization of fast ion conductors. In this paper we have briefly discussed applications which depend primarily on the dual ability of solid electrolytes to monitor their chemical environment as well as modify it by electrochemical means. Given these abilities, many intriguing and important feed-back control processes become possible.

ACKNOWLEDGEMENTS

The authors wish to thank the following organizations for their generous research support in the area of solid state ionics: The Ceramics and Electronic Materials Program, Division of Materials Science, National Science Foundation, under Contract DMR-87-20017; Lawrence Berkeley Laboratories under Subcontract 4548010; Basic Science Division, U.S. Department of Energy under Contract DE-FG02-86ER45261.

REFERENCES

1.   H.L. Tuller and P.K. Moon, Material Science & Engineering B2:in press.
2.   W. Weppner, these proceedings, pp.
3.   J. Maier, these proceedings, pp.
4.   Proc. Symp. on Chemical Sensors-Vol 87-9, ed. D.R. Turner, The Electrochem. Soc., Pennington, NJ (1987).
5.   Proc. 2nd Intl. Mtg. on Chemical Sensors, Bordeaux (1986).
6.   M. Barsoum, these proceedings, pp.
7.   W. Weppner, these proceedings, pp.
8.   I. Riess, these proceedings, pp.
9.   H.L. Tuller and M.W. Barsoum, Proc. Intl. Conf. Solid State Sensors and Actuators, IEEE, Pennington, NJ (1985) pp. 256.
10. S.F. Cogan, E.J. Anderson, T.D. Plante, and R.D. Rauh, Proc. 4th Conf. on Optical Materials Tech. for Energy Efficiency and Solar Energy Conversion, SPIE J. 562:23 (1985).

FAST ION CONDUCTION IN THE $Gd_2(Zr_xTi_{1-x})_2O_7$ PYROCHLORE SYSTEM

P.K. Moon and H.L. Tuller

Crystal Physics and Optical Electronics Laboratory
Department of Materials Science and Engineering
Massachusetts Institute of Technology
Cambridge, MA, USA

ABSTRACT

Electrical conductivity measurements as a function of temperature, oxygen fugacity and composition were made on the $Gd_2(Zr_xTi_{1-x})_2O_7$ solid solution system. A large increase in ionic conductivity with increasing Zr content was found and related to the intrinsic anionic disorder.

INTRODUCTION

Compounds with the pyrochlore structure encompass a wide variety of compositions and properties including fast ion conduction. Certain compositions such as $Tl_{1+x}Ta_{1+x}W_{1-x}O_6 \, nH_2O$ form fast cation conductors[1] while others such as $Gd_2Zr_2O_7$ are fast anion conductors. In this paper, we focus upon pyrochlore compounds in the $Gd_2Ti_2O_7$-$Gd_2Zr_2O_7$ solid solution system.

Pyrochlore may be considered as a superstructure of the defect fluorite structure (i.e., Y-stabilized ZrO_2) with twice the lattice constant. The two structures are distinguished by cation ordering, anion vacancy ordering and a slight distortion of the positions of the oxygen 48f sublattice in pyrochlore relative to fluorite.

Ionic disorder occurs easily in the pyrochlore structure because the enthalpy difference between fluorite and pyrochlore is often rather small[2]. This disorder, which can be considered as a partial reversion of the pyrochlore structure back towards fluorite, may occur on a scale of either atoms or tens of nanometers[3]. For example, van Dijk et al[4]. found a hybrid phase with ~50 nm. microdomains of pyrochlore in a fluorite matrix by TEM but thermodynamic considerations suggest the possibility of disorder on an atomic scale. We will treat the question of disorder using Kroger-Vink notation for convenience while acknowledging the likely presence of microdomains.

Cation disorder occurs in pyrochlores as cation anti-site defects:

$$A_A^x + B_B^x \longrightarrow A_B{}' + B_A{}^\bullet \qquad\qquad [1]$$

Anion disorder occurs as anion Frenkel defects:

$$O_o^x \longrightarrow V_o{}^{\bullet\bullet} + O_i{}'' \qquad\qquad [2]$$

The degree of cation and anion disorder in pyrochlores is related to the cation radius ratio (r_A/r_B). Indeed, the cation radius ratio must be within certain limits to form the pyrochlore structure at all[2,5,6]. This requirement is shown for $(A^{3+})_2(B^{4+})_2O_7$ compounds in Figure 1. As the cation radius ratio decreases, fluorite replaces pyrochlore as the stable structure. Even within the pyrochlore stability field, substantial increases in ionic disorder can occur as the cation radius ratio decreases as shown by IR and Raman spectroscopy[7,8,9,10] and electrical impedance measurements[11,12].

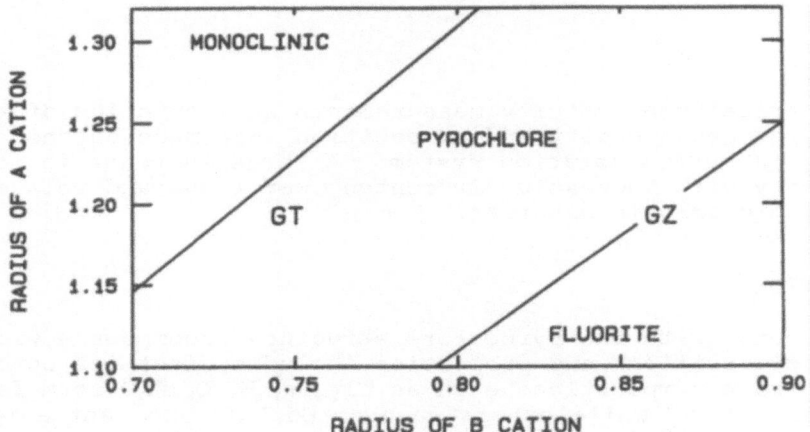

Fig. 1. Cation radius map for $(A^{3+})_2(B^{4+})_2O_7$; GT = Gadolinium Titanate and GZ = Gadolinium Zirconate.

In the present work, we exploit the sensitivity of the ionic disorder in pyrochlores to the cation radius ratio, by systematically varying the composition of a $Gd_2(Zr_xTi_{1-x})_2O_7$ solid solution and indirectly measuring the resulting ionic disorder with electrical conductivity measurements. In this manner, the relationships between fast ion conduction, composition and disorder can be conveniently studied in an isostructural system.

EXPERIMENTAL

Polycrystalline samples of compositions throughout the $Gd_2Ti_2O_7$-$Gd_2Zr_2O_7$ solid solution were formed via the Pechini[13] powder preparation technique followed by conventional sintering at ~1600 °C. Samples of $Gd_2(Zr_xTi_{1-x})_2O_7$ with x > 0.5 were annealed at 1300 °C for ~2 weeks in order to provide a reproducible thermal history. X-ray diffraction measurements showed that all samples had the pyrochlore structure and that the lattice constant varied linearly with composition. The sintered samples were polished and prepared as either discs (~13 mm diameter and 2 mm thickness) or parallelpipeds (~12 mm x 2 mm x 4 mm).

Two probe AC electrical impedance measurements were made using either a Hewlett Packard HP4192a impedance analyzer or a Solartron 1250 frequency response analyzer. The bulk impedance portion of the complex impedance spectrum was unambiguously determined by comparison of the permittivity at high frequency with the known bulk capacitance. Bulk conductivities were calculated from the low frequency intercept of the complex impedance curve after eliminating the effects of grain boundary and electrode impedances.

The bulk electrical conductivity was measured as a function of temperature, PO_2 and composition. The temperature range was 600 to 1400 °C with 50 °C increments and the PO_2 range was 1 to 10^{-20} atmospheres. The ionic conductivity at each temperature was determined by an iterative least squares best fit of the data as a function of PO_2 to the following equation.

$$\log \sigma = \log\{A + B(PO_2^{-1/4}) + C(PO_2^{1/4})\} \qquad [3]$$

in which A represents the ionic component of the conductivity and the second and third terms correspond to the electron and hole conductivities respectively.

Equation [3] is the general formula for the conductivity of an oxide compound when the electroneutrality condition is dominated by anion Frenkel defects and all the carrier mobilities are assumed to be PO_2 independent[14]. A, B and C are constants.

The bulk ionic conductivity for selected temperatures between 600 and 1000 °C is shown as a function of composition in Figure 2.

The activation energy (E^*) and pre-exponential constant (σ_o) for bulk ionic conduction were calculated from equation [4] for compositions with 30-100% Zr and temperatures of 600-1400 °C. The results are shown in Figures 3 and 4.

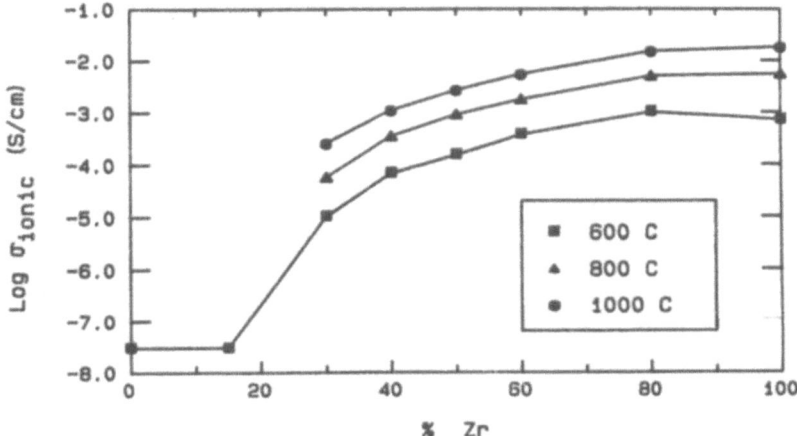

Fig. 2.   Log ionic conductivity in $Gd_2(Zr_xTi_{1-x})_2O_7$

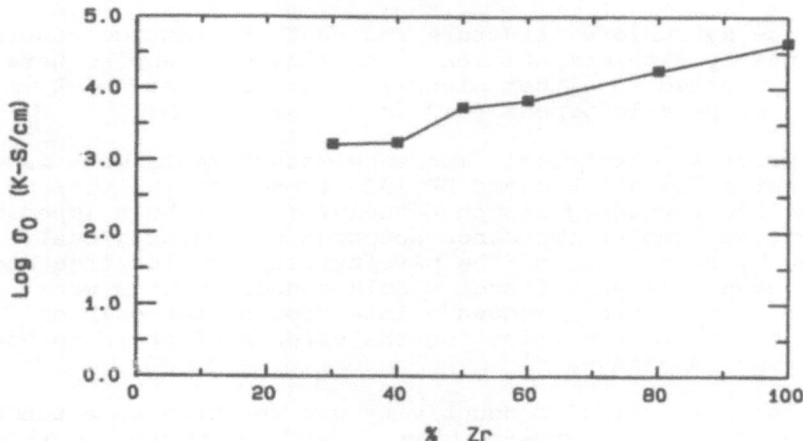

Fig. 3.   Log pre-exponential constant for ionic conduction in
$Gd_2(Zr_xTi_{1-x})_2O_7$

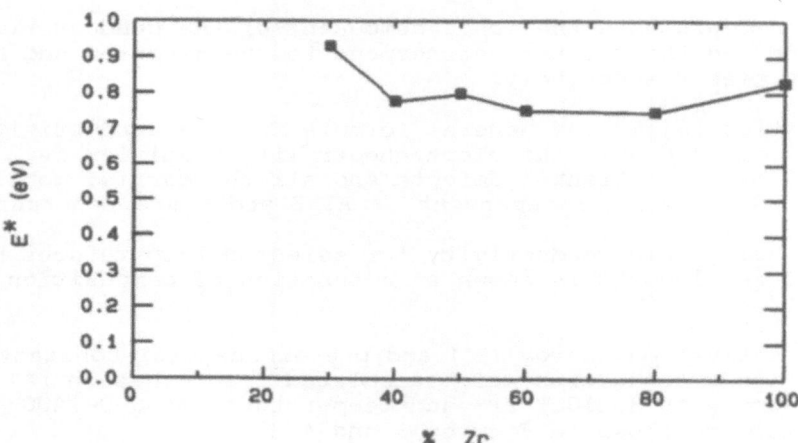

Fig. 4.   Activation energy for ionic conduction in
$Gd_2(Zr_xTi_{1-x})_2O_7$.

$$\sigma T = \sigma_0 \exp(-E^*/RT) \qquad [4]$$

The ionic conductivity of the low Zr content samples (x <
0.3) were masked by their electronic conductivity and could not
be determined for temperatures above 600 °C.   At temperatures below 600 °C some of the high Zr content compositions showed a
large increase in activation energy.   This increase in activation
energy was tentatively assigned to association between oxygen vacancies and cation anti-site defects.   An analysis of the low
temperature data is beyond the scope of this article and will not

be addressed. Therefore the reported conductivity data are for temperatures above 600 °C where the Arrhenius plots fit one straight line.

DISCUSSION

The ionic conductivity shows a large increase in magnitude as a function of Zr content for $x \geq 0.4$ in $Gd_2(Zr_xTi_{1-x})_2O_7$. The data in Figure 3 show that the pre-exponential constant also increases substantially as the Zr content goes from 40% to 100%. At the same time, the activation energy for ionic conduction is almost independent of composition ($E^* = 0.80 \pm 0.03$ eV) in this composition range. Thus the increase in ionic conductivity is primarily due to an increase in the pre-exponential constant which is a measure of the trend in carrier concentration. The identity of the charge carriers can be confidently assigned to oxygen vacancies on the basis of theoretical considerations[4] and aliovalent doping experiments[15]. The presence of a substantial oxygen vacancy concentration due to intrinsic disorder in $Gd_2(Zr_xTi_{1-x})_2O_7$ is confirmed by our quantitative measurement of the anion Frenkel constant in $Gd_2(Zr_{0.3}Ti_{0.7})_2O_7$, ($K_F \sim 10^{38}$ cm^{-6} at 1000 °C) as described in another paper[15]. Further confirmation comes from earlier conductivity[11,12], spectroscopic[7,8,9,10] and neutron diffraction[3] data obtained by other workers on similar pyrochlore systems. Therefore, the increase in ionic conductivity for $x \geq 0.4$ can be understood as an increase in the degree of intrinsic disorder with composition. Furthermore, $Gd_2(Zr_xTi_{1-x})_2O_7$, with $x \geq 0.4$ meets the criteria for an intrinsic fast ion conductor. It has a large intrinsic concentration of mobile charge carriers and a relatively low activation energy of conduction.

The increase in activation energy for $Gd_2(Zr_{0.3}Ti_{0.7})_2O_7$ relative to the compositions with more Zr suggests that reduced ionic mobility may play a role in the lower ionic conductivity of the $Gd_2(Zr_xTi_{1-x})_2O_7$, compositions with $x \leq 0.3$. The exact cause of the low ionic conductivity in these low Zr compositions is the subject of continuing research.

REFERENCES

1. J. Grins, M. Nygren and T. Wallin, Studies on Stoichiometry and Ionic Conductivity of the Pyrochlore type $Tl_{1+x}Ta_{1+x}W_{1-x}O_6 \cdot nH_2O$ System, **Electrochimica Acta** 24:803 (1979).

2. W.W. Barker, J. Graham, O. Knop and F. Brisse, Crystal Chemistry of Oxide Pyrochlores, in: "The Chemistry of Extended Defects in Non-Metallic Solids," L. Eyring and M. O'Keeffe, ed., North-Holland Publishing Co., Amsterdam (1969).

3. K.J. de Vries, M.P. van Dijk, G.M.H. van de Velde and A.J. Burggraaf, A Study of Defect Equilibria in $Nd_2Zr_2O_7$ with Pyrochlore Structure, in: "Solid State Chemistry 1982," R. Metselaar, H.J.M. Heijligers and J. Schoonman, ed., Elsevier Scientific Publishing Co., Amsterdam (1983).

4. M.P van Dijk, A.J. Burggraaf, A.N. Cormack and C.R.A. Catlow, Defect Structures and Migration Mechanisms in Oxide Pyrochlores, **Sol. St. Ionics**, 17:159 (1985).

5. M.A. Subramanian, G.Aravamudan and G.V Subba Rao, Oxide Pyrochlores- A Review, **Prog. Solid St. Chem**. 15:55 (1983).

6. R.A. McCauley, Structural Characteristics of Pyrochlore Formation, **J. Appl. Phys**. 51:290 (1980).

7. W.E. Klee and G. Weitz, Infrared Spectra of Ordered and Disordered Pyrochlore-type Compounds in the Series $RE_2Ti_2O_7$, $RE_2Zr_2O_7$, and $RE_2Hf_2O_7$, **J. Inorg. Nucl. Chem**. 31:2367 (1969).

8. B.E. Scheetz and W.B. White, Temperature-dependent Raman Spectra of Rare Earth Titanates with the Pyrochlore Structure: a Dipolar Order-Disorder Transition, **Opt. Eng**. 22:302 (1983).

9. B.E. Scheetz and W.B. White, Characterization of Anion Disorder in Zirconate $A_2B_2O_7$ Compounds by Raman Spectroscopy, **J. Am. Ceram. Soc**. 62:468 (1979).

10. D. Michel, M. Perez Y Jorba and R. Collongues, Etude de la Transformation Ordre-Desorde de la Structure Fluorite a la Structure Pyrochlore pour des Phases (1-x) ZrO_2 - x Ln_2O_3, **Mat. Res. Bull**., 9:1457 (1974).

11. J.D. Faktor, J.A. Kilner and B.C.H. Steele, A Study of the Effect of Order-Disorder on the Anion Conductivity in the System $Nd_2 Zr_2 O_7$ -$Nd_2Ce_2O_7$, <u>in</u>: "Solid State Chemistry 1982," R. Metselaar, H.J.M. Heijligers and J. Schoonman ed., Elsevier Scientific Publishing Co., Amsterdam (1983).

12. T. van Dijk, K.J. de Vries and A.J. Burggraaf, Electrical Conductivity of Fluorite and Pyrochlore $Ln_x Zr_{1-x}O_{2-x/2}$ (Ln = Gd,Nd) Solid Solutions, **Phys. Stat. Sol**. 58:115 (1980).

13. M.P. Pechini, "Method of Preparing Lead and Alkaline Earth Titanates and Niobates and Coating Method Using the Same to Form a Capacitor", U.S. Patent 3,330,697 (1967).

14. H.L. Tuller, Mixed Conduction in Nonstoichiometric Oxides, <u>in</u>: "Nonstoichiometric Oxides", O.T. Sorensen ed., Academic Press, New York (1981).

15. P.K. Moon and H.L. Tuller, Ionic Conduction in the $Gd_2Ti_2O_7$-$Gd_2Zr_2O_7$ System, to be published in "Proceedings of the Sixth International Conference on Solid State Ionics", Garmisch-Partenkirchen, FRG, September 6, 1987. **Solid State Ionics**.

PLASMA SPRAYED ZIRCONIA ELECTROLYTES

M. Scagliotti

Cise - Tecnologie Innovative SpA, Via Reggio Emilia, 39, 20090 Segrate (Milano), Italy

INTRODUCTION

Yttria stabilized zirconia is widely used as an electrolyte in electrochemical devices such as high temperature solid oxide fuel cells (HTSOFC) (1) and oxygen sensors (2). In order to reduce the ohmic resistance it is interesting to obtain electrolyte layers as thin as possible. Gas tight films of stabilized zirconia were prepared by chemical vapor deposition, plasma spraying and tape casting and successfully used in HTSOFC prototypes (1). As far as we know, however, the basic properties of these films, and in particular the correlation among preparation conditions, structure, microstructure and electrical properties, have not been extensively investigated. This fact has stimulated the present work on plasma sprayed zirconia based electrolytes (3-5).

Yttria stabilized zirconia films were prepared by plasma spraying and characterized by optical and scanning electron microscopies, X-ray diffraction, Raman scattering and complex impedance spectroscopy. The results are compared with those obtained on sintered pellets prepared with the same powders used for spraying and on commercial single crystals. In this paper the main results are summarized and briefly discussed.

SAMPLE PREPARATION

The powders used to prepare films and sintered pellets were obtained by mixing fine ZrO_2 and Y_2O_3 powders (Merck) in the composition range ZrO_2 - (8,10 and 12 mol%) Y_2O_3. In order to stabilize the zirconia cubic phase the powders were heated in air at 1550°C for 24 hours.

Films, 100-200 μm thick, were sprayed in air using argon as the plasma gas. To improve the density they were sintered in vacuum at 2000°C or 2100°C for 3 hours and then annealed in air at 1450°C for 3 hours to restore the oxygen stoichiometry. Hereafter the samples will be labelled 1 for sintering at 2000°C and 2 for sintering at 2100°C.

Pellets were sintered in air at 1700°C for 24 hours. Stabilized zirconia single crystals (ZrO_2 - 4.5, 12, 18 and 24 mol% Y_2O_3), grown by skull melting, were supplied by Ceres.

RESULTS AND DISCUSSION

The film stoichiometry was tested by proton induced X-ray emission (5). The actual composition are reported in Table 1. Slight deviations from the nominal composition, i.e. the composition of the powders, were found after spraying and vacuum sintering processes. However only in one sample (ZY8-2) was a large discrepance found between nominal and actual composition.

Table 1 - Composition, cubic lattice parameters and density of the samples (from ref. 4).

	Y_2O_3 (mol%)	latt.param. exper. (nm)	density calc. (g/cm^3)	density exp. (g/cm^3)	exp/calc %
sintered pellets	8	0.5139	5.95	5.50	92
	10	0.5141	5.93	5.50	92
	12	0.5143	5.91	5.53	93
films ZY 8-1	9.5	0.5132	5.97	5.85	98
ZY 10-2	10.7	0.5134	5.95	5.81	98
ZY 12-1,-2	11.1	0.5135	5.94	5.81	98
ZY 8-2	14.5	0.5150	5.86	5.79	98
single crystals	4.5	0.5123	6.04	6.05	100
	12	0.5150	5.89	5.90	100
	18	0.5164	5.79	5.82	100
	24	0.5177	5.71	5.72	100

The film microstructure was studied by optical microscopy and scanning electron microscopy. The sprayed films appear dense and without open porosity. Only a few large defects (10-20 micron) and some micron sized pores are observed on the scanning electron microphotographs of fractured samples. The average grain size, measured on polished and etched film cross sections, is about 20 microns.

The structure of the samples was carefully characterized by Raman spectroscopy, which is very sensitive to all the zirconia polimorphs. The films appear fully stabilized in the cubic phase (fig.1), even if some differences in the band shape are observed. They may be related to different defect structure.

The cubic lattice parameter a_0 was measured by using the (400) reflections in the X-ray diffractograms. It is found to increase with the yttria content in good agreement with the results obtained on sintered pellets and single crystals (Table 1). The a_0 values were used to calculate the theoretical density of the samples which are reported in Table 1 together with the experimental densities measured by the Archimede's technique (4).

The electrical properties of the sprayed films were studied by complex impedance spectroscopy (4). A Solartron 1174 frequency response analyzer equipped with a high impedance adaptator (10^{12} Ω , 6 pF) was used in the frequency range 0.01 Hz – 1 MHz.

The frequency dispersion measurements were analyzed according to the equivalent circuit reported in fig. 2a (6). This figure also shows typical impedance spectra of a sprayed film, a sintered pellet and a single crystal with the same nominal composition, i.e. ZrO_2-12 mol% Y_2O_3. The sprayed films and the single crystals are characterized by similar electrical responses. We note that the arc associated with grain boundary effects is not observed in sprayed film spectra.

The Arrhenius plots of the films are shown in fig. 3 while isothermal conductivities are reported in fig. 4 as a function of the Y_2O_3 content for films, pellets and crystals. The Arrhenius plots are characterized by two straight line portions with different slopes. Above 650°C the activation energies of the films range between 0.7 and 0.8 eV, while below 550°C they range between 1.1 and 1.2 eV. These results are in satisfactory agreement with those obtained on pellets and crystals with the same compositions (4).

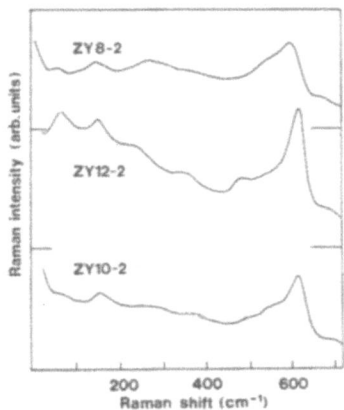

Fig. 1   Raman spectra of yttria stabilized zirconia sprayed films.

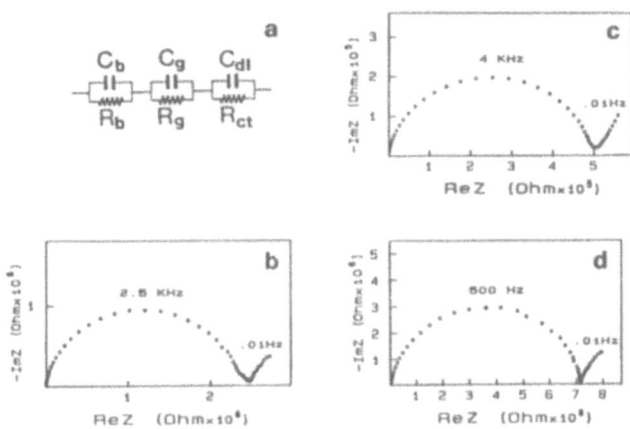

Fig. 2   Equivalent circuit for the electrical response of polycrystalline ceramic samples (a), typical complex impedance plots at T=220°C of a pellet (b), a sprayed film (c) and a single crystal (d).

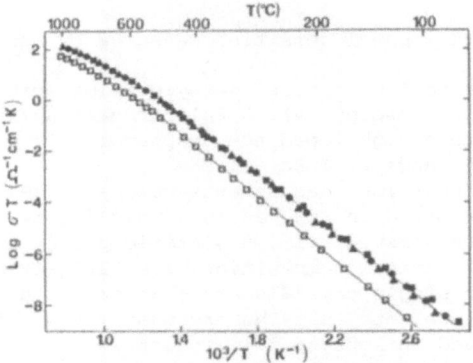

Fig. 3 Conductivities vs reciprocal T for different films (■ ZY8-1, ▲ ZY10-1,
●ZY12-1, □ ZY8-2).

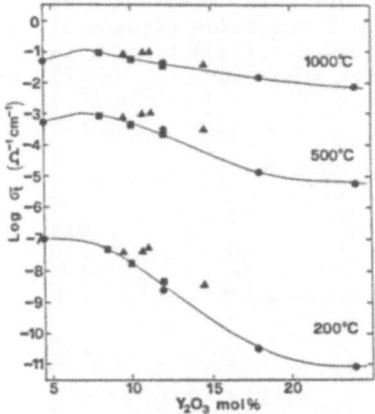

Fig. 4 Isothermal conductivities vs Y_2O_3 content of single crystals (●),
pellets (■) and plasma sprayed films (▲).

CONCLUSIONS

The present investigation shows that yttria stabilized zirconia
electrolytes prepared by plasma spraying can be made to be dense, fully
stabilized and to exhibit high ionic conductivity. In particular their
electric properties are similar to those of single crystals grown by
skull melting. This is due to the large grain size and high purity of
the plasma sprayed films. From the present data the plasma spray technique
appears suitable to prepare yttria stabilized zirconia electrolytes.

BIBLIOGRAPHY

1 - For a recent review on HTSOFC see for example B.C.H.Steele, High Tech
    Ceramics, ed. by P.Vincenzini, Elsevier 1987, p.105
2 - See for example E.C.Subbarao and H.S.Maiti, Solid State Ionics 11,317
    (1984)
3 - G.Lanzi, P.Milani,F.Parmigiani,D.Richon,G.Samoggia and M.Scagliotti
    in ref.1, p.1925
4 - G.Chiodelli, A.Magistris, M.Scagliotti and F.Parmigiani, J.Mater.
    Science 23, 1159 (1988)
5 - M.Scagliotti, F.Parmigiani, G.Samoggia, G.Lanzi and D.Richon, J.Mater.
    Science 23 (1988) (in press)
6 - J.A.Kilner and B.C.H.Steele in "Non Stoichiometric Oxides" ed. by O.T.
    Sørensen, Academic Press, London, 1981, p.233

FREE LITHIUM ION CONDUCTION IN LITHIUM BORATE GLASSES DOPED WITH Li_2SO_4 [*]

M. Balkanski[+], R. F. Wallis and I. Darianian[+]

Department of Physics
University of California
Irvine, California 92717, USA

I. INTRODUCTION

Extensive work has been performed in recent years on the structural and electrical properties of lithium borate glasses such as B_2O_3-Li_2O-LiX with X = F, Cl, Br, I. For B_2O_3-xLi_2O-yLiCl it has been found[1] that the ionic conductivity increases with increasing Li_2O and LiCl concentrations and that above a fixed value of x the increase is more rapid with y. For B_2O_3-xLi_2O-yLi_2SO_4 analogous results have been obtained.[2]

II. EXPERIMENTAL RESULTS

The electrical conductivities of glasses of the composition B_2O_3-xLi_2O-yLi_2SO_4 have been studied by the method of complex impedance in the range of 20 Hz to 1 MHz at temperatures ranging from 20 to 400°C. The activation energies were determined for the various compositions and were found to be 0.71, 0.71, 0.65 and 0.61 eV for x = 0.5 and y = 0.00, 0.05, 0.10, and 0.15, respectively. We see that there is a moderate reduction in the activation energy as y increases. At fixed temperature the conductivity increases rapidly with increasing values of y.

III. CALCULATION OF CONDUCTIVITY DUE TO FREE LITHIUM IONS

The electrical conductivity σ due to a concentration n_F of free Li^+ ions can be written as

$$\sigma = n_F e\mu \tag{1}$$

where e is the ionic charge and μ is the mobility. To calculate n_F we consider an ensemble of isolated Li_2SO_4 molecules dispersed in a borate glass assumed to be a continuous medium with static dielectric constant ε. Using the Boltzmann distribution the numbers of free Li^+ ions N_F and of bound Li^+ ions N_B are specified by

$$\left.\begin{matrix} N_F \\ N_B \end{matrix}\right\} = A \int d^3p \int d^3r\, e^{-E(p,r)/k_BT} \quad \text{for } \begin{cases} E(p,r) > 0 \\ E(p,r) < 0 \end{cases} \tag{2}$$

where the energy of a Li^+ ion, $E(p,r)$, is given by

$$E(p,r) = \frac{p^2}{2M} - \frac{e^2}{\varepsilon r} \quad ,$$ (3)

p is the momentum of the Li^+ ion, r is its distance from the negative $LiSO_4^-$ ion, and M is the Li^+ mass.

Evaluating the integrals in Eq. (2) and using approximations appropriate to the physical situation under discussion, one finds that

$$\frac{n_B}{n_F^2} = \frac{1}{n_\lambda} e^{\lambda/r_o} = K(T)$$ (4)

where $n_\lambda = \lambda/8\pi^{1/2}r_o^4$, $\lambda = e^2/\varepsilon k_B T$, and r_o is the distance of closest approach of a Li^+ ion to a $LiSO_4^-$ ion. Equation (4) is the law of mass action for the dissociation reaction and $K(T)$ is the equilibrium constant. Introducing the total concentration of Li^+ ions $n = n_F + n_B$ and using Eq. (1), we find to a good approximation that

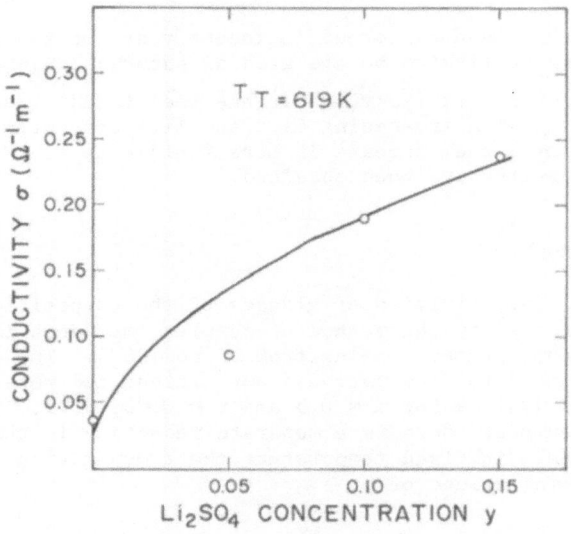

Figure 1. Variation of the conductivity σ as a function of the Li_2SO_4 concentration y in B_2O_3-$0.5Li_2O$-yLi_2SO_4. The circles are the experimental data and the solid curve is the theoretical result.

$$\sigma = e\mu[n_\lambda n_T(y+y_o)]^{1/2} \ e^{-E_{ad}/k_BT} \tag{5}$$

where n_T is the concentration of B_2O_3 molecules, $y = n/n_T$, y_o accounts for the source of free Li^+ ions when $y = 0$, and E_{ad} is the dissociation activation energy given by

$$E_{ad} = \frac{e^2}{2\varepsilon r_o} \ . \tag{6}$$

From Pauling's ionic radii[3] we estimate r_o to be 2.7A. Using $\varepsilon = 8.0$, we obtain $E_{ad} = 0.33$ eV. Taking 0.65 eV as representative of the values of the total activation energy given in Section II, we obtain for the mobility activation energy $E_{a\mu} = 0.32$ eV. Values of σ calculated from Eq. (5) for $T = 619K$ and taking $\mu = 1.69 \times 10^{-7} m^2/Vsec$, $n_T = 0.290 \times 10^{28} m^{-3}$ and $y_o = 2.0 \times 10^{-3}$ are plotted against y in Fig. 1 along with the experimental data. We see that the agreement is good.

[*]Work supported by DARPA Contract No. F33615-85-K-2501.
[+]Permanent address: Laboratoire de Physique des Solides, Université Pierre et Marie Curie, 4 Place Jussieu, 75252 Paris Cedex 05, France

REFERENCES

1. A. Levasseur, B. Cales, J. M. Réau and P. Hagenmuller, C. R. Acad. Sc. Paris 285-c, 471 (1977).
2. A. Levasseur, M. Kbala, J. C. Berthous, J. M. Réau and P. Hagenmuller, Solid State Commun. 32, 839 (1979).
3. L. Pauling, The Nature of the Chemical Bond, Second Edition (Cornell University Press, Ithaca, 1940).

EFFECTS OF HALIDE SUBSTITUTIONS IN POTASSIUM HALOBORATE GLASSES

F.A. Fusco and H.L. Tuller

Crystal Physics and Optical Electronics Laboratory
Massachusetts Institute of Technology
Cambridge, MA 02139

ABSTRACT

The effects of halide substitutions in potassium borate glasses are different from those observed in their lithium and sodium analogs. These results call attention to the important correlation between the activation energy for ionic conduction, E_A, and glass network rigidity.

INTRODUCTION

Fast ion conducting glasses possess properties which make them attractive materials for technological applications in high energy density, high performance batteries. In addition to their good ionic conductivities, isotropic, and grain boundary free structure, and ease of fabrication, they display wide compositional flexibility. We explore this last attribute and investigate how halide additivies affect the conductivity and physical properties of potassium diborate glasses. We interpret results of ionic conductivity studies in terms of the Anderson and Stuart[1] formalism for the strain energy barrier to potassium cation migration.

EXPERIMENTAL PROCEDURE

Glass samples were prepared by conventional glass melting and quenching techniques. Starting materials were anhydrous reagent grade potassium carbonate, anhydrous B_2O_3 and potassium halide salts. Charges of well blended powders were melted in air in an electric furnace in covered platinum crucibles. Melt times of approximately twenty to twenty-five minutes at temperatures between 950-1050°C were adequate to produce completely fused homogeneous melts. Chemical analyses verified that glasses retained the same chemical compositions as the starting batch powders.[2] The glass systems studied were $(0.33-y)K_2O-y(KX)_2-0.67B_2O_3$ where x=Cl or I and y ranges between 0.00 and 0.10 for Cl and 0.00 and 0.0165 for I. Glasses were quenched in air between room temperature graphite blocks, annealed at Tg-75°C for 1.5 hours and then allowed to furnace cool in air. Since these glasses are hygroscopic they were handled with tweezers or gloves and stored in dry environments.

Measurements of ionic conductivity were made on regular parallepipeds with shape factors between 10 and 20 cm^{-1} with sputtered Pt electrodes arranged in a two probe configuration. AC electrical measurements were performed on a computer controlled impedance spectrometer including a Solartron Frequency Response Analyzer (operating between 10^{-5} Hz and 65 KHz). Glass conductivity was measured over a temperature range of 100-350°C in an electrically shielded chamber flushed with dry nitrogen. All the potassium glasses exhibited nearly ideal responses, readily interpretable using complex plane analysis for extracting bulk conductivity values.

Glass density was obtained with the ASTM[3] test method. Measurements were made at room temperature using toluene as the immersion fluid. Density values were accurate to within 0.01 g/cm^3.

Glass transition temperatures, T_g's were determined using differential scanning calorimetry with a Perkin Elmer DSC 4 instrument operating at a scan rate of 10K/min between 500 and 800K.

RESULTS

The potassium ion conductivity at 150°C is presented in Figure 1 as a function of the chloride substitution. Also included for comparison are our earlier results for the corresponding lithium and sodium glasses.[4] Surprisingly, in contrast to the results for the smaller alkali ions, chloride substitutions appear to have little or no enhancing effect on the potassium ion conductivity. An examination of Figure 2 shows the corresponding activation energy for K ion motion to be, at best, very weakly dependent on chloride substitution while those for Li and Na motion are strongly depressed.

We have previously emphasized how halide additions to borate glasses dilute the glass structure. Figure 3, which shows the excess volume per mole of B as a function of halide substitution, confirms this observation for the K glasses as well. When we examine the dependence of T_g on chloride substitution in Figure 4 however, we again find some unexpected results. Rather than decreasing sharply as for the Li and Na glasses, T_g actually increases with chloride substitution after an initial small drop. For example, while a lithium glass with %(LiCl)$_2$/Li$_2$Z = 30 has a T_g roughly 40K lower than the chlorine free glass, a potassium glass with the same level of chloride substitution has a T_g nearly 10K greater than the corresponding chlorine free glass.

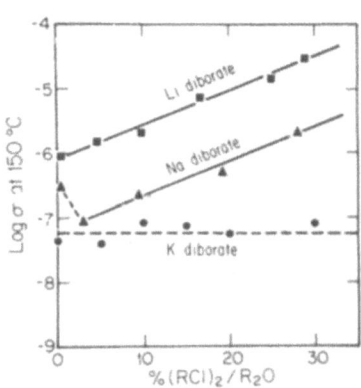

Fig. 1. Log σ versus chloride substitution for, Li, Na, and K diborate glasses

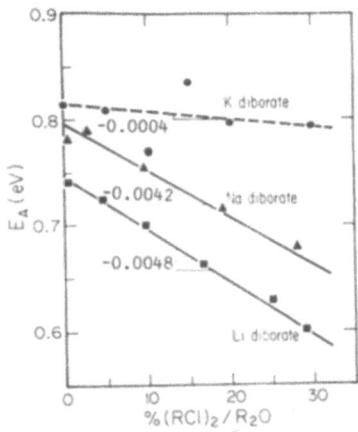

Fig. 2. Activation energy as a function of chloride substitution.

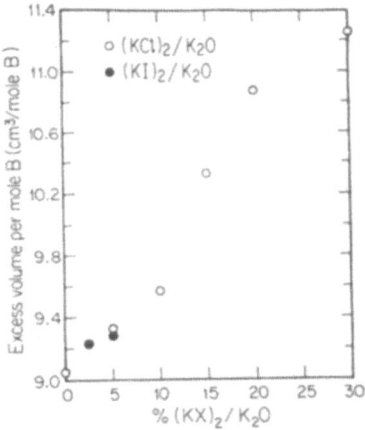

Fig. 3. Excess volume vs. halide content.

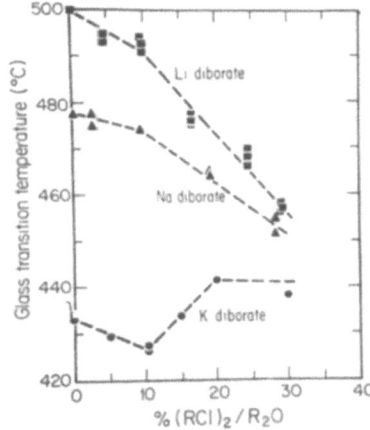

Fig. 4. Dependence of T_g on alkali halide content for alkali diborate glasses.

In Figure 5, we examine the effect of addition of the large iodide ion which in the lithium glasses was found to result in a greater enhancement of the ionic conductivity than that of chlorine.[5] For the potassium diborate glass we find the opposite trend. Figure 6 shows that the increase is

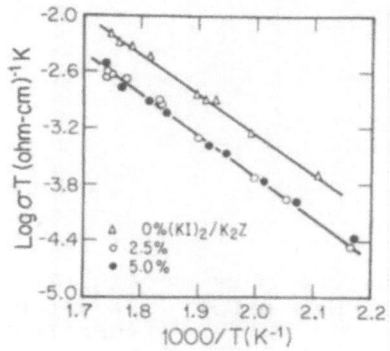

Fig. 5.   Arrhenius plot for potassium di-iodoborate glasses.

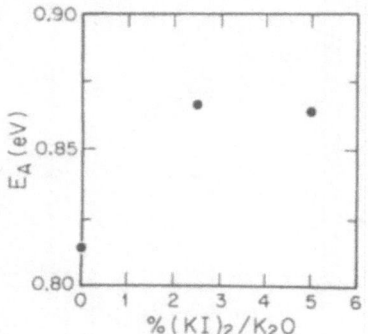

Fig. 6. E_A as a function of iodide substitution in glasses described in Fig. 5.

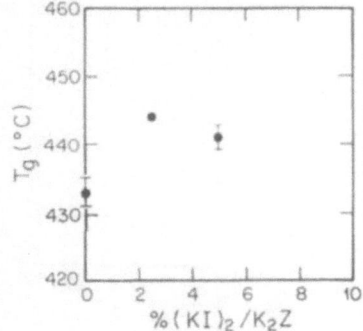

Fig. 7. T_g dependence on iodide substitution in glasses described in Fig. 5.

related to an increase in activation energy. The nearly equivalent results obtained for the 2.5 and 5.0% substitutions suggest that we may have exceeded the solubility limit of this large ion. Figure 7 shows the glass transition temperature to be increased as well.

DISCUSSION AND CONCLUSIONS

Halide substitutions in the lithium borate glasses enhanced σ monotonically. Button et al[4,5] clearly demonstrated that for the lithium and sodium dichloroborates, this conductivity improvement was associated with monotonic decreases in T_g and increases in excess volume. Thus, they concluded that Cl increases conductivity by plasticizing and dilating the glass network structure. In light of these conclusions, the results obtained for the potassium borate glasses are, at first, unexpected. However, after closer examination of expressions for ionic motion in glasses, we believe we can understand the source of this apparent discrepancy.

We recall from the Anderson and Stuart formalism,[1] as modified by McElfresh and Howitt,[6] that the strain energy barrier to cation migration depends directly on network rigidity through the parameter, G, the shear modulus and on the network doorway radius, r_D, which can be related to the excess volume. This strain energy term is given by Equation 1.

$$\Delta E_s = \frac{\pi G (r-r_D)^2 \ell}{2}$$ (1)

From the strain energy standpoint, both increasing excess volume and decreasing T_g (and correspondingly decreasing G) should result in decreased migration energies and increased ionic conductivities. Of these two factors (see Eq. 1) variations in excess volume are expected to play the dominant role in determining migration energies. We examine these expectations below.

For both the Li and Na di-chloroborates, the decrease in E_A with chloride is accompanied by increases in excess volume and decreases in T_g.[5] Here, both the volume and T_g variation work towards lowering E_A. For the K di-chloroborate, in the range 10-20%$(KCl)_2/K_2Z$, the excess volume and T_g play competing roles, resulting in an anomalous rise in E_A followed by a decrease once T_g saturates. Similarly, in the iodoborate glasses, both E_A and T_g rise with iodide additions confirming again the important influence of the network coherency on migration.

The results obtained for the potassium di-chloroborate glasses suggest that not only the number and size of "doorways" between sites but also the network rigidity (i.e., its resistance to dilation by carriers) controls the transport process. One may question why halide additions enhance network rigidity exclusively in the K glasses. To understand this, one needs a clearer vision of how the halide additions impact the glass structure in each of the alkali borate glass systems. We believe a productive approach may be to examine Krogh-Moe's model in more detail. This model proposes that a glass is composed of a random arrangement of the same short-range structural units found in crystals of similar chemical composition. Krogh-Moe's structure determinations for the Li, Na, and K diborate crystals[7-11] reveal that the boron-anion framework in these crystals is composed of different structural groupings: the Li diborate is composed of diborate groups exclusively; the Na diborate of triborate and di-pentaborate groups; and the K diborate of planar triangles, diborate groups, and triborate groups.

In conclusion, the role of halides in modifying σ differs depending upon the nature of the binary alkali borate glass network as determined by the identity of the alkali modifier cation. Excess volume increases normally contribute to σ enhancement. However, in already highly dilated glasses, such as the potassium di-chloroborates, network stiffness and coherency, as reflected in T_g, dominate the transport process.

ACKNOWLEDGEMENT

This research has been supported by Lawrence Berkeley Laboratory under subcontract to the US Department of Energy under contract #4548010. Discussions with Professor D.R. Uhlmann are appreciated.

REFERENCES

1. O.L. Anderson and D.A. Stuart, J. Amer. Ceram. Soc. 37: 573 (1954).
2. Chemical analysis were performed by Owens-Illinois Analytical Services, Toledo, Ohio.
3. ASTM "Standard Test Method for Density of Glass by Buoyancy" Test C693 Volume 15:02 (1983).
4. D.P. Button, P.K. Moon, H.L. Tuller, and D.R. Uhlmann, Glastechn. Ber. 56K Bd.2:856 (1983).
5. D.P. Button, Ph.D. Thesis, MIT, 1983.
6. D.K. McElfresh and D.G. Howitt, J. Amer. Ceram. Soc. 69: C-237 (1986).
7. J. Krogh-Moe, Ark. Kemi. 12:475 (1958).
8. J. Krogh-Moe, Acta Cryst. 15:190 (1962).
9. J. Krogh-Moe, Acta Cryst. B24:179 (1968)
10. J. Krogh-Moe, Acta Cryst. B30:578 (1974).
11. J. Krogh-Moe, Acta Cryst. B28:3089 (1972).

EVIDENCE OF BORON-OXYGEN NETWORK MODIFICATIONS IN ALKALI

BORATE GLASSES

M. Massot, A. Ayyadi, and M. Balkanski

Laboratoire de Physique des Solides de
l'Universite Pierre et Marie Curie
Paris Cedex France

ABSTRACT

In recent investigations of alkali borate glasses
B_2O_3-Li_2O-LiX (X=Cl, Br, or I) by light scattering, we find
that glasses in the binary and in the ternary system with
common O/B ratios do not exhibit identical boron-oxygen
networks. A very simple model, based on electrostatic
interactions is proposed.

INTRODUCTION

Because of their high conductivities, alkali borate
glasses are of technological interest as solid electrolytes in
solid state batteries. The ionic conductivity of these
materials increases as the content of lithium oxide and
lithium salt is increased. Furthermore, in the ternary
systems with fixed O/B ratio, for similar halogen
concentrations, the ionic conductivity increases as one goes
from a fluoride to an iodide glass.[1]

In attempting to understand the cause of the greatly
enhanced lithium ion conductivity upon the addition of alkali
halides, we have examined the effect of the halides on the
glass network by Raman spectroscopy.

RAMAN SCATTERING EXPERIMENTS

The Raman spectra were recorded on a PHO Coderg double
monochromator equipped with Jobin Yvon holographic gratings.
The emission line at 514.5 nm of a argon-ion laser was used
with an incident power of about 200 mW. Detection was made
with a cooled ITT FW 130 photomultiplier coupled with a Racal
Dana 9500 photoncounting system. An Apple II micro-computer
was used to drive the spectrometer and accumulate the Raman
data.

The Raman spectra of ternary systems with the same O/B ratio and different lithium halide concentrations are shown in Figure 1 and compared with the spectrum of the binary glass B_2O_3-0.57 Li_2O.

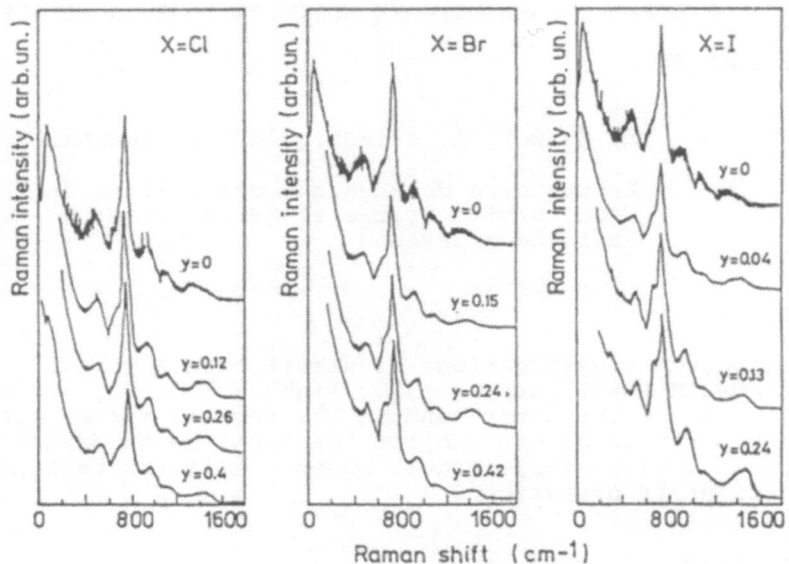

Fig. 1. Raman spectra of B_2O_3-0.57Li_2O-yLiX with X=Cl, Br, and I for different lithium halide concentrations.

The spectrum of the binary glass is characterized by a very sharp line at about 770 cm^{-1}. This peak, insensitive to the ^{10}B --> ^{11}B isotopic substitution is attributed to the breathing vibration of six-membered borate rings with one or two BO_4 tetrahedra.[2,3] Because of the exceptional internal compensation of force constants acting on the boron atoms, these latter groups do not contribute to the ring vibration. The Raman line at 520 cm^{-1} is analogous to those, observed at 450 cm^{-1} in v-B_2O_3[4] in that it represents in-phase motion of the bridging oxygen atoms with little or no boron motion.

In the ternary glasses, the modifications of the boron-oxygen network with the halide addition are evidenced for the first time in Raman spectroscopy by the shift to lower frequency of the shoulder observed at 720 cm^{-1} in the glassy matrix spectrum (Fig. 1 and Fig. 2) At the same time, the vibration frequency of the bridging oxygen increases and this later result may be understood as a pressure effect induced by the halide addition. We note also that the frequency of the borate ring vibration seems to be unchanged upon halide doping.

A second characteristic of the present result is its sensitivity to the nature of the halide. The frequencies shift more rapidly as one goes from a chloride to an iodide glass. We also observe, for glasses with fixed O/B ratio and similar halide concentrations, that the intensity of this new line increases and becomes maximum for the lithium iodide doped glass (Fig. 3).

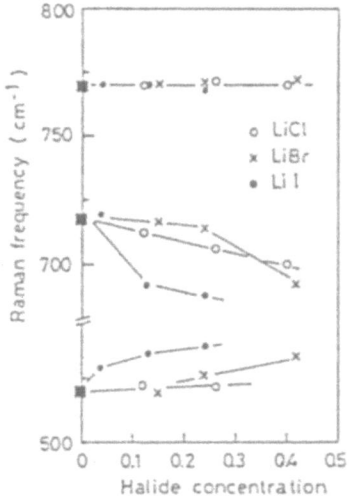

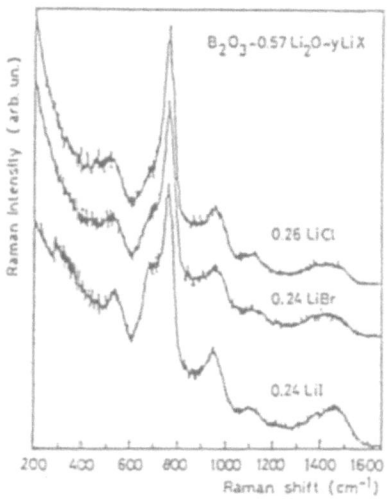

Fig. 2. Raman frequency of the shifted peaks vs. halide concentration for the ternary glasses $B_2O_3-0.57Li_2O-yLiX$ (X = Cl, Br, or I).

Fig. 3. Raman spectra of the ternary glasses with fixed O/B ratio and similar halide concentrations.

DISCUSSION AND CONCLUSION

Our present results show that structural modifications of the boron-oxygen network are induced by the doping salt. Presently, we are able to formulate two hypotheses.

The first one is based on the work of Konijnendijk and Stevels.[3] These authors have attributed the 720cm^{-1} peak to the existence of metaborate groups in the glass. These groups are composed of chains with non-bridging oxygens of negative charge which can easily trap the mobile lithium ions. Such a mechanism does not contribute to the increase of conductivity observed in this system.

In the second one, it is expected that this new peak is an internal mode of new borate group formed during the "doping" process. So the nature of this new line is the same

as those of the 786cm^{-1} peak in the binary. NMR results have
shown that tetraborate and diborate groups are present in the
B_2O_3 -0.57Li_2O glass.[3] During the "doping salt" addition,
tetraborate groups are transformed into diborate groups by the
formation of triangle BO_3. Now the existence of internal
isolated modes for the diborate rings may be possible.
Probably electrostatic interactions between the negative
charged BO_4 units and the halide ions are responsible for this
transformation of the glass network.

REFERENCES

1.   P. Hagenmuller, A. Levasseur, C. Lucat, J.M. Reau, and
      G. Villeneuve, "Fast Ion Transport in Solids," eds.
      P. Vashihta, J.N. Mundy, and G.K. Shenoy, Elsevier
      North Holland, Inc., p. 637, (1979).
2.   J. Lorosch, M. Couzi, J. Pelous, R. Vacher, and A.
      Levasseur, J. Non-Cryst. Solid. 69:1 (1984).
3.   W.L. Konijnendijk and J.M. Stevels, J. Non-Cryst. Solids
      18:307 (1975).
4.   F.L. Galeener and A.E. Geissberger, J. de Phys. C9 43:343
      (1982).
5.   P.J. Bray, A.E. Geissberger, F. Bucholtz, and I.A.
      Harris, J. Non-Cryst. Solids 52:45 (1982).

CHARACTERIZATION OF ALKALI-OXIDE ELECTROLYTE

GLASS IN THIN FILMS

P.Dzwonkowski, C.Julien and M.Balkanski

Laboratoire de Physique des Solides, associé au CNRS
Université Pierre et Marie Curie
4, place Jussieu, 75252 Paris Cedex 05, France

INTRODUCTION

Thin film alkali-oxide glasses could be used as electrolytes in solid state batteries. Recent studies on the B_2O_3-Li_2O family have been reported on bulk material[1,2]. Some of these glasses have a conductivity of the order of 10^{-2} S/cm at 600 K and the conductivity increases as the lithium halide increases[3]. Levasseur et al.[4] have reported the elaboration and characterization of lithium conducting thin film glasses prepared from binary glassy material.

In this paper, we present the electrical conductivity and structural properties of alkali-oxide electrolyte glass in thin film prepared from two different starting materials : (i) binary B_2O_3-Li_2O powder of electrolyte glass, (ii) lithium metaborate $LiBO_2$. Using both the structural and electrical measurements, the correlation between composition and ionic conductivity is discussed.

EXPERIMENTAL

Sample preparation

B_2O_3-Li_2O thin films have been prepared using a quasi-flash evaporation process in a vacuum chamber with 1-10 mPa gas pressure conditions. Thin films of 0.1-1.8 μm thick have been formed on a pyrex substrate heated below 400 K. The starting materials were : (i) binary alkali-oxide electrolyte B_2O_3-xLi_2O with 0.1< x<0.7 in form of powder, (ii) 99.9% purity powdered lithium metaborate $LiBO_2$ (Aldrich-Chimie). The films were formed by evaporating small amounts of the mixtures (0.3 g) charged in a molybdenum boat which is designed with a special chimney. Its temperature was measured by means of an optical pyrometer while the substrate temperature was controlled by a thermocouple. Silicon wafers with (111) orientation having a low electrical resistivity were used as a transparent substrate for the infrared absorption measurements.

Electrical and optical measurements

A sandwich structure has been obtained by successive deposition of metal and electrolyte as shown in figure 1. The geometrical arrangement was realized using several masks. The Au, Al or Ni metals have been utlized as the electrode materials. The distance between electrode planes

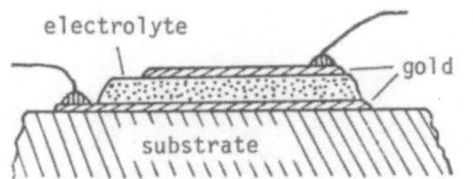

Fig.1 Structure of the Au/glass/Au thin
film cell

equal to the electrolyte thin film thickness has been measured by means
of the double beam interferometer method.

Ionic conductivity was measured by a classical complex impedance
method[5] using a Hewlett-Packard 4192A frequency pulse analyser in the
frequency range of 5 Hz up to 10 MHz. Activation energy has been determined by heating the samples up to 400–450 K in the inert argon atmosphere.
The contacts with the sample electrodes have been prepared by use of small
gold balls. The value of the conductivity was determined by employing the
complex impedance plane method at the time of data analysis. In the low
frequency region, the impedance plot was a close circular arc with the
classical electrode contribution.

The infrared absorption spectra of the films deposited onto silicon
wafers were measured over a 400–2000 cm^{-1} range at room temperature using
a Bruker IFS 113V FTIR spectrometer. In this region, the substrate transmission was better than 50%, and the experimental resolution was taken
at 2 cm^{-1}.

RESULTS AND DISCUSSION

The typical conditions used in the experiments to prepare thin films
are the following : substrate temperature T_s=350.K, source temperature
T_i=1100–1600 K, film thickness e=1μm, deposition rate D_e=10 nm/s. The films
were found to have a smooth and compact surface.

Figure 2. shows the infrared absorption spectra of $LiBO_2$ powder (a),
$LiBO_2$ thin film prepared with T_i=1200 K (b) and $LiBO_2$ thin film prepared
with a thermal procedure starting at 1100 K and finishing at 1600 K during
100 s.(c). As it has been reported previously[6], in the infrared spectra

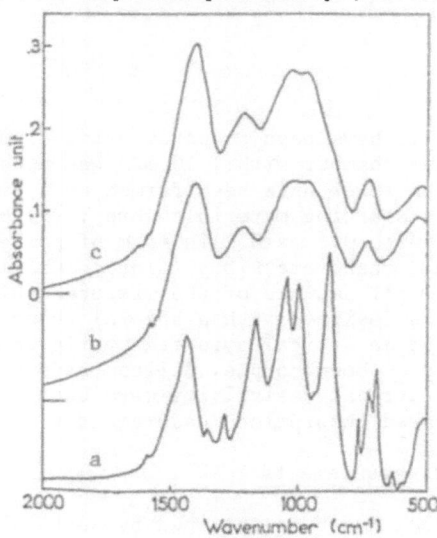

Fig.2 Infrared absorption spectra of starting lithium metaborate (a), thin
film prepared at T_a=1200 K (b) and at 1100–1600 K (c).

absorption bands were generally observed at about 1400, 1200, 1000 and 750 cm^{-1}. It can be seen in the figure that these absorption bands shift continuously in relative intensity, this has been attributed with the changes in Li content in the film. In particular, the broad band around 1000 cm^{-1} decreases progressively in intensity with increasing Li content.

A systematic study has been carried out on thin films using the starting material B_2O_3-xLi_2O where $0.1 < x < 0.7$. In figure 3., a plot of the intensity ratio of the two main bands A=1400 cm^{-1} and B=1000 cm^{-1} versus x lithium oxide content is given. This ratio increases continuously and for interpretation of the spectra of amorphous phases, comparison with appropriate compounds whose structures are known is generally attempted.

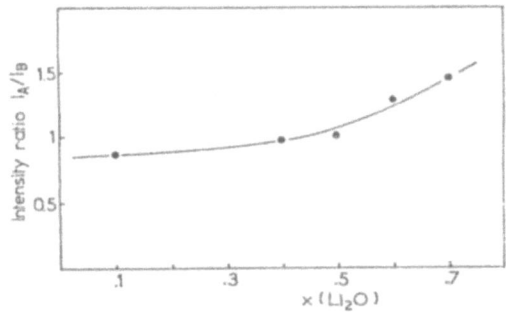

Fig.3 Intensity ratio between the absorption peaks at 1400 cm^{-1} and 1000 cm^{-1} versus Li_2O content in thin films.

The infrared spectra of B_2O_3-Li_2O thin films show a close similarity to those of the polycrystalline material. The presence of BO_4 units in diborate groups can be recognized by the appearance of an absorption band near 1000 cm^{-1}. As was already demonstrated[7] an increase of Li content leads to a decrease of the absorption band near 1000 cm^{-1}, thus showing the transformation of boroxol rings into triborate cycles or di-triborate groups.

Figure 4. shows the variation of the ionic conductivity for $LiBO_2$ thin films versus inverse absolute temperature using three kinds of cell electrodes. Ni electrodes give the rather unstable samples which resistance can be changed 2-3 times when the amplitude of the sinusoidal voltage signal is changed. The Au and Al electrode sample curves are composed of the two parts. Low frequency linear joined with the higher frequency semi circular part. Conductivity values obtained for thin films are in good agreement with those determined for bulk material, the ionic conductivity varies exponentialy with temperature corresponding to the classical Arrhenius law. Thus, the activation energy is in the range 0.64 eV-0.81 eV which is similar to the bulk material.

CONCLUSION

This paper describes the preparation conditions of alkali-oxide electrolyte thin films and their structural and electrical properties.

The structural analysis is in good agreement with the work of Ito et al.[6] where the chemical analysis showed no significant composition change between thin films and starting materials.

Electrical measurements seem to show that the thin film conductivity is similar to that observed in bulk material.

Use of alkali-oxide glass as electrolyte in micro-solid-state batteries is in progress.

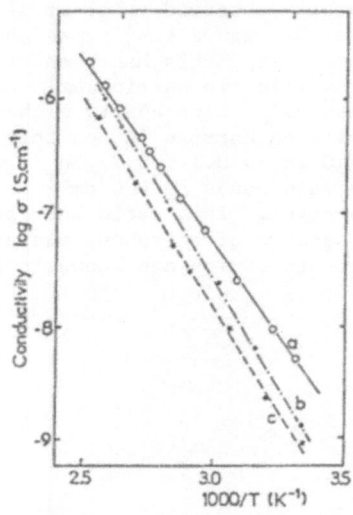

Fig.4 Conductivity of LiBO$_2$ thin films
versus inverse absolute temperature using
electrodes in Au (a), Al (b), Ni (c).

REFERENCES

1. A.Levasseur, M.Kbala, J.C.Brethous, J.M.Reau, P.Hagenmuller and
   M.Couzi, Solid State Commun., 32:839 (1979).
2. H.L.Tuller, D.P.Button and D.R.Uhlmann, J.Non-Crystalline Solids
   40:93 (1980).
3. J.C.Brethous, A.Levasseur, J.P.Bonnet and P.Hagenmuller, Solid
   State Ionics, 6:97 (1982).
4. A.Levasseur, M.Kbala, P.Hagenmuller, G.Couturier and Y.Danto,
   Solid State Ionics, 9-10:1439 (1983).
5. J.E.Bauerle, J.Phys.Chem.Solids, 30:2657 (1969).
6. Y.Ito, K.Miyauchi and T.Oi, J.Non-Crystalline Solids, 57:389 (1983).
7. T.Minami, Y.Ikeda and M.Tanaka, J.Non-Crystalline Solids, 52:159
   (1982).

GLASS SOLID ELECTROLYTE THIN FILMS

C. Julien

Laboratoire de Physique des Solides, associé au CNRS
Université Pierre et Marie Curie
4, place Jussieu, 75252 Paris Cedex 05, France

An important recent development concerns the research on lithium ion solid electrolytes in thin film form. Shrink of power source to miniature size is an urgent necessity for batteries to electronic devices developed recently.

We highlight in perspective gained as results of recent works which concern mainly the elaboration of lithium film cells using glass electrolyte system.

INTRODUCTION

In the last decade, a large variety of solid electrolyte thin films have been elaborated which are mainly silver, lithium, sodium or oxygen fast ionic conductors. All of them have been developed directly for constructing solid state ionic devices (SSID) such as micro-batteries, gas sensors or micro-coulometric systems.

Recent studies on glasses used as solid electrolyte have shown that the lithium ion conductivity, the large electrochemical stability range and the very low electronic conductivity give good performances which incited us to include such materials in solid state batteries[1].

Solid electrolyte thin films offer two main advantages : (i) thinning of layers gives a lower ionic resistance in transversal direction, (ii) thin films are well adapted for devices design using advanced microelectronic technology. However, a major constraint for building one SSID electrochemical cell is obtaining a good quality interface between electrolyte and electrodes. The use of glasses in form of thin films as electrolyte could allow to minimize this difficulty.

The subject of SSID thin films has been discussed in the scientific literature for many years. The review by Kennedy[2] is a good source for work prior to 1977 while that by Julien[3] reviews recent studies.

The aim of this paper is twofold. First, we present a reminder of what we know about glass thin films which can be obtained by different techniques. Second, we deal the growth and the characterization of alkalioxide electrolyte glass in thin films prepared by the Paris University group.

GLASS THIN FILM FORMATION

One can employ a large variety of deposition techniques (electrolytic, sputtering, evaporation, CVD, MBD, spray, etc.) to obtain thin layers of many different materials. The experimental conditions of vacuum-deposition techniques, such as sputtering, can be well defined, yielding reproducible deposits and by varying the experimental parameters (source or cathode composition, deposition rate, substrate temperature, etc.) deposits of various composition or structure can be obtained.

For example, in the case of the B_2O_3-Li_2O system the relationship between the mole ratio of Li and B in deposited thin films and that of the starting material has been discussed by the determination of the different parameters[4] and explained by using thermodynamic calculations[5]. For the binary oxide, the vapor pressure of a complex oxide $LiBO_2$ is greater than that of any other vapor species in a wide temperature range and the flux ratio of lithium to boron at 1200 K remains approximately at the composition $0.5B_2O_3$-$0.5Li_2O$ even if the composition of the evaporation source is changed. But, at higher temperature the change of the ratio with temperature depends on the composition of the evaporation source. In fact Zhang et al.[4] claim first that no linear relation exists between the thin film composition and the starting material one as it has been reported by previous authors[6-7], and second, that the lithium content in the thin films is independent of the temperature of the evaporation source but is dependent on the temperature of the substrate.

The different glass thin films and their main characteristics reported in the literature are compiled in Table 1.

Table 1. Characteristics of different lithium glass thin films. TVE:Thermal Vacuum Evaporation, QFE:Quasi Flash Evaporation, RFS:R.F Sputtering

Compound	Evaporation technique	Conductivity at RT ($\Omega^{-1}cm^{-1}$)	Activation energy (eV)	Other investigations	Reference
B_2O_3-Li_2O	TVE	10^{-11}	1.0	Raman, IR	(6)
B_2O_3-$2Li_2O$	TVE	10^{-8}	0.7	Raman, IR	(6)
$0.8B_2O_3$-Li_2O-$0.2SiO_2$	TVE	5×10^{-7}	0.8	SEM	(7)
B_2O_3-Li_2O	TVE	-	-	X-rays	(4)
B_2S_3-Li_2S	TVE	10^{-4}	0.18	X-rays	(8)
$BLiO_2$	QFE	10^{-8}	0.64	IR	(9)
$Li_{3.6}Si_{0.6}P_{0.4}O_4$	RFS	5×10^{-6}	0.5	-	(10)
Li_2O-SiO_2-ZrO_2	RFS	10^{-6}	0.54	-	(11)

For comparison, the experimental results reported by the japanese group[10-11] are also shown. Glass thin films obtained by evaporation of B_2S_3-Li_2S systems where substituting oxygen by sulfur enhances strongly the conductivity, is promising, but necessary controlled thermal treatment seems difficult. Perhaps the most promising example of a glass thin film electrolyte yet reported is due to Kanehori et al.[10]. $Li_{3.6}Si_{0.6}P_{0.4}O_4$ electrolyte was amorphous, ionic conductivity was 5×10^{-6} $\Omega^{-1}cm^{-1}$ and its Li^+ transference number was unity.

ALKALI-OXIDE GLASS THIN FILMS

Alkali-oxide glass thin film can be prepared by different techniques. We have obtained alkali-oxide films by using both the quasi-flash evaporation (QFE) technique which has been described elsewhere[9] and the molecular beam deposition (MBD) technique in ultra high vacuum (UHV) conditions. In the QFE method the source temperature is a sensitive parameter to increase the Li content in the film. The substrate temperature is less sensitive and it is possible to add different lithium salts in the same boat to increase the ionic conductivity of the films.

The MBD technique appears an elegant method. By using individual beam effusion cells, the different component fluxes i.e $LiBO_2$, Li_2O,

LiCl, etc, can be monitored with a movable ion gauge.

Figure 1. shows ionic conductivities parallel to the film surface which were evaluated by complex impedance analysis using interdigital blocking electrodes for $BLiO_2$ (a) and $BLiO_2+0.1LiCl$ (b) thin films. In comparison the experimental results reported by Ito et al.[6] (c) and Levasseur et al.[7] (d) are also shown in figure 1. Incorporating few amount of lithium salt enhances strongly the conductivity by one order of magnitude while the activation energy decreases from 0.74 eV in the metaborate film to 0.63 eV in the LiCl doped film.

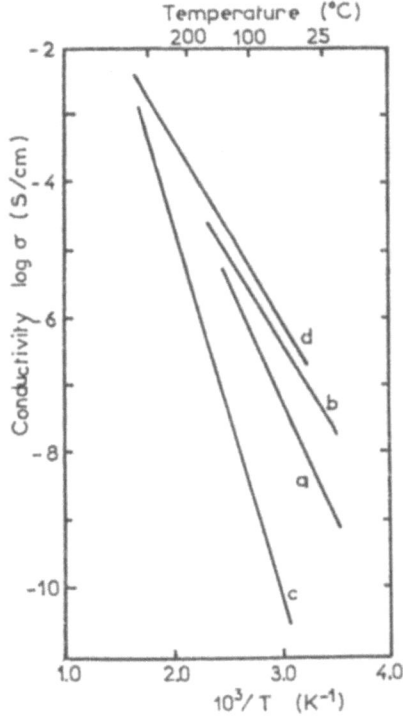

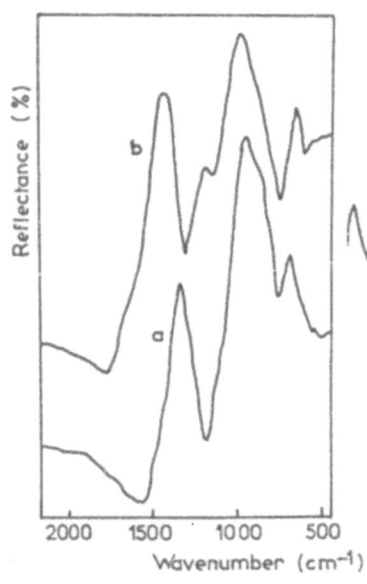

Fig.1 Ionic conductivities of different glass thin films. $LiBO_2$(a), $LiBO_2+0.1LiCl$(b) from ref.6(c), from ref.7(d)

Fig.2 Infrared reflectance spectra of bulk $B_2O_3-Li_2O$ glass (a) and $LiBO_2$ thin film (b).

Infrared or Raman spectroscopies are well adapted tools for the characterization of glass thin films. Using infrared measurements in absorption and reflection modes provide structural informations on the substance and if, the frequency domain is large enough give experimental measure of the dielectric response of the glass.

Figure 2. represents the infrared reflectance spectra of $B_2O_3-Li_2O$ bulk (a) and $LiBO_2$ thin film (b) prepared by MBD. These spectra show different reflectance peaks in the range 2000-500 cm^{-1} and the spectrum of $LiBO_2$ thin film is similar to that of the bulk. The shift of the band at 1400 cm^{-1} is attributed to the effect of increase of the Li content and to the transformation of boroxol rings into triborate cycles or di-triborate groups.

CONCLUSION

This paper describes that the glass thin film formation is not only function of the experimental conditions but also to the adopted

technique. Results given on the $B_2O_3-Li_2O$ system show that the MBD technique appears one of the best method for producing alkali-oxide glass thin
films. By control of the different fluxes, the Li/B ratio can be optimized
for obtain higher ionic conductivity.

ACKNOWLEGDMENT

The author thanks the European Research Office (London) for
partial financial support.

REFERENCES

1. H.L.Tuller and M.W.Barsoum, J.Non-Cryst.Solids, 73:331 (1985)
2. J.H.Kennedy, Thin Solid Films, 43:92 (1977)
3. C.Julien, Micro-Solid-State Batteries, Univ.of Paris Rept.,France (1987)
4. L.W.Zhang, M.Kobayashi and K.S.Goto, Solid State Ionics, 18-19:741 (1986)
5. L.W.Zhang, M.Yahagi and K.S.Goto, Solid State Ionics, 18-19:1163 (1986)
6. Y.Ito, K.Miyauchi and T.Oi, J.Non-Cryst.Solids, 57:398 (1983)
7. A.Levasseur, M.Kbala and P.Hagenmuller, Solid State Ionics, 9-10:1439
   (1983)
8. M.Kbala, M.Makyta, A.Levasseur and P.Hagenmuller, Solid State Ionics,
   15:163 (1985)
9. P.Dzwonkowski, C.Julien and M.Balkanski, this volume, p.
10. K.Kanehori, K.Matsumoto, K.Miyauchi and T.Kudo, Solid State Ionics,
    9-10:1445 (1983)
11. K.Miyauchi, K.Matsumoto, K.Kanehori and T.Kudo, Solid State Ionics,
    9-10:1469 (1983)

ELECTRONIC STRUCTURE, BONDING, AND LITHIUM MIGRATION EFFECTS INVOLVING THE SURFACE OF THE MIXED CONDUCTOR β-LiAi

I. M. Curelaru [a], K.-S. Din [a], G.-E. Jang [a], E. E. Koch [b], K. Horn [b], J. Ghijsen [c]*, R. L. Johnson [c], S. Susman [d], T. O. Brun [d], and K.J. Volin [d]

[a] Department of Materials Science and Engineering, University of Utah, Salt Lake City, Utah 84112, USA
[b] Fritz-Haber-Institut der Max-Planck-Gesellschaft, Faradayweg 4 - 6, D-1000 Berlin 33, FRG
[c] Max-Planck-Institut für Festkörperforschung, Heisenbergstrasse 1, D-7000 Stuttgart 80, FRG
[d] Materials Science and Technology Division, Argonne National Laboratory, Argonne, Illinois 60439, USA

Abstract

Detailed experimental studies of the electronic structure of the valence and conduction bands of the mixed conductor β-LiAl indicate that a quasi-gap opens at the Fermi level, and the conduction states are highly localized, as opposed to the theoretical band structure calculations that predict predominant metallic behavior. Evidence for complex lithium migration effects involving the surface of LiAl, induced by particle (electron or ion) bombardment, and mechanical treatment, has been obtained as a by-product of these experiments.

1. *Introduction*

The β-LiAl intermetallic compound is successfully used as the negative electrode in lithium-based high energy density batteries at both ambient and elevated temperatures /1 - 3/. The material has the advantage to possess mixed (electronic and ionic) conductivity, a stable electrode potential in repeated cycles of charging and deep discharging, and a lower reactivity with the molten electrolyte than lithium metal. On charging and discharging of a LiAl electrode, lithium is transported through the electrolyte due to concentration gradients, and should be able to exit and re-enter easily the electrode matrix with a desired efficiency as close as possible to 100 %. Accurate determination of diffusion co-efficients, the mechanism of ionic and electronic conductivity, and the surface-interface properties of this material are therefore problems of interest from both practical and basic viewpoints.

Beta-LiAl is a rather intriguing material. Its physical properties resemble both those of metals and non-metals. The phase is stable over a stoichiometry range of 48 - 54 at % Li /4/ and forms a highly ordered intermetallic compound with B32 (Zintl) structure /5/. It consists of two interpenetrating diamond-like sublattices, one for lithium and one for aluminum. The aluminum lattice is rigid, while

---

*Present address: Technical Physics Laboratory, University of Groningen, Nijenborgh 18, NL-9747 AG Groningen, The Netherlands.

lithium is highly mobile, with a diffusion coefficient of about 10^{-5} cm^2/sec and an activation energy of 1.5 eV /6/. The crystal has a complex defect structure consisting of lithium vacancies (that dominate at low lithium content) and lithium anti-site defects (that dominate at high lithium content) /7, 8/. The electronic conductivity is very low, and the positive Hall coefficient indicate that the majority charge carriers are holes /9, 10/. The room temperature resistivity varies between 20 and 50 $\mu\Omega$.cm over the composition range of the beta phase /9, 10/.

Band structure and cluster model calculations performed in various approximations /11/ have indicated close similarities with the band structure of silicon. The valence band with a total width of about 9.5 eV is composed of three subbands almost completely separated from each other. The lowest subband is mainly of s-type, the next one is an s-p hybrid, and the highest valence subband is dominated by p-states originating from both lithium and aluminum. This suggested significant s-to-p charge promotion at both components, as opposed to earlier models that assumed Li-to-Al charge transfer /12/. There is a charge accumulation between the Al atoms in the aluminum sublattice, and a remarkable charge depletion along Li-Li directions. Thus, the Li-ionic cores are in a "non-bonding" state, this explaining the easy lithium migration through the solid. The low electronic conductivity is due to the very low density of states at the Fermi level. Bonding in LiAl has been described as polarized covalent (in the lower part of the valence band) and metallic (in the higher electronic states).

The experimental and theoretical treatments mentioned above refer to an admittedly infinitely large bulk material. In real, finite systems, surfaces are an integral component of the objects to be studied, and in particular in batteries surface/interface electrode processes may determine the conductivity properties of the cell. We have recently initiated a systematic study of β-LiAl by using a number of complementary X-ray and electron excited techniques shown in Table 1. This synergistic approach provided a wealth of information on the electronic structure and related properties of this material and permitted important correlations to be made between its structure at the atomic scale and its macroscopic behavior.

Table 1. Basic information provided by various spectroscopic techniques

Method	Information
PES [a]	- core level binding energies - structure of the valence band - nature of bonding - dynamics of photoelectron emission (resonances) - absorption edges → structure of the conduction band - surface/subsurface composition
EELS [b]	- collective (plasmon) oscillations of valence electrons - interband transitions → structure of valence and conduction bands - effects of non-stoichiometry and lattice defect structure
AES [c]	- AES characteristic energies - surface/subsurface composition - effects of electron and ion bombardment
APS [d]	- APS binding energies - structure of conduction band - localization of conduction states

[a] Photoelectron Spectroscopy (NSLS-Brookhaven, BESSY-Berlin, DESY-Hamburg)
[b] Electron Energy Loss Spectroscopy (MRL-Urbana)
[c] Auger Electron Spectroscopy (MRL-Urbana)
[d] Appearance Potential Spectroscopy (University of Utah)

2. Experimental Conditions

The measurements were performed on polycrystalline sample materials of composition $Li_{0.513}Al_{0.487}$ synthesized by solidification from the melt. Photoemission experiments were carried out at the synchrotron radiation facilities NSLS-Brookhaven, BESSY-Berlin, and DESY-HASYLAB-Hamburg. The energy resolution in all these set-ups was estimated to be 0.3 eV. Calibration of the energy scale was done by reference to gold. Electron energy loss and Auger spectra were measured with a conventional PHI-595 Auger microbrobe at the Materials Research Laboratory, Urbana-Champain. Appearance potential experiments used a home-made instrument constructed at the University of Utah /13/. The instrument was operated with electron excitation (100 - 1000 eV), and soft X-ray total yield detection.

3. Structure of Valence and Conduction Bands

Figure 1 shows a set of valence band photoelectron spectra of LiAl measured at three different photon energies, compared with the total density of states (DoS) distribution calculated by Hafner and Weber /11/. In general, the theory predicts correctly the position of the bands, but the intensity ratios are significantly different. The leading photoemission peak (B) corresponds to an envelope of bands I and II in the DoS curve. Structure C has no counterpart in the theoretical DoS distribution, and its origin is not yet understood. The photoemission intensity from the uppermost valence band (band III) is

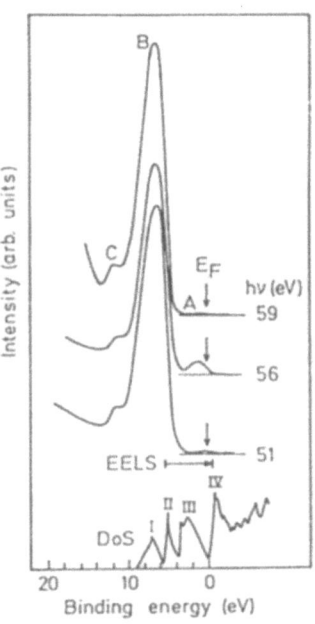

Fig. 1. A set of photoemission spectra of β-LiAl at various incident photon energies compared with the theoretical DoS curve (after /11/). An interband transition observed in EELS is marked by a horizontal arrow.

remarkably low. In fact this band is only measurable at incident photon energies close to 56 eV (Figure 1) and 74 eV (not shown), which correspond to the excitation edges of Li-1s and Al-2p core electrons, respectively. Since band III is dominated by p-type states originating from both lithium and aluminum, we conclude that the population of these states below the Fermi level is very low, and the s-to-p charge promotion, if at all present, is apparently much less significant than that predicted by theory. The resonant enhancement of the photoemission intensity from band A at the Li-1s and Al-2p edges is clearly illustrated in Figure 2. We explain these resonances as the added contribution of three possible processes shown schematically in Figure 3. Process (a) represents direct photoemission from band III. Processes (b) and (c) involve formation of an excitonic state (an excited state consisting of an electron lifted from a core level to an empty conduction band state, coupled to the core hole left behind) and its subsequent decay via either direct recombination (b) or Auger electron emission

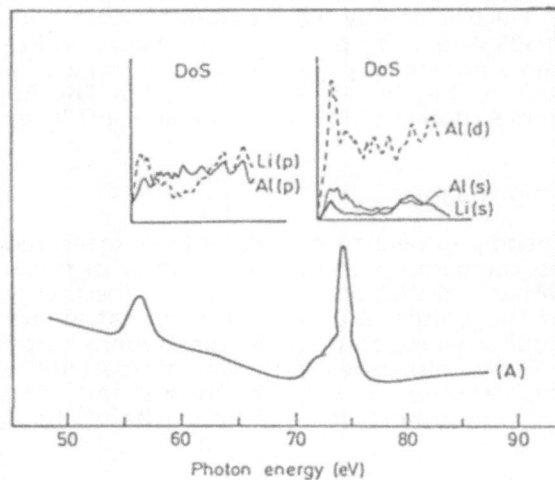

Fig. 2. Resonant photoemission from bands A and B at the Li-1s and Al-2p excitation edges.

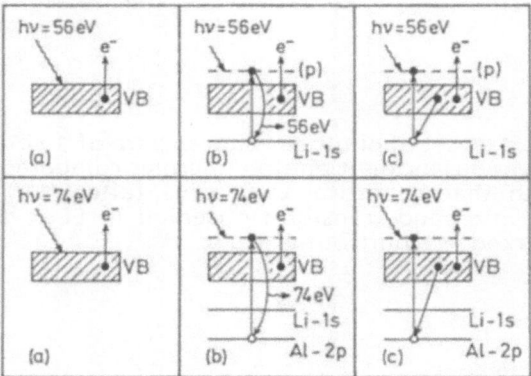

Fig. 3. Schematic diagram of processes that contribute to resonant photoemission: (a) direct photoemission; (b) exciton formation and decay by direct recombination; (c) exciton formation and decay by Auger electron emission.

(c). Formation of excitons (quasi-bound electron-hole pairs) implies a high degree of localization of the conduction band states. In the case of LiAl, the very low population of the electronic states in band III closest to the Fermi level resembles the existence of a band gap. Our electron energy loss experiment /14/ identified a loss structure of 5.8 eV, which we interpret as an interband transition between the valence band II and the first conduction band of high density IV. This transition, which is shown in Figure 1 by a horizontal arrow labeled EELS, confirms the position of the lowest conduction band at about 0.8 - 1.0 eV above the Fermi level, in agreement with the theory /11/. The distance between the low binding energy edge of the leading photoemission peak and the lowest conduction band (band IV) is about 5 eV. It is therefore understandable that, with such a wide "quasi-band-gap", formation of excitons is highly **favored**. A detailed analysis of the photoemission data will be published elsewhere /15/.

4. Lithium Migration Effects Involving the Surface of LiAl, Induced by Mechanical Treatment and Particle (Ion or Electron) Bombardment

For the experiments discussed in this paper, the high chemical reactivity of LiAl imposed stringent requirements on surface sample preparation. In various experimental facilities used, this was attempted by fracturing, scraping with a diamond file, or cutting with a bolt-cutter. Argon ion bombardment was also used. Electron bombardment was either a component part of the spectroscopic processes used (e.g. EELS, AES, APS, and also soft X-ray emission spectroscopy, which will be presented in a separate publication /16/), or was purposely associated with an Auger measurement as discussed below.

We have observed that all the methods used for preparation of clean surfaces resulted in a surface enrichment in lithium, in some experiments up to a complete coverage. The best case appeared to be fracturing, when the signal from both lithium and aluminum components could be measured, and the worst cases were cutting by a bolt-cutter and scraping, when aluminum was completely masked by lithium. This segregation effect caused by mechanical damage was presumably due to heating, or lowering of the crystal surface energy by the induced lattice strain. Ion bombardment resulted in mass removal from the surface of LiAl with preferential sputtering of the light component. In different experiments electron bombardment had opposite effects. In X-ray emission spectroscopy, an extended exposure to high energy electrons (10^3 eV) evaporated lithium from the surface /16/. Attempts to measure the appearance potential spectra of LiAl failed because of massive surface accumulation of lithium, apparently caused by the absorbed dose of low energy electrons (10^2 ev) /17/.

Figure 4 shows the Auger spectra of a fresh LiAl sample fractured in ultrahigh vacuum (a) and of the same sample after argon ion bombardment (800 eV, 30 mA, 5 minutes) (b) and subsequent exposure to high enery electrons (10 keV, 15 µA, 20 minutes) (c). Also shown are the Auger spectra of lithium and aluminum taken from literature /18/.

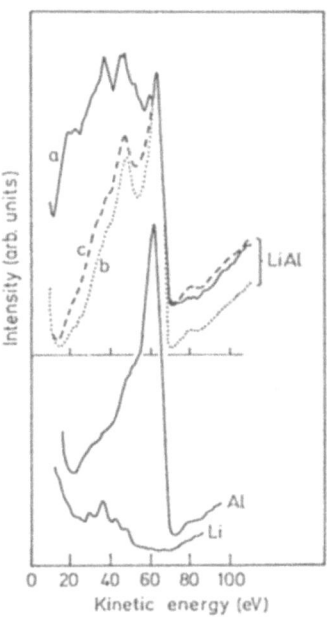

Fig. 4. Auger electron spectra of β-LiAl in the "as fractured" state (a); after ion bombardment (b); and after subsequent electron bombardment (c). Also shown are the Auger spectra of Li and Al.

A clear effect of ion bombardment (spectrum b) was the removal of most of the low energy fine structure which in spectrum (a) corresponded to Auger transitions involving lithium atoms. The total band width of the Auger spectrum was also significantly reduced. Based on the known sputtering yield data for various elements, it is understood that lithium was preferentially sputtered away from the LiAl surface. Spectrum (c) measured after electron bombardment indicates a tendency toward recovery of the low energy part, which demonstrates an increased contribution from lithium atoms. Here again, electron bombardment apparently caused migration of lithium atoms from the bulk towards the surface. Neither the exposure to electrons during measurement of the Auger spectra, nor the final prolonged bombardment with high energy electrons resulted in any shifts or fluctuations of the Auger structures. This indicates that the electron bombardment had no significant effects on the chemical state and phase stability of the LiAl intermetallic compound. Also, the electronic conductivity of the sample was sufficiently high to prevent any charging effects even under impact of a massive electron dose.

All these experimental observations have at this stage only a documentary value. The dynamics of lithium migration through the lattice of LiAl is complex. More studies are needed to elucidate the mechanisms of this process and its effects on the macroscopic behavior of the Zintl intermetallic compounds.

Acknowledgements

This work was supported by the U.S. Department of Energy under Grant DE-FG02-84ER 45143 (University of Utah) and Contract W-31-109-Eng-38 (Argonne National Laboratory), and the Bundesministerium für Forschung und Technologie under Projects 05390 CA B and 05250 CA. The expert technical assistance of the NSLS, BESSY, DESY-HASYLAB, and MRL staff is gratefully acknowledged. One of us (IMC) is grateful to the Max-Planck-Gesellschaft for a fellowship and to the Fritz-Haber-Institut der Max-Planck-Gesellschaft for hospitality and assistance during completion of parts of this work.

References

1. D. L. Barney: US DOE Report ANL-80-128, Argonne National Laboratory, February 1981.
2. W. Fischer: Solid State Ionics 3/4, 413 (1981).
3. B. N. McNicol and D. A. J. Rand (eds.): Power Sources for Electric Vehicles, Elsevier, Amsterdam-New York, 1984.
4. R. Thümmel and W. Klemm: Z. Anorg. Allg. Chem. 376, 44 (1970).
5. E. Zintl and G. Brauer: Z. Phys. Chem. B20, 245 (1933).
6. C. J. Wen, B. A. Boucamp, R. A. Huggins and W. Weppner: J. Electrochem. Soc. 126, 2258 (1979).
7. K. Kishio and J. O. Brittain: J. Phys. Chem. Solids 40, 933 (1979).
8. L. N. Hall, T. O. Brun, G. W. Crabtree, J. E. Robinson, S. Susman and T. Tokuhiro: Solid State Commun. 48, 547 (1983).
9. M. Yahagi: Phys. Rev. B24, 7401 (1981).
10. K. Kuriyama, T. Nozaki and T. Kamijoh: Phys. Rev. B26, 2235 (1982).
11. J. Hafner and W. Weber: Phys. Rev. B33, 747 (1985) and the references therein.
12. W. Hückel: Structural Chemistry of Inorganic Compounds, Elsevier, Amsterdam, 1951.
13. K.-S. Din, R. Chin and I. M. Curelaru: MSE-UofU Report, 1985.
14. I. M. Curelaru, K.-S. Din, S. Susman, T. O. Brun and K. J. Volin: submitted for publication.
15. I. M. Curelaru, G.-E. Jang, E. E. Koch, K. Horn, J. Ghijsen, R. L. Johnson, S. Susman, T. O. Brun and K. J. Volin: to be published.
16. I. M Curelaru, G.-E. Jang, T. A. Callcott, K. L. Tsang, C. H. Zhang, S. Susman, T. O. Brun and K. J. Volin: to be published.
17. I. M. Curelaru, K.-S. Din and G.-E. Jang: unpublished.
18. Handbook of Auger Electron Spectroscopy, JEOL, Tokyo, 1981.

DEFECTS WITH VARIABLE CHARGES: INFLUENCE ON CHEMICAL DIFFUSION

AND ON THE EVALUATION OF ELECTROCHEMICAL EXPERIMENTS

Joachim Maier

Max-Planck Institut für Festkörperforschung
Heisenbergstr. 1
D-7000 Stuttgart-80 (FR-Germany)

ABSTRACT

It is emphasized that the concept of chemical diffusion holds formally if and only if the deviations from stoichiometry (more generally: conservative ensembles) are considered, since the source terms in the equations of continuity cancel under these conditions. It is shown that, if partially ionized or neutral defects play a role in the transport processes, non-trivial modifications appear with respect to the chemical diffusion coefficient, to the cell voltage, as well as to the evaluation formulae of electrochemical or chemical polarization experiments. The importance is highlighted particularly for oxygen concentration cell experiments and for electrochemical polarizations with blocking electrodes as characteristic examples. In the first case, EMF values may be obtained that are greater than the Nernst EMF expected for fully ionized defects only. In the latter case internal dissociations of (or associations to) ionic-electronic defect-complexes can counteract the blocking effects. Particularly in the steady state (Wagner-Hebb-polarization) the apparent partial conductivities are not identical with the values in the non-polarized state.

INTRODUCTION

Recently experimental progress has been made and different publications appeared with respect to the relevance of charge carriers that are not fully ionized or even neutral[1-7] (e.g. $V_o^{\cdot}$, V_o in addition to $V_o^{\cdot\cdot}$), particularly in the case of transition metal oxides. The author, in cooperation with G. Schwitzgebel, formulated the necessary extension of the basic theory.[8,9] We could show in which manner the chemical diffusion coefficient, the cell voltage and the evaluation formulae for electrochemical experiments must be modified. This extension enables us, moreover, to treat the subject in a generalized manner (complex defect chemistry, mixed conductors, electronic and ionic electrodes).[9] The evaluation formulae have been applied to electrochemical measuremens on PbO, BaTiO$_3$, and SrTiO$_3$.[10] In the case of PbO, it could be concluded by comparing results of concentration cell experiments and electrochemical polarization experiments, that the presence of an overwhelming concentration of neutral oxygen defects (as proposed by Heyne et al.[11]) can be excluded as long as defect chemical equilibrium can be assumed.

Because of the recent interest in this theme, a short outline of the basic results of our theory will be given in a descriptive manner. We will first tackle the problem of chemical diffusion by approaching the decisive point (step IV) step by step and then discuss the implications for appropriate measurments. As an example we consider a binary oxide with ionic disorder only in the O-sublattice. (For background, the reader is also referred to I. Riess' article in this book).

CHEMICAL DIFFUSION

Step I ($V_o^{\cdot\cdot}$; e$'$)

In the case that only fully ionized oxygen defects (here vacancies) and conduction electrons are present the flux equations are (neglecting cross-coefficients)[12,13]

$$j_{o2-} = - \frac{\sigma_{o2-}}{4F^2} \frac{\partial \tilde{\mu}_{o2-}}{\partial x} \tag{1}$$

(where $j_{o2-} = -j(V_o^{\cdot\cdot})$ and $\mu_{o2-} = -\mu(V_o^{\cdot\cdot})$) and

$$j_{e-} = - \frac{\sigma_{e-}}{F^2} \frac{\partial \tilde{\mu}_{e-}}{\partial x} \tag{2}$$

where $j_{e-} = j(e')$, $\sigma_{e-} = \sigma(e')$ and $\tilde{\mu}_{e-} = \tilde{\mu}(e')$. The usual procedure is to split the electrochemical potential ($\tilde{\mu}$) into the chemical potential (μ) and the electrical potential (Φ), and to eliminate ($\delta/\delta x)\Phi$ by making use of the condition of local electroneutrality (internal electric currents must vanish or, in the case of a polarization, equalize the external current). One obtains, e.g. for the electronic current (i:external current)

$$j_{e-} = - \left(\frac{1}{F} \frac{\sigma_{e-}}{\sigma} \right) i - \tilde{D} \partial C_{e-} / \partial x \tag{3}$$

(σ = total conductivity and $\tilde{D} = (8F^2)^{-1} (\sigma_{o2-}\sigma_{e-}/\sigma) d\mu(O_2)dc_{o2-}$ = chemical diffusion coefficient). The relation for j_{o2-} is analogous. Since in this case the (time) increase of the defect concentration is simply given by the (negative) divergence of the corresponding flux (equation of continuity), Fick's second law appears, e.g.

$$\partial C_{e-} / \partial t = \tilde{D} \partial^2 C_{e-} / \partial x^2 \tag{4}$$

Step II ($V_o^{..}$; e', $h^.$)

If we have to consider additionally the presence of holes, source terms become important in the equation of continuity, i.e. a time change of the electronic concentration can occur by internal defect-chemical reactions (Null = e' + $h^.$). (For a more detailed explanation of the consistency of non-zero source terms, deviations from transport equilibrium and local defect -chemical equilibrium, see Ref. 8). Obviously, the source terms cancel if we consider $c(e')$ - $c(h^.)$, i.e. $2c(V_o^{..})$ (proportional to degree of non-stoichiometry). We find by combinations of the individual flux equations that the above relationships are still valid with j_{e^-} = $j(e')$ - $j(h^.)$; δc_{e^-} = $\delta c(e')$ - $\delta c(h^.)$ and σ_{e^-} = $\sigma(e')$ + $\sigma(h^.)$.

Step III ($V_o^{..}$, $o_i{}''$; e', $h^.$)

As in Step II, the new defect equilibrium (Const = $V_o^{..}$ + $O_i{}''$) can be taken account of by considering $\delta(c(O_i{}'') - c(V_o^{..}))$ = δc_{o2-} = $1/2$ $\delta(c(e') - c(h^.))$ = $1/2$ δC_{e^-} = which is proportional to degree of nonstoichiometry), i.e. we have j_{o2-} = $j(O_i{}'')$ - $j(V_o^{..})$ and σ_{o2-} = $\sigma(o_i{}'')$ + $\sigma(V_o^{..})$ and thus qualitatively the same diffusion coefficient. So far no exciting changes appeared since we did not have any coupling between electronic and ionic defects (see separation of indices o^{2-} and e^- in the above equations).

Step IV ($V_o^{..}$, $V_o^{.}$ ...; e', $h^.$)

This is no longer true if partially ionized ($V_o^.$, O_i') and/or neutral defects (V_o^x, O_i^x) must be considered. It can be shown[8] that the above equations can be formally maintained if the non-stocihiometry (i.e. c_{ion}^* = $c(O_i'')$ + $c(O_i')$ + $c(O_i^x)$ - $c(V_o^{..})$ - $c(V_o^.)$ - $c(V_o^x)$) is considered as quasi-species. The transport quantities, e.g. σ_{ion}^*, can no longer be interpreted in a simple way (e.g. as σ_{ion}). Qualitative changes occur. Details will not be given here (see Refs. 8,9). Just a few examples may highlight its importance*.

CONSEQUENCES OF ION-ELECTRON COUPLING

Diffusion Equation and Boundary Conditions

The form of Eq. 4 is still maintained if c_{ion}^* (or c_e^*) is used. The chemical diffusion coefficient now is given by

$$\tilde{D} = (8F^2)^{-1} \{2\sigma_1 + 4S$$

$$+ (\sigma_{ion}+\sigma_1)(\sigma_e-\sigma_1)/\sigma\} \; d\mu(O_2)/dc_{ion}^* \qquad (5)$$

We see that the transport coefficients of the non-fully ionized defects (σ_1 = σ (onefold ionized defect); S α

---

*Complications with the measurement of the total conductivity do, of course, not arise, since formally speaking σ_{ion}^* + σ_e^* = σ_{ion} + σ_e = σ (i.e. the perturbations cancel in the sum)[8].

diffusivity of the neutral defects) enter Eq. 5 in a non-trivial manner ($\sigma_{ion} = \sigma_2 + \sigma_1$; $\sigma_2 = \sigma(V_o^{\cdot\cdot}) + \sigma(O_i'')$; $\sigma_e = \sigma(e') + \sigma(h^\cdot)$). As far as the evaluation of dynamical electrochemical experiments[9] (chemical or electrochemical polarization) is concerned, an additional point is important, namely that the boundary conditions can only be easily formulated within the concept of conservative ensembles.[8,9]

Cell Voltage of a (Electro-)Chemically Polarized Cell

The general relation for the cell voltage at any electrochemically or chemically polarized cell[12,13] must be extended[8] with respect to the appearance of σ_1 to:

$$U_{e,ion} = i \int \frac{dx}{\sigma} \pm \frac{1}{4F} \int \frac{\sigma_{ion,e} \pm \sigma_1}{\sigma} \, d\, \mu(O_2) \quad (6)$$

U_e holds if electronic electrodes are used, U_{ion} if ionic electrodes are applied.

Concentration Cell Experiment (Steady State Chemical Polarization)

A very important special case is met in an (oxygen) concentration cell experiment (i = 0, electronic electrodes). It is obvious that the classical Wagner-formula[12] must be generalized[9] to (Po_2:oxygen partial pressure)

$$E = \frac{RT}{4F} \left\langle \frac{\sigma_{ion}+\sigma_1}{\sigma} \right\rangle \ln \frac{Po_2(l.h.s)}{Po_2(r.h.s)}$$

$$= \frac{RT}{4F} \left\langle \frac{\sigma_2+2\sigma_1}{\sigma} \right\rangle \ln \frac{Po_2(l.h.s)}{Po_2(r.h.s)} \quad (7)$$

σ_1 appears in a non-trivial way, viz. twice: on the one hand included in σ_{ion} and on the other hand, as an extra contribution. For an extreme case, let $\sigma_1 \gg \sigma_2$, then $\sigma_{ion} + \sigma_1 = 2\sigma_1$. Hence, the factor of 2 cancels against 4 in the denominator and we obtain the classical Wagner formula that would be expected for monovalent oxygen (i.e. 2 $E(\sigma_2 \gg \sigma_1)$). The evaluation formula for measuring with ionic probes in a chemical potential gradient (see U_{ion} in Eq. (6)) is analogous.

Wagner-Hebb Polarization

In the case of a polarization experiment, an internal supply of lacking charge carriers is now possible. (In a true steady state a "titration" of complexes should be possible.) If the reservoir of defects is large compared to changes due to the polarization, the equations of Ref. 8,9 are obtained. In such an approximation, it can be shown that in the extreme cases of a large excess of neutral or partially ionized defects (which are mobile in a chemical or electrochemical gradient) no polarization occurs since all blocking effects are cancelled by internal dissociaton. For details and applications on other examples (e.g. impedance measurements) see Ref. 9.

CONCLUDING REMARKS (DEFECT DESCRIPTION VS. COMPONENT DESCRIPTION)

At present, we are looking for examples to check the equations in a quantitative manner. So far, we were only able to obtain negative information by their application.[10] In any case, if experiments are performed suggesting complex defect chemical behavior, one should be aware that modified evaluation formulae must be used in order to collect consistent information.

The above results have been obtained by considering the defects and not the components. Two further examples where the consideration of different levels of description (defects/ions) causes misunderstandings, or at least difficulties in communication, are worth being considered:

1) Since $|\mu_{ion}| = |\mu_d|$ in the space charge regions, both $(\delta/\delta x)\tilde{\mu}_{ion}$ and $(\delta/\delta x)\tilde{\mu}_d$ vanish. We assume here a fully ionized ionic defect, d (see e.g. Schottky[14]). A detailed derivation through minimizing the total Gibbs' energy[15] is not necessary.

2) Different authors with electrochemical background use the concept of the thermodynamic factor of the ions (T_{ion}) for describing the differences between tracer and chemical diffusion coefficient, whereas most semiconductor physicists (starting from an atomistic viewpoint) ascribe this discrepancy to the low defect concentration (c_d/c_{ion}). Applying the relation between μ_{ion} and μ_d, we obtain immediately the connection ($\delta \ln a_{ion} = \pm \; \delta \ln a_d$; $\delta c_{ion} = \pm \; \delta c_d$)

$$T_{ion} \equiv \frac{\partial \ln d_{ion}}{\partial \ln c_{ion}} = \frac{\partial \ln C_d}{\partial \ln C_d} \left(\frac{C_d}{C_{ion}} \right)^{-1} \tag{8}$$

with $T_d = \delta \ln a_d/\delta \ln c_d$ being of the order of unity for dilute solutions.

REFERENCES

1. N. Ait-Jounes, F. Millort, and P. Gerdanian, Solid State Ionics 12:431 (1984); 12:437 (1984).
2. F.A. Kroger, Solid State Ionics, 15:39 (1985).
3. M. Martin and H. Schmalzried, Solid State Ionics 20:75 (1986).
4. L. Dufour, personal communication (1987).
5. R. Dickman, personal communication (1985).
6. I. Riess, this book (lecture on mixed conductors).
7. J. Mizuski and K. Fueki, Solid State Ionics 6:85 (1982).
8. J. Maier, G. Schwitzgebel, Phys. Stat. Sol.(b), 113:535 (1982).
9. J. Maier, Z.Phys. Chem. N.F., 140:191 (1984).
10. J. Maier and G. Schwitzgebel, Mater. Res. Bull. 18:601 (1983); Mater.Res. Bull. 17:1061 (1982); J. Solid State Chem., 58:1 (1985).
11. L. Heyne, N.M. Beckmans, and A. de Beer, J. Electrochem. Soc. 119:77 (1972).

12. C. Wagner, Z. Elektrochemie 60:4 (1956); Progr. Solid State Chem., 10:3 (1975).
13. I. Yokota, J. Phys. Soc. Japan 16:2213 (1961).
14. W. Schottky, in: "Halbleiterprobleme," Vol. IV and Vol. VII, W. Schottky, ed. Vieweg, Baunschweig, 1958.
15. K.L. Kliewer and J. Kohler, Phys. Rev. A, 140:1126 (1965).

ATTENDEES

Dr. Ingvar Albinsson
Ms. G. Allitsch
A. Ayyadi
Dr. Kumara Bandaranayake
Mr. Marco Bernasconi
M. Robert Creus
Prof. I.M. Curelaru
I. Darianian
Jose M. Delgado
Dr. V.K. Despande
P. Dzwonkowski
Ms. Rosana Z.D. Fernandes
Sten Frostang
Florence Fusco
Mr. S.J. Golden
James Harvie
C. Julien
Dr. W. Kernler
Dr. C. Leach
A. Lobert
Mandy Mackenzie
Dr Gianluigi Marra
Steve W. Martin
M. Massot
A. Menne
Peter Moon
G. Meunier
G. Albert Popson, Jr.
Mr. M.A. Priestnall
Kenneth Pytlewski
R. Safadi
Johan Sandahl
Dr. M. Scagliotti
G. Schafer
Mark Sheldon
Graham Sorrie
Dr. Jaime Valderrama-N.
Richard F. Wallis

LECTURERS

Dr. Michel Armand
Professor M. Balkanski
Professor Michael W. Barsoum
Professor C.R.A. Catlow
Dr. Michel Kleitz
Dr. Joachim Maier
Dr. Ilan Riess
Professor Harry L. Tuller
Dr. Werner Weppner
Professor B.J. Wuensch

ATTENDEES/LECTURERS (by country)

Ms. G. Allitsch
Institut fur Technische
Elektrochemie
Getreidemarkt 9/158
A-1060 Wien AUSTRIA

Ms. Rosana Z.D. Fernandes
Laboratoire d'Ionique et
d'Electrochimie du solide
LIES - ENSEEG Domaine Univ. BP 75
38402 Saint Martin d'Heres CEDEX FRANCE

G. Meunier
Laboratoire de Chimie
du Solide du CNRS
351, Cours de la Liberation
33405 Talence CEDEX FRANCE

Dr. V.K. Despande
Laboratoire de Chimie Minerale C
ERA 314 Chimie Des Materiaux
Place E. Bataillon
34060 Montpellier CEDEX FRANCE

M. Robert Creus
Laboratoire de Chimie Minerale C
ERA 314 Chimie Des Materiaux
Place E. Bataillon
34060 Montpellier CEDEX FRANCE

M. Massot
Laboratoire de Physique du Solide
Universite Pierre et Marie Curie
Tour 13, 75230 Paris Cedex 05,FRANCE

A. Ayyadi
Laboratoire de Physique du Solide
Universite Pierre et Marie Curie
Tour 13, 75230 Paris Cedex 05,FRANCE

I. Darianian
Laboratoire de Physique du Solide
Universite Pierre et Marie Curie
Tour 13, 75230 Paris Cedex 05,FRANCE

P. Dzwonkowski
Laboratoire de Physique du Solide
Universite Pierre et Marie Curie
Tour 13, 75230 Paris Cedex 05,FRANCE

C. Julien
Laboratoire de Physique du Solide
Universite Pierre et Marie Curie
Tour 13, 75230 Paris Cedex 05,FRANCE

Mr. S.J. Golden
Imperial College of Science & Technology
Royal School of Mines
Prince Consort Road
London, SW7 2BP ENGLAND

Mr. M.A. Priestnall
Imperial College of Science & Technology
Royal School of Mines
Prince Consort Road
London, SW7 2BP ENGLAND

Dr. C. Leach
Imperial College of Science & Technology
Royal School of Mines
Prince Consort Road
London, SW7 2BP ENGLAND

Mark Sheldon
School of Chemistry
Leicester Polytechnic
P.O. Box 143, Leicester
LE1 9BH ENGLAND

Mandy Mackenzie
University of Aberdeen
Department of Chemistry,
Meston Walk, Old Aberdeen
AB9 2UE, SCOTLAND

James Harvie
University of Aberdeen
Department of Chemistry,
Meston Walk, Old Aberdeen
AB9 2UE, SCOTLAND

Graham Sorrie
University of Aberdeen
Department of Chemistry,
Meston Walk, Old Aberdeen
AB9 2UE, SCOTLAND

R. Safadi
Technion
Israel Institute of Technology
Department of Physics
Technion City, 32 000 Haifa ISRAEL

Dr. M. Scagliotti
CISE-Tecnologie Innovative
Segrate Via Reggio Emilia, 39
Casella Postale 12081
20134 Milano ITALY

Dr Gianluigi Marra
Universita degli Studi di Milano
Dipartimento di Fisica
Sezione Stati Aggregati e Biomolecole
Via Celoria 16, I-20133 Milano ITALY

Mr. Marco Bernasconi
Universita degli Studi di Milano
Dipartimento di Fisica
Sezione Stati Aggregati e Biomolecole
Via Celoria 16, I-20133 Milano ITALY

Dr. Ingvar Albinsson
Dept. 5, Physicum
Chalmers University of Gothenburg
412 96 Gothenburg SWEDEN

Dr. Kumara Bandaranayake
Dept. 5, Physicum
Chalmers University of Gothenburg
412 96 Gothenburg SWEDEN

Dr. Jaime Valderrama-N.
Dept. 5, Physicum
Chalmers University of Gothenburg
412 96 Gothenburg SWEDEN

Sten Frostang
Dept. of Inorg. Chemistry
Arrhenius Laboratory
University of Stockholm
S-106 91 Stockhlom - SWEDEN

Johan Sandahl
Dept. of Physics
Chalmers Univ. of Technology
Gottenburg S-41296 SWEDEN

Jose M. Delgado
MIT Rm 13-4069

Florence Fusco
MIT RM 13-4010

Peter Moon
MIT RM 13-4006

Steve W. Martin
Iowa State University
Dept. Materials Science & Eng.
1110 Engineering Annex
Ames, Iowa 50011

G. Albert Popson, Jr.
Kinard Laboratory of Physics
Clemson University
Clemson, SC 29634-1911

Richard F. Wallis
Department of Physics
University of California-Irvine
Irvine, CA 92717

Prof. I.M. Curelaru
University Of Utah
Dept. of Materials Science & Eng.
Salt lake City, UT 84112

Kenneth Pytlewski
Department of Materials Engineering
College of Engineering
Drexel University
Philadelphia, PA 19104

Dr. W. Kernler
MPI fur Festkorperforschung
Heisenbergstrasse 1, Postfach 80 06 65
D-7000 Stuttgart, WEST GERMANY

A. Lobert
Max-Planck-Institut fur Festkorperforschung
Heisenbergstrasse 1 Postfach 80 06 65
7000 Stuttgart 80 WEST GERMANY

A. Menne
Max-Planck-Institut fur Festkorperforschung
Heisenbergstrasse 1 Postfach 80 06 65
7000 Stuttgart 80 WEST GERMANY

G. Schafer
Max-Planck-Institut fur Festkorperforschung
Heisenbergstrasse 1 Postfach 80 06 65
7000 Stuttgart 80 WEST GERMANY

LIST OF LECTURERS
NATO ASI
The Science & Technolog of Fast Ion Conductors
ERICE, ITALY July 1-15, 1987

Dr. Michel Armand, Laboratoire D'Ionique
et d'Electrochimie du Solid
ENSEEG/CNRS 1213 INPG
Universite de Grenoble, B.P. 75,
38402 St. Martin d'Heres, FRANCE

Professor M. Balkanski
Laboratoire de Physique du Solide
Universite Pierre et Marie Curie
Tour 13, 75230 Paris Cedex 05,FRANCE

Professor Michael W. Barsoum
Department of Materials Engineering
Drexel University
Philadelphia, PA   19104

Professor C.R.A. Catlow
University of Keele
Keele, Staffordshire
ENGLAND ST5 5BG

Dr. Michel Kleitz
ENS Electronchimie
Domaine Universitaire
B.P. 75, 38402 Grenoble, FRANCE

Dr. Joachim Maier
Max-Planck-Institut fur Festkorperforschung
Heisenbergstrasse 1 Postfach 80 06 65
7000, Stuttgart 80, West Germany

Dr. Ilan Riess
Technion-Israel Institute of Technology
Department of Physics
Technion City,32 000 Haifa,ISRAEL

Professor Harry L. Tuller
Director, NATO ASI
Science & Technology of Fast Ion Conductors
MIT Room 13-3126

Dr. Werner Weppner
Max-Planck-Institut fur Festkorperforschung
D-7000 Stuttgart 80, West Germany

Professor B.J. Wuensch
TDK Professor of Materials Science & Engineering
MIT Room 13-4037

INTERNATIONAL SCHOOL OF MATERIALS SCIENCE AND TECHNOLOGY
12th course on FAST ION CONDUCTORS
Held July 1-15, in Erice, Italy

INDEX